Springer Proceedings in Business and Economics

More information about this series at http://www.springer.com/series/11960

Albertina Dias • Bror Salmelin • David Pereira •
Miguel Sales Dias
Editors

Modeling Innovation Sustainability and Technologies

Economic and Policy Perspectives

Springer

Editors
Albertina Dias
Business and Economics
New Atlantic University
Porto Salvo, Portugal

Bror Salmelin
DG Communications Networks,
Contents and Technology
European Commission
Brussels, Belgium

David Pereira
Faculdade de Ciências e Tecnologia
Universidade Nova de Lisboa
Lisboa, Portugal

Miguel Sales Dias
ISCTE-Instituto
Universitario de Lisboa
Lisboa, Portugal

ISSN 2198-7246 ISSN 2198-7254 (electronic)
Springer Proceedings in Business and Economics
ISBN 978-3-319-67100-0 ISBN 978-3-319-67101-7 (eBook)
https://doi.org/10.1007/978-3-319-67101-7

Library of Congress Control Number: 2017963636

Printed on acid-free paper

This Springer imprint is published by Springer Nature
The registered company is Springer International Publishing AG
The registered company address is: Gewerbestrasse 11, 6330 Cham, Switzerland

Foreword

This is a most refreshing set of stories from the front line of innovation.

While the EU institutions review current policy and, I hope, increase the focus on pro-innovation policies, the wide range of initiatives reported in these pages show very clearly how fast open innovation practice spreads already across regions and across sectors.

The cases coming from services promotion and from environmental policy are particularly inspiring: whatever a community's challenge, more and more creative innovation will be part of the answer.

The focus on modeling reported here is a proof of the increasing maturity of our understanding of innovation systems. But the self-developing (autopoietic) nature of innovation systems means that modeling them is in itself a pioneering challenge, where brute force and established practice must evolve toward new data analysis approaches. More innovation is needed.

Oxford
April 2017

Robert Madelin

Foreword by the European Commission

MIST stands for Modeling Innovation Sustainability and Technologies, which is of growing importance. The innovation landscape has fundamentally changed from the rather well-defined, even predictable "old" economic behaviors to the new, connected and networked world where the interdependencies together with the dynamics present a real challenge.

What I see as the challenge for this book too is to keep the strong focus on modeling itself, based on an understanding of modern innovation. We are no longer in linear, predictable innovation processes, since much of the innovation is parallel, networked, and multidisciplinary.

For policy making, the innovation modeling can bring valuable hints on which kinds of processes best foster the success and speed of innovation and how to build enabling technologies but increasingly more relational infrastructures for success. According to the work by Lin and Edvinsson, the structural intellectual capital is the key factor for successful innovation policy, i.e., how the various elements in the whole innovation system interact! It is not about creating innovation or knowledge islands but to create the interactions, reflections, and processes.

Also noteworthy is discovering that in the Western world knowledge is seen very much as an object one can possess, but in Eastern cultures the knowledge is seen as interaction, as a reflective process between the individuals.

Innovation is to make things happen, and innovation ecosystems at their best catalyze collisions igniting new ideas and bringing them into innovations. Infrastructures, i.e., open innovation environments including all stakeholders in the Open Innovation 2.0 perspective, make rapid prototyping and experimentation in real-world settings possible, creating early indications on the successful paths to go forward and of course those to be killed as well.

Modeling this complexity and these interdependencies can help policy makers discover the successful paths for building connected innovation ecosystems. I hope that this first MIST book will develop a strong driving force behind innovation modeling for policy development in Europe. We surely have interesting elements and emerging infrastructures in Europe, but having a holistic vision on the needed actions for impact is very much what is now needed.

Advisor for Innovation Systems
Directorate General for
Communications Networks
Content and Technology
(DGCONNECT)
European Commission
Brussels
Belgium

Bror Salmelin

Brief Description

Today, we are witnessing an extraordinary fast changing world with multiple disruptive technologies. The growing power of artificial intelligence, machine learning, cloud computing, big data and the so called IoT (Internet of the Things) have brought the need to adapt for many individuals and organizations, which is resulting in radical new forms of business modeling, policies, societal systems and society itself, as well.

This book focuses, not exclusively but mainly, on a set of real cases across several European regions where innovation modeling is needed urgently. By documenting the analysis on the new creative ideas and on the concrete bottlenecks to implement innovations capable of satisfying market needs and its challenges we attempt to distil the wisdom from the ongoing reality.

Albertina Dias

Contents

About the Editors

Albertina Dias Among several academic positions, in different universities in Portugal and Spain, Albertina Dias qualified as a Coordinating Professor of Economics and Industrial Management (2009–2012). Between 2014 and 2015, Albertina Dias was the Director of the Department of Business Science at New Atlantic University in Portugal. In 2015, she founded a new association for Modeling Innovation Sustainability and Technologies (MIST). She is the MIST Association Chairman and Business Scientific Director as well (www.mist.pt). With more than 15 years of work experience, in the areas of Economics, Finance, Innovation Strategic Management, and Information and Communication Technologies (ICT), Albertina has been working actively in the "knowledge triangle" (research, education and innovation) having run her own company on innovative ICT between 2005 and 2013 (SbH, Ltd., developing multimedia interactive books, based on augmented reality technology and computer-aided eBook interaction systems). In the past, she has assumed several degrees of responsibilities in companies such as Paulo Branco Holding (CEO, film industries), Oracle (Financial Analyst), Allianz (Financial Control), and Cap Gemini (Project Manager). Under the (EU funded) Program Equal e-Change, she participated since 2005 in the presentation, demonstration, and validation of innovative products in the areas of flexible security on employment and audiovisual and multimedia tools. She was responsible for the audiobook library of the Economists' College in PT and has produced several audiobooks of economic content such as the "Portuguese Governmental Budget, 2006," "2nd Annual Conference of Economists, 2007," "SCUTS—An Economic Analysis, 2005," among others. Since 2008, she is an Independent Expert for the European Commission, acting as a Reviewer/Evaluator for the implementation of indirect actions under specific programs. Since 2005, she is a fellow researcher of GITICE/ Universidad de Huelva in Spain (scientific domain of economics and research line of ICT). In this institution, she is developing research on Innovative Tools toward

Regional Economic Development. Previous academic background includes three principal scientific degrees (pre-Bologna):

- Ph.D. on Business and Economics (Huelva University, Spain, registered in Nova University, Lisbon, Portugal);
- Sc. Master in Statistics and Information Management (Nova University, Lisbon, Portugal);
- Degree in Economics (Nova University, Lisbon, Portugal).

She is the author of a considerable number of academic theses, articles, papers, and books. Her recent research activity involved studies on the Economics of Innovation (international comparative studies), Information and Communication Technologies (ICT), and the development of innovative business models for several environments. Currently, her research interests focus on the political economic models and public policies.

Miguel Sales Dias José Miguel Sales Dias holds a B.Sc. (1985), a M.Sc. (1988), both in Electrical and Computer Engineering (IST-UTL, Portugal), and a Ph.D. (1998) in Computer Graphics and Multimedia from ISCTE-IUL, Portugal, where he was an Associated Professor until 2005, currently holding an Invited Associated Professor position, teaching and conducting research in Computer Graphics, Virtual and Augmented reality, Ambient Assisted Living, and Human–Computer Interaction (HCI). He coordinates the Digital Living Spaces group of ISTAR-IUL, an R&D unit of ISCTE-IUL (http://istar.iscte-iul.pt/index.php/Digital_Living_Spaces). From early 2017, he is as a member of the Board of Directors of ADENE, the Portuguese Energy Agency. He has been involved in entrepreneurship initiatives. He is past director of the first European R&D Centre in Speech and natural HCI technologies of Microsoft Corporation, in Portugal (Microsoft Language Development Center, MLDC), from 2005, and during 11 years, where he collaborated closely with global product groups (USA, China, Europe) and local teams of marketing, sales, and services. He is regularly commissioned by the European Commission and Portuguese Ministry of Economy for R&D project evaluations and reviews. He is the author of one patent and author, coauthor, or editor of 11 scientific books or journal editions, 12 indexed papers in international journals, 26 chapters in indexed international books, and 144 other publications, workshops, or keynotes in international conferences. Since 1992, he participated or participates in 33 International R&D projects (ESPRIT, RACE, ACTS, TELEMATICS, TEN-IBC, EUREKA, INTERREG, FP5 IST-IPS, FP6 IST, ESA, Marie Curie, AAL, ACP) and 15 Portuguese (FCT, QREN, NITEC, POSC, POCTI, POSI, ICPME, TIT). He obtained five scientific prizes. He is member of ACM SIGGRAPH, Eurographics, ISCA, and IEEE; editorial boards of several journals; several Program Committees of National and International conferences in Computer Graphics, Virtual and Augmented Reality, Speech technologies, Accessibility, and Ambient Assisted Living. He is past President of ADETTI, an ISCTE-IUL R&D research center. He was past Vice-president of the Portuguese Group of

Computer Graphics, Eurographics Portuguese Chapter. Born on December 1, 1961, he is the happy father of three children, two girls and one boy.

https://www.researchgate.net/profile/Miguel_Dias2/
http://orcid.org/0000-0002-1445-2695
https://scholar.google.com/citations?hl=pt-PT&user=48Zq8ecAAAAJ

David Pereira Civil and Sanitary Engineer, Ph.D., Born in 1956, with a professional career in very practical technology application and bestseller of Portuguese innovative technical software, since 1977. He had a parallel career in both the academic and the industry sides, being partner of a consulting engineering conglomerate and an environmental sustainability company. Core competencies include water, wastewater, and solid waste systems modeling (simulation and optimization); technology application in real life (water/wastewater transport and treatment, solid waste management); and technical and economic planning of those systems.

Bror Salmelin is the adviser for Innovation Systems at the European Commission, Directorate General for Communications, Network, Content, and Technology (DG CONNECT) where he is responsible for Open Innovation and Modern Innovation systems.

He is currently managing the activities of the Open Innovation Strategy and Policy Group (OISPG), an industry-led group which advises on strategic priorities for open and service innovation.

As a head of unit, he developed the concept of European Network of Living Labs, which is grown through EU presidencies to a 150+ sites innovation network for ICT intense services. Previously, he held the position of Deputy of the ICT Section in the Technology Development Centre and served as the Finnish representative at the ESPRIT/IST program of the EU.

Bror Salmelin is a member of the New Club of Paris and the Advisory Board for the Innovation Value Institute in Ireland. He has expertise in intangible economy and value creation, related to policies like innovation policy, productivity, and creativity in particular focused on new service innovation.

Editorial Introduction

Albertina Dias, Bror Salmelin, David Pereira, and Miguel Sales Dias

Abstract This book is about Modeling Innovation Policies focusing on a wide variety of topics, aiming to encourage interdisciplinary and comparative approaches, while bringing together researchers and public institutions interested in Sustainable Economics and Technological Development issues—including all stakeholders and all the important factors that influence Open Innovation development, diffusion and its practical use. Modeling Innovation Sustainability and Technologies (MIST) has a strong focus on new research and innovation methodologies, following the Experimentation and Application Research (EAR) approach as well as Open Innovation 2.0 principles. This book aims to summarize new ideas and to provide insights into new approaches for research and innovation, planting the seed for new collaborative projects across boundaries and with practical industrial applications, in a context of sustainable economics and technologies.

A. Dias (✉)
GITICE-Reseach Group of Information Technology and Communication in Business, Universidad de Huelva, Huelva, Spain

New Atlantic University, R. Nuno de Bragança, 6, 1°B, 2740-282 Porto Salvo, Portugal
e-mail: tina.melo.dias@gmail.com

B. Salmelin
Innovation Systems, DG Communications Networks, Contents and Technology, European Commission, Avenue de Beaulieu 25, 1160 Brussels, Belgium
e-mail: bror.salmelin@ec.europa.eu

D. Pereira
FCT-UNL- Faculdade de Ciências e Tecnologia da Universidade Nova de Lisboa, Caparica, Portugal

MARE-Centro de Ciência do Mar e do Ambiente, Rua João de Freitas Branco, 32, 4° Direito, 1500-359 Lisboa, Portugal
e-mail: djp@fct.unl.pt

M.S. Dias
ISCTE-IUL - Instituto Universitário de Lisboa, Lisboa, Portugal

Digital Living Spaces Lab, ISTAR-IUL/ISCTE-IUL – Information Sciences, Technologies and Architecture Research Center, Av.ª das Forças Armadas, 1649-026 Lisboa, Portugal
e-mail: miguel.dias@iscte.pt

A. Dias et al. (eds.), *Modeling Innovation Sustainability and Technologies*, Springer Proceedings in Business and Economics,
https://doi.org/10.1007/978-3-319-67101-7_1

Keywords Systems' modeling • Innovation • Economic and environment sustainability • Real systems' management and technological application • International investment law • Human rights law • Agricultural foreign direct investments • Developing countries • Land grabbing • Food security

This book is about Modeling Innovation Policies focusing on a wide variety of topics, aiming to encourage interdisciplinary and comparative approaches, while bringing together researchers and public institutions interested in Sustainable Economics and Technological Development issues—including all stakeholders and all the important factors that influence Open Innovation development, diffusion and its practical use.

Modeling Innovation Sustainability and Technologies (MIST) has a strong focus on new research and innovation methodologies, following the Experimentation and Application Research (EAR) approach as well as Open Innovation 2.0 principles.

This book aims to summarize new ideas and to provide insights into new approaches for research and innovation, planting the seed for new collaborative projects across boundaries and with practical industrial applications, in a context of sustainable economics and enabling information and communication technologies. The purpose is to give insight to new research and innovation approaches within cross-disciplinary interaction providing a possible shift in the formation of policy-making. By not following solely a traditional academic approach, we are specially determined to create the time and space for interaction and creativity for the future actions and thus, not so much focused on reporting from the past.

The book is structured in five parts:

Part I Innovation Economics and Policy Perspectives This part comprises three chapters addressing Open Innovation 2.0 conceptual innovation model, the dynamic interdependencies across territories and the influence on the innovative performance of regions.

Part II Innovation Management and Policy Perspectives: Case Studies in the Central European Countries This part focuses on a set of real case studies aiming to offer new insights and practical lessons on the issues of innovation management, both in the Central European countries, and in the context of the EU global internal market.

Part III Modeling Innovation and Sustainability of Future Tourism Destinations This part explores the multidimensional nature of tourism phenomena and discusses the role of tourism in spatial development and the innovation processes in places that do not belong among well-known tourist destinations, based on case studies from four different countries, namely, Italy, Poland, Slovakia and Russia.

Part IV Scalable and Sustainable Information and Communication Technologies This part of the book addresses some challenges associated with the implementation of ICTs and highlights the importance of the adoption of such technologies by consumers, corporations and business, with a focus on long term adoption.

Part V Environmental Sustainability Innovation Management: Case Studies in the Southern European Countries These selected chapters focus on a critical tradeoff between environmental and energy sustainability and the requirements for large amounts of capital needed for such investments. Other important discussions are made when analyzing the spatial, demographic and socioeconomic characteristics of concrete regions and its common fiscal and regulatory framework for the covered southern European countries.

Part I
Innovation Economics and Policy Perspectives

In essence, innovation is the ability to generate and manage knowledge creatively in response to market-articulated demands and other social needs. Enterprises are the main source of innovation. Their performance depends on incentives provided by the economic and regulatory environment, their access to critical inputs (via factor markets or through interactions in networks and clusters of knowledge-based organizations) and their internal capacity to seize market and technological opportunities.

Several trends combine to change the conditions for successful innovation: innovation increasingly relies on effective interaction between the science base and the business sector; more competitive markets and the accelerating pace of scientific and technological change force firms to innovate more rapidly; networking and collaboration among firms are now more important than in the past and increasingly involve knowledge-intensive services; small and medium-sized enterprises, especially new technology-based firms, have a more important role in the development and diffusions of new technologies; the globalization of economies is making countries' innovation systems more important. In sum, innovation performance depends not only on how specific actors (e.g. enterprises, research institutions, universities) perform, but also on how they interact with one another as elements of an innovation system, at local, regional, national and international levels.

In recent years, the systemic dimension of the generation and distribution of technological knowledge and innovation at large is placed at the center of the economics of innovation. Conceptual and empirical research underlines that interactions and learning efforts are needed in order to strengthen the generation and accumulation of knowledge, the emergence of innovation systems, and also the growth performances of firms and economies. In particular, the conditions of knowledge interactions emerge as a central feature characterizing the systemic dynamics of innovation and knowledge.

Living Labs and Open Innovation in European Context

Albertina Dias and Bror Salmelin

Abstract The article elaborates the background thinking and path for Open Innovation 2.0 conceptual innovation model. It is based on virtual enterprises, Holonic enterprises and fractal enterprises theory, combined with MIT Living Lab concept developed by Bill Mitchell (Me++: the cyborg self and the networked city. MIT Press; 2003). Combining this with the internet/connectivity revolution the need to have faster pace and more successful innovation rate led to the thinking of the quadruple helix, including the citizens as active agents in the innovation process, not only as verificators as they were used to be in the previous triple helix thinking.

Based on the work of New Club of Paris (Lin and Edvinsson. National intellectual capital: a comparison of 40 countries. Springer; 2011) the structural intellectual capital (IC) is a key for national prosperity. Open innovation integrating the crowd into the innovation process seamlessly seems to increase the structural IC. Hence, integrating all these components: quadruple helix, non-linear innovation, fractal and dynamic organizations into innovation processes in real world with real market creation with the users who become co-creators seem to be the key for future success.

The new open innovation 2.0 paradigm seems to be serving the innovation needs very well in time—if we dare to take it on board.

Keywords Innovation ecosystems • Open innovation • Living labs • Innovation models • Policy modelling • Quadruple helix • Complex systems

---

A. Dias (✉)
European Commission, Brussels, Belgium
e-mail: tina.melo.dias@gmail.com

B. Salmelin
Innovation Systems, European Commission, DG Communications Networks, Contents and Technology, Brussels, Belgium
e-mail: bror.salmelin@ec.europa.eu

A. Dias et al. (eds.), *Modeling Innovation Sustainability and Technologies*,
Springer Proceedings in Business and Economics,
https://doi.org/10.1007/978-3-319-67101-7_2

1 Introduction

European Union has set innovation as high priority as part of the Europe 2020 strategy. Europe is focusing on jobs and growth through innovation. Innovation Union is one of the key flagships to target this ambitious goal for Europe to become a leading region in the world of modern innovation.

By focusing on both quantitative (3% of Gross Domestic Product) and qualitative goals in innovation policy this has led to a good mix of instruments supporting modern innovation systems.

In the Horizon 2020 (H2020) framework research and innovation are seamlessly integrated, and entirely new instruments for funding are created. In the text we will describe those in the context of European Innovation Ecosystem thinking, linking that to the experiences we already have from Living Labs and Open Innovation, since 10 years. This article also describes the background thinking and the developed Open Innovation 2.0 perspective on modern innovation Systems.

2 Living Labs in European Context

The origin of Living Labs thinking was in Massachusetts Institute of Technology (MIT) where the approach was to construct test and verification environments in laboratory settings to develop and experiment different technology solutions with real users invited to visit those environments. This led to early prototyping with real users again with the probability to have faster scale-up of the results. Bill Mitchell was one of the key drivers in this new research and prototyping approaches. Since then, the concept has been widely spread among different concepts within new projects and programs in different parts of the world. Thus, there is no unique definition for the Living Labs' concept; each concrete definition approach is defining its major and specific aspects related to the objectives of a program or project. It is not the purpose of this article to carry out a review of the state of the art of the concept and definition of Living Lab; still we recall some of the latest or most generally known definitions:

- according to Niitamo, V. P., Kulkki, S., Eriksson, M. & Hribernik, K. A. (2006, pp. 26-28) "Living Labs are an emerging Public Private Partnership (PPP) concept in which firms, public authorities and citizens work together to create, prototype, validate and test new services, businesses, markets and technologies in real-life contexts, such as cities, city regions, rural areas and collaborative virtual networks between public and private players";
- Based on the components and principles of a Living Lab, a Living Lab is a citizen-centric innovation milieu built on every-day practice and research, with an approach that facilitates user influence in an open innovation environment engaging all relevant stakeholders—business, academia partnership, citizens and government—in real-life contexts, aiming to create sustainable values (Bergvall-Kåreborn and Ståhlbröst 2009, pp. 356–358);

- Yet the opinion for the contextual definition and purpose of a Living Lab within the South African context (rural area) is foreseen as "a real-time experimental environment that enables different role players with some or other common interest within a domain to collaborate in the use and development of innovative ideas to solve current and real world problems in a unique and integrated way" (van der Walt, J. S., Buitendag, A. A. K., Zaaiman, J. J. & van Vuuren, J. C. J., 2009, p. 430)
- more recently Lucassen, I., Klievink, A. J., & Tavasszy, L. A. (2014, p. 5) suggest that a Living Lab consists of a "Test environment for cyclical development and evaluation of complex, innovative concepts and technology, as part of a real-world, operational system, in which multiple stakeholders with different background and interest work together towards a common goal, as part of medium to long-term study".

When discussing such variety of approaching Living Lab from European perspective it soon became evident that from innovation system perspective end-user involvement could be THE key factor for renewing European Innovation System. We have the most demanding but also very diverse user communities for our products and services; the question stands for how to harness that to increase success rate and speed of the innovation processes in Europe.

The work of Niitamo et al. (2006, pp. 26-28) cannot be enough appreciated when developing the strategy but also practicing it in large and small scale, and again also in practice. This vision still recalls to the linear innovation policy understanding instead of more recent debate on Europe's understanding and practice of an holistic view of the innovation policy (Edquist 2014).

At the same time "Democratization of Innovation" driven by Von Hippel (2005) triggered the thinking of co-creation and user involvement in the innovation processes.

The industrially led think-tank for Living Labs strategy in Europe was established in liaison with European Commission, DG Information Society and Media Directorate (DG INFSO, currently DG CONNECT) in 2003 to conceptualize the European approach. Further this Living Labs think-tank focused on Open Innovation becoming the Open Innovation Strategy and Policy Group (OISPG).

Soon it became evident that the European approach should be focusing on creation of innovation hubs which would build on the quadruple helix innovation model, i.e. strong and seamless interaction of the industry, public sector, research institutions and universities, and finally also the "people".

The target was to create attractive environments, which would be attractive for industrial and research investment due to better innovation dynamics. This dynamics would be supported by the public sector and one of the focus areas would be public sector service, which could be co-developed with the user communities, in real world settings. Part of this thinking was based on the idea to stretch the boundaries of societal behavior as well, as we saw the connectivity and ICT shared environments (with emerging social media) to change the society as well. The quest was to push the boundaries with real world projects including strong technological development too. Only by doing the research and development with citizens we

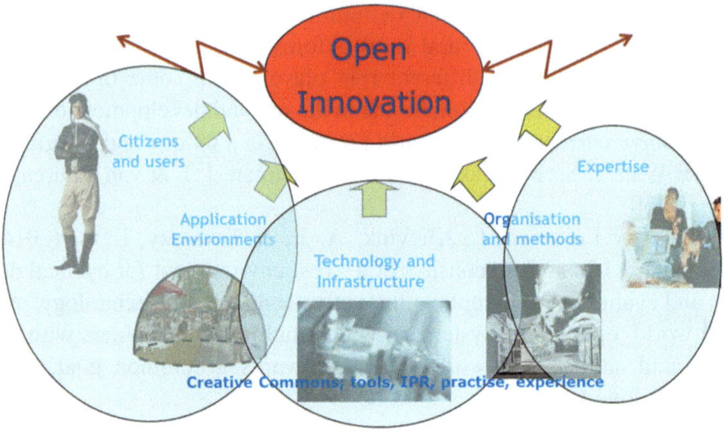

Fig. 1 Living Labs as European concept

could see what finally would be acceptable and thus scalable to products and services.

This led to the first concept of Living Lab in European context; a real world site, not an extension of laboratory. Important was also the scale as it was seen that for scalability we needed the "sample users" to be large enough, at least in hundreds.

In Fig. 1 we have all the components needed for European Living Labs: Citizens, application environments, technology infrastructure, organizations and experts. Important to see is the later addition of societal capital into the picture as functioning Living Labs build strongly on the idea of spill-over effects back to the society, giving motivation for all of the stakeholders, including citizens to contribute to the common goal, making Living Lab a winning game.

Based on these conceptual thoughts European Commission and the Finnish EU presidency launched in 2006 the first wave of European Living Labs which built a network—European Network of Living Labs—which became later the ENoLL movement. From the first wave the network grew fast under the following EU presidencies to the substantial scale it has now, 340 sites even beyond European borders (ENoLL 2015). And, the network is still growing. What we can say that the Living Labs have now a strong foothold in all European regions, and is being applied as important component in regional innovation systems too.

On European level the networking of Living Labs is of utmost importance. Using Living Labs methodology to find common, scalable solutions with different user environments is essential when driving to common European services based on common architectural approaches. I am happy to see that the thematic cross border networking of the sites is speeding up, enabling the most interesting Living Labs to collaborate as partners e.g. in the H2020 projects, especially in smart city or public services context.

3 Open Innovation as Part of Living Labs' Thinking

As starting point when developing Living Labs in the European way was openness; in sharing platforms for services but also open mind set for collaboration amongst all stakeholders. The thinking stems from the early 1990s when the hot topic was virtual and holonic enterprises, which were as group creating both agile and scalable structures for operations; by sharing common operating architectures and by collaborating strongly on task-driven basis (Leitão 2004). Good examples of holonic/fractal/virtual enterprise theory was developed e.g. in the IMS (Intelligent Manufacturing Systems) initiative among the leading industrial economies in the 1990s (Tharumarajah, Wells & Nemes, 1996). Scaling up this thinking we come very close to the foundations of Living Labs by adding the public and societal components to it.

Combining the approach by von Hippel about the user-driven and co-creativity in innovation processes with the approach Chesbrough introduced in 2003—open innovation—we come to the two fundamental of modern innovation theory. The definition of open innovation by von Hippel focuses on creation of public goods whilst the one by Chesbrough builds on sharing, cross licensing and in that way being a market and product driven approach.

Open platforms, sharing and seamless interaction of all stakeholders is essential in Living Labs. Quadruple helix has thus been central as innovation model from the very first beginning onwards.

Open Innovation Ecosystems are increasingly becoming the synthesis of Living Labs and open innovation processes. We see real new paradigm evolving when combining these. Open innovation has become much more than cross fertilization of ideas between organizations, it has become a flow of colliding ideas, raising sparks for new innovations in real world settings.

4 Open Innovation 2.0 and Ecosystems

Following the research of Lin and Edvinsson (2011) there are clear indications that intellectual capital, and especially structural intellectual capital drives competitiveness and innovation. This means in turn that from innovation policy perspective the interaction fluidity is a critical feature of any successful innovation system.

Fluidity in this context means frictionless interaction, experimentation in real world, and a lot of unexpected, non planned collisions of ideas, problems and of course competencies to collide, giving the spark. It is not only about single excellent components in the system, it is centrally about collisions and connectivity.

It was already shown in 2004 that the diversity of research teams increases significantly the probability of breakthroughs, and actually we can also say that mediocre inventions are not enough. We need to combine the best. Cross-fertilization of ideas is nothing new as such, but what ecosystem thinking does is embedding diversity and serendipity in the innovation process more systematically than ever before.

It is important to move from clusters to ecosystems in our innovation system design. It's nothing wrong with clusters, but they tend to be rather monolithic focusing on one sector only. Of course the clusters reinforce the sector they work in, but the tendency more towards improving, extrapolating than to create something new. Hence the emphasis on modern innovation systems need to be increasingly on the "in-between" areas where creation of new is likely, and as consequence also the fast growth.

To substantiate the potential for new market creation the end-users need to turn to be active drivers together with the other stakeholder in jointly creating the new. Quadruple helix innovation model gives clear roles to all stakeholders, including the users as active agents from the first beginning. Earlier the users were objects in the process, not co-creators. By taking the users actively on board we see immediately which solutions can be scaled up and which will fail due to various reasons. Scaling up fast the emerging successes is key to maintain the dynamics in the innovation system. There are also indications that those organizations which cut failing projects at earlier stage will be more successful in the longer run.

Again, we come to the ecosystem when we think about where the experimentation and early prototyping is to be done. In real world settings one can at early stage see the potential and also identify the paths for fast enter into the full scale markets. Seamless user involvement is thus essential. It is important also to understand that properly designed innovation ecosystems provide a safety net from the ideation to the market. Failing fast means also often failing small, and experimentation and early prototyping in turn means faster results to be brought to the market, even incrementally.

Business model experimentation in these open innovation ecosystems is also essential. Due to the dynamics in the economy and technology it rarely is possible to write the old fashioned extensive business plans. Often it is enough to have a business model idea and develop it continuously further in the real world settings, to finally see what works and what not. Fast adjustment and experimentation is the way forward.

Here legislation can play also a remarkable role if it is a catalyzing one. Restrictive legislation again is a strong hinder for business model innovation. Proper legal framework is one of the important factors for the fluidity of the innovation space we spoke about earlier.

Innovation has moved from linear processes to mash-up processes where diversity, speed and experimentation are the fundamentals.

We have moved from closed innovation to open innovation and further towards open innovation 2.0 which highlights the interaction, fluidity and mash-up nature of innovation processes, including all stakeholders in quadruple helix innovation (Fig. 2).

Open Innovation ecosystems can be regional or thematic, or both. They are built on strong interaction between the competencies illustrated in the picture by different coloured dots. The ecosystem itself has tens or more projects (funnels in

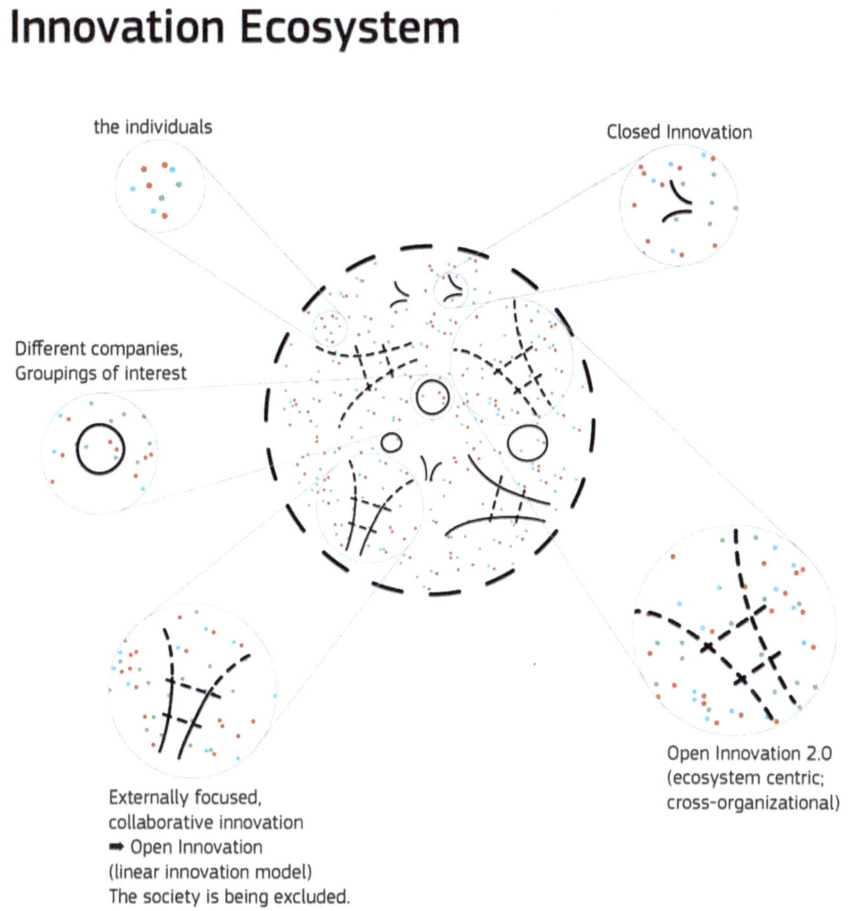

Fig. 2 Open Innovation 2.0 view on open innovation ecosystems (OIE)

Chesbrough sense) which can be more or less leaking to broaden the competence base of each action. The funnels in this context represent development projects, not organizational boundaries. Spill-over effects to the whole ecosystem by the projects should increase the societal but also knowledge structural capital enabling continuous rise of the value proposition of the new activities.

Sharing infrastructures but also experience and knowledge is a key of the trust to be built within the ecosystem itself. The trust is of very high importance because of the interdependence of all stakeholders in this mutual win-win process. Cross-fertilization and sharing does not happen without trust.

In ecosystems it is, as previously stated, very important to allow collisions to spark the real innovations, even disruptive ones. Hence the creation process requires courage to design a governance structure for the ecosystem to let it grow organically. Prototyping and experimentation of policies is one of the important components in this development too.

5 Open Innovation 2.0 in 20 Snapshots

In the white paper from 2013 manifesting the Open Innovation 2.0 paradigm Martin Curley and Bror Salmelin highlight 20 key elements as the transformative factors for the modern innovation approach.

The OI2 approach emphasizes the importance of Quadruple Helix innovation where the private, public and research institutions collaborate seamlessly and in which from the very beginning the user(s) communities co-create the new products and services. This leads to win-win approaches as the users get products and services they need, and the suppliers get scalable products and services. If this co-creativity and prototyping in real world settings would not take place there would be a real risk that the development work would lead to a win-lose setting between the existing players in the market, and no new markets would be created either.

Cross-disciplinary innovation together with prototyping and experimentation is bringing forward the required dynamics. Failing fast and getting directions to potentially successful solutions at early stage is essential. Traditional piloting or test bed approaches are not sufficiently scalable to verify the market potential of the inventions.

In this palette of 20 drivers for Open Innovation 2.0 (Fig. 3) one needs to highlight both societal and technological innovation which enable business model (more generally value creation model) innovations. The area of business model innovation together with the new markets emergence is clearly dimensions/realms in which we Europeans can do/perform much better.

How to achieve the fluidity and frictionless environments for multi-stakeholder trials, including legal and policy elements is the key to root in the European mindset. We need to speak about openness for innovation, innovation 2.0 culture, to complement the view.

New types of leadership, new processes and new approach to ecosystems—paradigm change is real.

The paradigm has changed. Table 1 illustrates some aspects to concretize this change and illustrate its drivers. Of course these factors are interrelated in complex systemic manner and lead to the need of looking at successful innovation ecosystems and innovation processes together.

Closed innovation reflects the traditional linear paradigm, often based on brilliant individuals or performing industrial labs. Open Innovation, as introduced by Henry Chesbrough, is a move towards collaborative innovation structures, where those ideas not used by oneself can be seen as tradable assets to those who might have need for specific technologies.

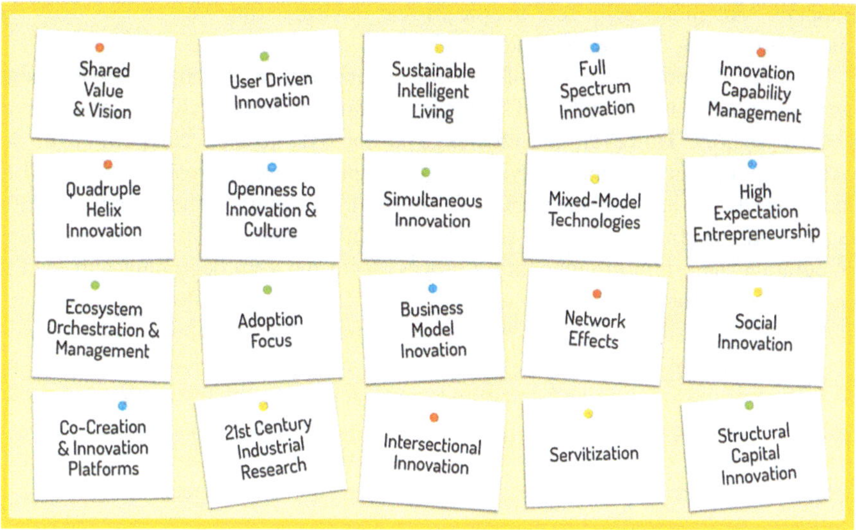

Fig. 3 20 drivers for Open Innovation 2.0

Table 1 The change and drivers of the innovation paradigm

Closed innovation	Open innovation	Open innovation 2.0
Dependency	Independency	Interdependency
Subcontracting	Cross-licensing	Cross-fertilization
Solo	Cluster	Ecosystem
Linear	Linear, leaking	Mash-up
Linear subcontracts	Triple helix	Quadruple helix
Planning	Validation, pilots	Experimentation
Control	Management	Orchestration
Win-lose game	Win-win game	Win more-Win more
Box thinking	Out of the box	No boxes!
Single entity	Single discipline	Interdisciplinary
Value chain	Value network	Value constellation

When we began to analyse the innovation processes and the success closer, we realized that one of the critical elements is the scalability of the work, which naturally results in increased success rate. But how to achieve this?

We need to break out from the traditional linear models; we need to dare to do more experimentation in real world settings as only then we learn very fast what is scalable, successful, as opposed to what is simply not worth going forward with. Traditionally we see pilots and validation in many projects, but. . .often they come too late to have any influence of the project work itself. This triple helix approach which excludes end-users from the actual innovation process is by far too slow. Only by moving to the quadruple helix model where the innovation process happens "out there" with real people in real environments we can speed up the successful results and kill the bad ones in time.

Another dominant element of the open innovation traditional cross-licensing process is the cluster thinking. Cluster operations reinforce well the competitiveness of sectors. However, the challenge is not only to stay competitive in the existing field, but also to find entirely new areas for value creation. We need to have interdisciplinary manner actions between the clusters in the open innovation ecosystems to strengthen cross-fertilization. And, taking the users on board and integrating them into the innovation process from the very beginning will lead to the creation of new markets. If we target only traditional clusters and traditional industries we easily end up with a win-lose game.

Organizational changes and collaboration changes are also clearly moving towards this mash-up, mixed disciplines approach. Value chains with subcontractors highlight the linearity in innovation processes together with control approach which is typical for the manufacturing and traditional industry era. When products integrate into services and get more complex, we have seen networking between suppliers to be established, e.g. in the automotive sector, where independent component manufacturers deliver to many brands simultaneously, based on their special competencies. In open innovation 2.0 we go even further into dynamic value constellations where the links are not a priori determined, but more task driven. Competencies and resources are combined based on the tasks, not as earlier when the services were determined by organizational structures. In turn, this also means that the end users will be much more dominant in the innovation process for modern products and services, especially on their functional level.

The innovation process change affects also radically the management styles of successful companies. We have plenty of examples where an authoritarian control-type of management is replaced by strong leadership. However, we need to go into even further metaphors when we move to open innovation 2.0. The successful leadership will be mentoring, catalyzing, inspiring: it will be orchestration of fluid resources to perform their best. And, what makes all interesting is that the orchestration conducts not only the known players, but also the audience to create fantastic joint experiences with the interaction internally AND externally; Like in a successful concert where the ambience and success is all about the interaction and not just the play, even professionally.

6 Conclusion

Open Innovation 2.0 is a new mind-set; it is openness for innovation. It is the courage to experiment and prototype. It is the courage to fail and scale. And, as a consequence, it builds up a growing spiral of performance built on success and motivation.

Living Labs, networked society, democratizing innovation, open innovation, disruptive innovation. . .many words, which are fluently used without often thinking about the reality behind. The reality is however in the courage to change the behavior, including the governance structures to create something new. The reality is also to turn these buzzwords into a functioning innovation ecosystem with new dynamics.

ICT provides connectivity and the shared space of knowledge, meaning that the new paradigm of open innovation ecosystem is more doable than ever before. In the rich connectivity we need to see the new role of all players in the spirit of quadruple helix innovation, and move due to the dynamics needed to an experimentation and prototyping culture. This shows the options for success earlier and significantly reduces the risk for big failures too.

The fundamentals are developing in Europe. Our challenge is to make these fundamentals to work together, to fully use the potential we have as the single biggest market in the world.

The paradigm has changed.

References

Bergvall-Kåreborn B, Ståhlbröst A (2009) Living Lab: an open and citizen-centric approach for Innovation. Int J Innov Reg Dev 1(4):356–358

Chesbrough HW (2003) Open innovation: the new imperative for creating and profiting from technology. Harvard Business Press, Boston, MA

Edquist C (2014) Striving towards a holistic innovation policy in European countries – but linearity still prevails! STI Policy Rev 5(2), Lund University, Lund. https://charlesedquist.files.wordpress.com/2014/11/step-2014-v5-no2_01.pdf

ENoLL – European Network of Living Labs (2015) http://www.openlivinglabs.eu/. Accessed 26 Aug 2015

Leitão Paulo JP (2004) An agile and adaptive holonic architecture for manufacturing control. PhD Thesis, University of Porto

Lin C, Edvinsson L (2011) National intellectual capital: a comparison of 40 countries. Springer, New York

Lucassen I, Klievink AJ, Tavasszy LA (2014) A living lab framework: facilitating the adoption of innovations in international information infrastructures. In: Proceedings of transport research arena 2014, Paris, France, p 5

Mitchell WJ (2003) Me++: the cyborg self and the networked city. MIT Press, Cambridge, MA

Niitamo VP, Kulkki S, Eriksson M, Hribernik KA (2006) State-of-the-art and good practice in the field of living labs. In: Proceedings of the 12th international conference on concurrent enterprising: innovative products and services through collaborative networks (Italy: Milan), pp 26–28

Tharumarajah A, Wells AJ, Nemes L (1996) Comparison of emerging manufacturing concepts. CSIRO manufacturing science & technology, Preston, Victoria, Australia. Int J Comput Integr Manuf 9(3):217–226. www.fractal.org/Fractal-Research-and-Products/Fractalfactory.pdf

van der Walt JS, Buitendag AAK, Zaaiman JJ, van Vuuren JCJ (2009) Community living lab as a collaborative innovation environment. Issues Inform Sci Inf Technol 6:430

Von Hippel E (2005) Democratizing innovation. MIT Press, Boston

Albertina Dias Among several academic positions, in different universities in Portugal and Spain, Albertina Dias qualified as a Coordinating-Professor of Economics and Industrial Management (2009–2012). Between 2014 and 2015 Albertina Dias was the Director of the Department of Business Science at New Atlantic University in Portugal. In 2015 she founded a new association for Modeling Innovation Sustainability and Technologies (MIST). She is the MIST Association Chairman and Business Scientific Director as well (www.mist.pt). With more than 15 years of work experience, in the areas of Economics, Finance, Innovation Strategic Management and Information and Communication Technologies (ICT), Albertina has been working actively in the knowledge triangle having run her own company on innovative ICT between 2005 and 2013 (SbH, Ltd., developing multimedia interactive books, based on augmented reality technology and computer aided eBook interaction systems). In the past she has assumed several degrees of responsibilities in companies such as Paulo Branco Holding (CEO, film industries), Oracle (Financial Analyst), Allianz (Financial Control), and Cap Gemini (Project Manager). Under the (EU funded) Program Equal e-Change she participated since 2005 in the presentation, demonstration and validation of innovative products in the areas of flexible security on employment and audiovisual and multimedia tools. She was responsible for the audiobook library of the Economists' College in PT and has produced several audiobooks of economic content such as the "Portuguese Governmental Budget-2006", "2nd Annual Conference of Economists-2007", "SCUTS-An Economic Analysis, 2005", among others. Since 2008 she is an Independent Expert for the European Commission, acting as a Reviewer/Evaluator for the implementation of indirect actions under specific programs. Since 2005, she is a fellow researcher of GITICE/Universidad de Huelva in Spain (scientific domain of Economics and research line of ICT). In this institution she is developing research on Innovative Tools towards Regional Economic Development. Previous academic background includes three principal scientific degrees (pre-Bologna):

- Ph.D. on Business and Economics (Huelva University-Spain, registered in Nova University, Lisbon-Portugal);
- Sc. Master in Statistics and Information Management (Nova University, Lisbon-Portugal);
- Degree in Economics (Nova University, Lisbon-Portugal).

She is the author of a considerable number of academic theses, articles, papers and books. Her recent research activity involved studies on the Economics of Innovation (international comparative studies), Information and Communication Technologies (ICT) and the development of innovative business models for several environments. Currently, her research interests focus on the political economic models and public policies.

Bror Salmelin is the adviser for Innovation Systems at the European Commission, Directorate General for Communications, Network, Content, and Technology (DG CONNECT) where he is responsible for Open Innovation and Modern Innovation systems.

He is currently managing the activities of the Open Innovation Strategy and Policy Group (OISPG), an industry-led group which advises on strategic priorities for open and service innovation.

As a head of unit he developed the concept of European Network of Living Labs, which is grown through EU presidencies to a 150+ sites innovation network for ICT intense services. Previously, he held the position of Deputy of the ICT Section in the Technology Development Centre and served as the Finnish representative at the ESPRIT/IST program of the EU.

Bror Salmelin is a member of the New Club of Paris and the Advisory Board for the Innovation Value Institute in Ireland. He has an expertise in intangible economy and value creation, related to policies like innovation policy, productivity and creativity in particular focused on new service innovation.

The Consequences of Tax Base Rules on Enterprise Innovation in the European Union

Žaneta Lacová and Ján Huňady

Abstract Traditionally, there is an important role that external conditions such as establishment of tax rules can play in fostering innovation process in companies. When considering Innovation Union in the European Union context, we need to take into consideration the fact that companies meet twenty-eight different tax systems. While the differences concerning the nominal tax rates are obvious, another aspect comprising tax base rules differences is less visible, although they can play a relevant role in stimulating innovation activity. In some countries, the tax base composition is affected by the existence of R&D tax incentives concerning the company's income tax, but the situation differs according to the EU member state.

Our study questions the existence of the link between the above-mentioned aspects of national tax regulation and a country's innovation performance with a special emphasis on the entrepreneurial innovation activity. In accordance with a broader definition of innovation activity, both the R&D expenditure and the non-R&D innovation expenditures in the business sector are taken into account in our analysis. For empirical testing, the Granger causality methodology and panel fixed-effect regression analysis are applied.

Our results find evidence that countries proposing more generous possibilities in the statutory or effective tax rates don't meet more suitable performances in entrepreneurial innovation activity. Similar results can be found in estimating the impact of different tax base rules, approximated by the difference between the statutory and the effective tax rate. Another important aspect of our study concerns testing of correlation between different forms of R&D tax incentives and enterprise innovation activity. Our results indicate a positive relationship between R&D tax incentives and enterprise R&D expenditures while a negative relationship between such incentives and enterprise non-R&D innovation activities can be identified as well. It seems that tax incentives affecting the income tax base composition (enhanced allowances and accelerated depreciation) do not indicate considerably different results from those proposed by the tax incentives affecting the income tax rate (tax credits and patent boxes). In conclusion, the results we have identified are

Ž. Lacová (✉) • J. Huňady
Faculty of Economics, Matej Bel University, Banská Bystrica, Slovak Republic
e-mail: zaneta.lacova@umb.sk; jan.hunady@umb.sk

A. Dias et al. (eds.), *Modeling Innovation Sustainability and Technologies*, Springer Proceedings in Business and Economics,
https://doi.org/10.1007/978-3-319-67101-7_3

interpreted in the context of the European Commission initiative of the rebirth of the Common and Consolidated Corporate Tax Base (CCCTB) proposal, announced in 2016. Thus, the chapter tries to contribute to the renewal debate concerning the consequences of CCCTB from the perspective of business innovation activities.

Keywords Enterprise innovation • European Union • Common and Consolidated Corporate Tax Base • Tax incentives

1 Introduction

As a part of the Europe 2020 strategy, fostering the conditions for innovation plays an important role among priorities of the policy-making in Europe. What concerns the tax rules conditions for companies as a factor influencing enterprise innovation process in Europe, a heterogeneity among the European Union (EU) member states persists. The existence of 28 tax systems means that enterprises need to adapt to a country-specific tax conditions when making all kind of decisions, including the innovation strategy decisions. In such circumstances, the development of two phenomena is specifically not desirable: (1) companies innovation strategies can become limited by tax system borders, and (2) multinational companies are motivated to waste their innovation capacity for tax planning strategies (so-called tax innovation) instead of using it for innovation activities in the areas with the potential of growing productivity and efficiency (e.g. the core-business activities). To avoid the occurrence of such phenomena, the projects in order to some forms of standardisation or harmonisation in this area are highly welcomed (see Uramová et al. 2016). The directive proposal of Common and Consolidated Company Tax Base (CCCTB) of the European Commission (EC) from 2011 was presented as an initiative with two principal objectives: to make the corporate tax framework in Europe to be more simple, and to reduce opportunities for multinational companies to avoid income tax payments. The main idea was that companies operating within the EU would have to comply with only one system for computing its taxable income, rather than different sets of rules in each member state in which they operate. It is important to underline that according to this project, each member state will keep its right to apply its own corporate tax rate. However, this proposal hasn't met a necessary political support within the European Council yet.

As the priority is to harmonize the national corporate income tax base and establish a Common Company Tax Base (CCTB), the current debate of types of common rules which would best foster the innovation activity of companies is necessary. Our paper tries to contribute to this debate by identifying the potential causalities between tax rules and innovation activities in the EU member states (especially the business innovation activities) at three levels. Firstly, we focus on the links between innovation activity and nominal tax rates in order to confirm the relevance of the approach "tax base harmonisation only" for the innovation process in the EU. Afterwards, we directly concentrate on the links between innovation activity and tax base rules in the EU member states. Finally, we look to the potential causalities between country-specific tax incentives driving the innovation process in companies and the enterprise innovation performance in these countries. Our ambition in such testing was to identify the role of these incentives for innovation process in Europe from the perspective of an eventual impact of the CCTB proposal.

This research paper originated in partial fulfilment, and with support of, the project ITMS 26110230082 Mobility—Support of Science, Research and Education at Matej Bel University in Banská Bystrica (Mobility—podpora vedy, výskumu a vzdelávania na UMB) under the Operational Program Education co-financed by the European Social Fund within the bounds of financial subsidy contract No. 018/2012/1.2/OPV.

The paper has been also supported by the Scientific Grant Agency of the Ministry of Education of the Slovak Republic and Slovak Academy of Science (VEGA) under the contract No. 1/1009/16 Innovation potential of the regions of Slovakia, its measurement and innovation policy at the regional level.

2 Literature Review

Mulgan and Albury (2003) define innovation as the successful implementation of a new or significantly improved product, service, marketing strategy or new organization method that will bring substantial improvement to the economy, efficiency or quality of the outputs. According to this definition of innovation activity, firm's investment in innovation comprises the R&D investments as well as the non-R&D investments, and both of these aspects should be taken into consideration. Concerning the R&D innovation only, Zemplinerová and Hromádková (2012) argue that private companies invest in this innovation less than would be socially desirable.

The literature is proposing several external factors which can drive entreprise innovation processes focusing mainly on enterprise R&D activities. In this context, the public financial support as a factor promoting R&D in business sector is usually tested. As reported by Hunady et al. (2014) the public financial support for R&D and innovation is one of the most important factors affecting the firm's innovation activities. The authors also found many other determinants of innovation such as market competition, type of the industry as well as export focus of the firm. Based on the date from OECD countries, Falk (2005) found that there are two important political instruments supporting R&D in firms: special tax treatment for companies that invest in R&D and direct financial support.

What concerns the tax incentives efficiency, the evidence in literature is ambiguous. Based on the data from Canadian firms, Czarnitzki et al. (2011) conclude that R&D tax credit increase the R&D engagement at firm level. Similarly, Cappelen et al. (2012) found that use of tax credit often lead to successful developed of new production processes and products in the case of Norwegian firms. On the other hand, Tassey (2007) stated that R&D tax credits when applied in US, were ineffective. He proposes the changes that should be made to increase its effectiveness. He argues that a flat rate applied to all R&D is the most effective way to promote R&D. In the context of the European Union, Ientile and Mairesse (2009) also conclude that the impact of R&D tax credits on R&D investment is quite heterogeneous, likely sensitive to the country analysed and methodology used. The authors identified that while the R&D tax incentives appear to be efficient in Norway and France, evaluations for Spain and The Netherlands provide less convincing results.

Several characteristics of the existing literature appear to be relevant for our research: (1) the authors concentrate on R&D expenditures and the impact of tax conditions on innovation (e.g. the non-R&D innovation) is missing, (2) most of the studies focus on R&D tax credits and other forms of tax incentives (enhanced allowances and accelerated depreciation) are not analysed, (3) there is no one 'perfect' way how to assess the effectiveness of tax conditions and the results depend on data and methods applied. When choosing methodology of our research, we were trying to reflect this characteristics.

3 Data and Methodology

In order to reach the potential causalities between chosen aspects of the tax system and the innovation activities in the EU member states, diverse data sources were used (Table 1). Most of the data were retrieved from the following EC publications: European Innovation Scoreboards (2007–2009) and Innovation Union Scoreboards (2010, 2011, 2013–2015), Taxation trends in European Union (European Union 2014b), Tax reform in EU Member states 2015 (European Union 2015) and EC Study on R&D Tax Incentives (European Union 2014a).

We obtained panel data for the first six variables. In the case of SII, STR, EATR, TB and Firm's R&D expenditures, we used the data for 28 EU countries in the period of 2007–2014. Thus, we gained 196 observations, but this number has been slightly decreased by the application of first difference in the models. Due to the several missing observations, the number of observation for non-R&D innovation expenditure was lower and included 162 observations.

As our approach took into consideration the potential impact of tax base composition, we needed to choose a quantitative indicator to capture this phenomena. For this purpose, we used the effective corporate tax rates (the third variable) which implicitly contains the effect of the tax base composition as well as the effect of a statutory tax rate level. Furthermore, we also calculated the difference between effective and statutory tax rate (the fourth variable) in order to approximate only the potential effect of tax base (without a rate dimension).

Concerning the tax incentive score (the seventh variable), data for a certain period (year 2014) were available for 26 EU member states. Data for Germany and

Table 1 Description and data sources of variables used in the analysis

Variable's abbreviations	Description	Source
SII	Summary innovation index—the composite indicator published in European Innovation Union scoreboards	European Union (2015)
STR	Top statutory corporate tax rate (%)	European Union (2014b)
EATR	Effective average corporate tax rate (%)	European Union (2014b)
TB (EATR–STR)	A proxy of corporate tax base calculated	Own calculation
Firm's R&D expenditures	All R&D expenditure in business sector (as % of gross domestic product)	European Union (2015) and previous reports
Firm's non-R&D innovation expenditures	Sum of total innovation expenditure in business sector (as % of total turnover)	European Union (2015) and previous reports
Tax incentives score	Average of scores for all existing tax incentives in order to facilitate R&D in a specific country	European Union (2014a)

Estonia are missing due to the fact that these two countries haven't implemented a specific tax incentives to facilitate enterprise R&D activity in their tax systems.

Different types of analysis have been conducted in this dataset in order to test assumed correlations or causal relationships: correlation analysis, panel Granger causality tests and panel fixed-effect regression analysis. In the first two parts of our analysis, we applied the panel data analysis to search for potential dependencies between indicators of innovation performance and tax system specificities (corporate tax rates and corporate tax bases). In this case, we were able to capture the dynamic aspect as well as to test the lagged dependencies between variables. All the variables have been tested for the stationarity with various panel stationarity tests. Most of the tests indicated that all variables appeared to be non-stationary at level, but stationary at their first difference. In accordance with these results, we decided to use differenced data in order to avoid the potential problem of spurious regression, which seemed to be very high.

In the third part of our analysis, the correlation analysis based on the cross-section data were used. As this part of our analysis focused on examination of link between R&D tax incentives and innovation activity in EU countries, we put under the question the assumed correlation between R&D tax incentives (by country and by tax incentive type) and firm's R&D and non-R&D innovation expenditures.

Table 2 Results of Granger causality tests

Hypothesis	Lags = 1 F-stat	Lags = 2 F-stat
Δ Effective average tax rate (EATR) does not Granger Cause Δ Summary innovation index (SII)	2.64	2.25
Δ Summary innovation index (SII) does not Granger Cause Δ Effective average tax rate (EATR)	0.20	0.02
Δ Statutory tax rate (STR) does not Granger Cause Δ SII	4.47**	4.42**
Δ SII does not Granger Cause Δ Statutory tax rate (STR)	0.90	0.71
Δ (EATR-STR) does not Granger Cause Δ Firm's R&D expenditure	0.07	0.83
Δ Firm's R&D expenditure does not Granger Cause Δ (EATR-STR)	0.50	0.76
Δ Statutory tax rate (STR) does not Granger Cause Δ Firm's R&D expenditure	0.09	0.63
Δ Firm's R&D expenditure does not Granger Cause Δ Statutory tax rate (STR)	1.23	0.41
Δ Effective average tax rate (EATR) does not Granger Cause Δ Firm's R&D exp.	0.33	0.20
Δ Firm's R&D exp. does not Granger Cause Δ Effective average tax rate (EATR)	0.11	0.72
Δ EATR does not Granger Cause Δ Firm's non-R&D innovation expenditure	0.05	0.51
Δ Firm's non-R&D innovation expenditure does not Granger Cause Δ EATR	0.0001	0.35
Δ STR does not Granger Cause Δ Firm's non-R&D innovation expenditure	0.01	1.33
Δ Firm's non-R&D innovation expenditure does not Granger Cause Δ STR	0.03	0.06
Δ (EATR-STR) does not Granger Cause Δ Firm's non-R&D innovation exp.	0.10	0.37
Δ Firm's non-R&D innovation exp. does not Granger Cause Δ (EATR-STR)	0.17	0.85

** represents statistically significant results at 5% level of significance

4 Results

In our analysis structure, three different approaches could be identified. Firstly, we were trying to focus on the relationships between the enterprise innovation activity (both R&D and non-R&D expenditures) on one side and the corporate statutory tax rates on the other side. To identify an eventual existence of innovation transfer between companies and other groups of economic subjects (like the spillovers effects of large companies), we proceeded to enlarge our analysis by taking the Summary innovation index into account. Secondly, we were trying to test the potential causality between the existing tax base rules (represented both by the effective tax rate and by the numerical difference between the statutory and effective tax rate) and the innovation activity (enterprise R&D expenditures, enterprise non-R&D expenditures and overall SII index) in the EU Member states.

Finally, the links between chosen features of tax incentives and enterprise innovation activity as well as between tax incentives' ranking and the innovation activity were tested.

As a first step of our analysis, we tested the Granger causalities between selected pairs of variables. The results of the tests are summarized in Table 2. In vast majority of cases, no significant Granger causalities between the observed variables can be identified. However, it seems that there is a significant Granger causality arising from statutory corporate tax rates to summary innovation index. This could represent a kind of causality in Granger sense between the level of corporate tax rates and the innovation performance of the whole economy. Surprisingly, no analogical significant evidence for statutory tax rates and enterprise innovation activity represented by R&D and non-R&D expenditures can be identified. Although it seems that level of corporate tax rate can have a positive impact on innovation activity in a specific country, there is no evidence that this impact passes through the innovation activity of the all companies sector.

What concerns the effective corporate tax rates and the difference between statutory and effective tax rate (approximations of tax base), the Granger causality between these variables and firm's R&D and non-R&D innovation expenditures appear to be insignificant. So it seems that different rules of tax base composition in the EU member states don't influence the innovation activity in these countries, at least for the analysed period.

As a next step of our analysis, we decided to explore potential causalities using simple panel fixed-effects regression models. To keep it simple, each model

Table 3 Results of panel regressions with Summary innovation index as dependent variable

Dependent variable: Δ Summary innovation index (SII)						
Regression no.	Independent variable	Coef. (t-stat)	Fixed effects	No. of observations	R-squared	Akaike criterion
1.	Δ EATR	0.0005 (1.06)	Cross-section	168	0.0876	−5.463
2.	Δ EATR	0.0005 (1.148)	Period	168	0.0607	−5.696
3.	Δ EATR (lag = 1)	−0.0011* (−1,723)	Cross-section	140	0.1389	−5.363
4.	Δ EATR (lag = 1)	−0.0014* (−1.879)	Period	140	0.0713	−5.616
5.	Δ STR	0.0003 (0.672)	Cross-section	168	0.0858	−5.462
6.	Δ STR	0.0002 (0.318)	Period	168	0.0573	−5.693
7.	Δ STR (lag = 1)	−0.0013* (−1.966)	Cross-section	140	0.143	−5.368
8.	Δ STR (lag = 1)	−0.002*** (−2.783)	Period	140	0.085	−5.631

Symbols */**/*** denotes statistically significant at the 10/5/1 percent level

Table 4 Result of panel regressions with firm's R&D expenditures as dependent variable

Dependent variable: Δ Firm's R&D expenditures						
Regression no.	Independent variable	Coef. (t-stat)	Fixed effects	No. of observations	R-squared	Akaike criterion
1.	Δ EATR	−0.0024 (−0.768)	Cross-section	168	0.223	−1.302
2.	Δ EATR	−0.0041 (−0.679)	Period	168	0.018	−1.330
3.	Δ EATR (lag = 1)	−0.0026 (−0.885)	Cross-section	140	0.225	−1.117
4.	Δ EATR (lag = 1)	−0.0049 (−0.976)	Period	140	0.021	−1.272
5.	Δ STR	0.0004 (0.142)	Cross-section	168	0.222	−1.301
6.	Δ STR	−0.0012 (−0.401)	Period	168	0.016	−1.328
7.	Δ STR (lag = 1)	0.0006 (0.161)	Cross-section	140	0.224	−1.176
8.	Δ STR (lag = 1)	−0.0023 (−0.437)	Period	140	0.018	−1.269

contained one dependent and one independent variable. All variables have been used at their first differences, thus the number of observation have been redacted by one period for each country. Furthermore, the White diagonal robust standard errors have been applied in all the models. We alternated the cross-section and period fixed-effects in each model. The outcomes of the first models are shown in Table 3. In this case, the Summary innovation index is used as a dependent variable.

In most cases, the outcomes of regression analysis are in line with the results of Granger causality tests. On one hand, there is no evident relationship between tax rates and innovation index, when using the variables from the same period. However, the negative effect of tax rates becomes significant at 10% level, once we lag the tax rates variables by one period. Moreover, the impact of statutory tax rates in period fixed-effect model seems to be significant even at 1% level of significance. Hence, there is some evidence that higher nominal corporate tax rates can have a negative effect on overall innovation performance of the country.

Furthermore, we continued in proceeding analogical regression analysis, but with the Firm's R&D expenditure and non-R&D innovation expenditure as a dependent variable. The outcomes of the models are summarized in the Tables 4 and 5, respectively.

Based on the results, we can say that there is no significant relationship between the firm's R&D expenditures and effective or statutory tax rates. The same is true for the firm's non-R&D innovation expenditure (Table 5). While performing 16 fixed-effect regressions with different specifications, we failed to find any statistically significant causality.

Table 5 Result of panel regressions with firm's non-R&D innovation expenditures as dependent variable

Dependent variable : Δ Firm's non-R&D innovation expenditures						
Regression no.	Independent variable	Coef. (t-stat)	Fixed effects	No. of observation	R-squared	Akaike criterion
1.	Δ EATR	−0.0042 (−0.951)	Cross-section	135	0.163	0.339
2.	Δ EATR	−0.0079 (−0.597)	Period	135	0.104	0.082
3.	Δ EATR (lag = 1)	−0.0017 (−0.313)	Cross-section	108	0.237	0.506
4.	Δ EATR (lag = 1)	−0.0001 (−0.006)	Period	108	0.102	0.244
5.	Δ STR	−0.0058 (−0.738)	Cross-section	135	0.163	0.340
6.	Δ STR	−0.0081 (−1.082)	Period	135	0.103	0.083
7.	Δ STR (lag = 1)	0.0149 (1.277)	Cross-section	108	0.240	0.503
8.	Δ STR (lag = 1)	−0.0020 (−0.130)	Period	108	0.102	0.244

To sum it up, we can say that probably, there is an impact of corporate tax rate on innovation performance of the country as whole. However, this effect is delayed by at least 1 year. On the other hand, any comparable causality was not found in the case of firm's R&D expenditures and non-R&D innovation activities.

In the context of innovation fostering, the existence of various tax incentives supporting the R&D activities in almost all EU member states can eventually represent an efficient channel. To test this assumption, we decided to study the impact of R&D tax incentives on enterprise innovation activity. As described in details by European Union (2014a), different types of R&D tax incentives as well as other tax rules and tax administrative features (eventually beneficial for the tax payer innovation activities) are applied by EU member states. From this point of view, Belgium and the United Kingdom are the leading member states with relatively more suitable tax rules for R&D and innovation. On the other hand, the tax system of Germany and Estonia do not use any specific initiative to focus on innovation activity. What concerns the form of the most widely used tax incentive, the tax credit for R&D expenditures are the most represented—this instrument which is not affecting the tax base rather decreasing the corporate tax rate, is applied in sixteen EU member states.

To find an evidence concerning eventual efficiency of different tax incentives, we decided to proceed the correlation analysis between selected features of R&D tax incentives and firm's R&D and non-R&D innovation expenditures. Firstly, we calculated standard Pearson correlation coefficient for all selected variables in the sample and we found a positive, but weak correlation between most of the R&D tax incentives and firm's R&D expenditure (Table 6). The total number of R&D tax

Table 6 Pearson correlation coefficients for selected variable (cross-sections)

	Firm's R&D expenditures	Firm's non_R&D innovation expenditures	Summary innovation index
Total number of R&D tax incentives	0.279	−0.510	0.303
Tax credits	0.166	−0.318	0.221
Enhanced allowances	−0.088	−0.040	−0.150
Accelerated depreciation	0.178	−0.310	0.020
Patent box	−0.207	−0.330	0.085

Table 7 Tetrachoric correlations for binary variables (cross-sectional data)

	Firm's R&D expenditures	Firm's non_R&D innovation expenditures	Summary innovation index
Tax credits	0.353	−0.415	0.131
Enhanced allowances	0.232	0.131	−0.216
Accelerated depreciation	0.201	−0.705	0.204
Patent box	−0.131	−0.482	−0.131

incentives, calculated as the sum of tax incentives used in certain country, correlates positively with firm's R&D expenditure, but this correlation is rather weak. The same evidence is true for tax credits and accelerated depreciation. On the other hand, there is rather significant negative correlation between total number of R&D tax incentives and firm's non R&D innovation expenditure. Moreover, all tax incentives are negatively correlated with non R&D innovation expenditures.

According to these findings, the firms in the tax environment with more R&D tax incentives prefer to spend more on R&D. But this readiness to invest in R&D seems to have a negative impact on other forms of innovation activities (represented by non-R&D innovation). This could indicate that tax incentives could have more effect on the structure of innovation activities (the share between R&D and non-R&D innovation expenditures), rather than on the total volume of R&D and non-R&D innovation expenditure.

One can argue that the method we applied is not appropriate for the analysis of discrete binary variables, which are mostly used in the sample. Thus, we also decided to apply the tetra choric correlation, suitable only for binary variables. Therefore, the continuous variables had to be recoded to binary ones. The average value of each variable has been used as the threshold between zero and one. The results we obtained are to some extent similar to those concerning the Pearson correlation coefficients. However, the negative correlation between non-R&D innovation expenditure and tax credit, accelerated depreciation as well as patent box appears to be even stronger. This observation is especially true for the form of

Table 8 Pearson correlation coefficients (cross-sectional data)

	Firm's R&D expenditures	Firm's non_R&D innovation expenditures	Summary innovation index
R&D tax incentives score	0.538	−0.538	0.413

accelerated depreciation, where the correlation seems to be very strong, as indicated in Table 7.

Since different types of tax incentives ensure different conditions for enterprise innovation activity, we also decided to apply the results of ranking of tax incentives in respect to R&D activities in European countries, calculated by EC Study on R&D tax incentives (European Union 2014a). The latter study takes into account three categories of features of the R&D tax incentives: (1) scope of the policy, including the type of R&D tax incentive and costs covered, (2) targeting of specific groups of firms, according to their size, age, region, etc. (3) organization, including administrative practices and evaluation (European Union 2014a, p. 73). According to this ranking, Denmark and Ireland seem to have the most suitable R&D tax incentives among the EU member states. On the other hand, the results of this ranking indicate that the least appropriate R&D tax incentive can be found in Malta, Cyprus and Greece.

The results of the correlation analysis between the tax incentives scores and other selected variables are presented in Table 8. These results are in compliance with the previous results gained by testing different forms of tax incentives independently. They indicate that there is a positive correlation between a country's R&D tax incentives score and enterprise R&D expenditures. In addition, the positive correlation between better-scored country's tax incentives and higher values of country's summary innovation index can be found. In accordance to our previous results, the correlation between R&D tax incentives score and non-R&D innovation expenditure is again negative in our sample of the 28 EU member states.

5 Discussion and Conclusions

The objective of our study was to find an evidence about the relationship between chosen features of national tax system related to tax base composition and the innovation activity in EU member states with a special emphasis on enterprise innovations. The empirical testing of such relationship is interesting by itself. However, as our research tries to contribute to the renewal debate concerning the CCTB proposal, we proceed in interpretation of our findings from this point of view.

The first characteristics of the CCTB directive proposal is related to the fact that harmonisation of tax base rules doesn't need any harmonisation of corporate tax rates in the EU. As we found no evidence of the impact of statutory tax rate on enterprise innovation activity, we can support this approach of not bringing such politically difficult topic into consideration. However, one should notice, that a possibility of positive impact of the level of corporate statutory tax rate on overall innovation activity in a specific country may exist thanks to effects of spill overs between a certain groups of economic subjects (e.g. large companies) and other groups of economic subjects (e.g. innovation activity in the public sector). The probability of an existence of such effects seem to increase with the identification of the link between the statutory tax rates and Summary Innovation Index.

Further, an EC initiative towards harmonisation of tax base rules should have neither positive nor negative impact on enterprise innovation activity as the latter seems to be unaffected by the composition of tax base rules. At least in the analysed period, the approximation of tax base composition by two measurable variables—the effective tax rate and the difference between statutory and effective tax rates—seems not to be able to explain the differences in enterprise or overall innovation activity in EU member states. Eventually, other variables representing the tax base differences can be taken into account for further research, but in this case a firm-level data approach should be appropriate.

Although it seems that the enterprise innovation activity is not influenced by the tax base as a whole, our results indicate it might be affected by a certain part of tax base related to corresponding tax incentives effect. The use of R&D tax incentives to wider extent in some EU member state seems to lead to higher R&D innovation activity in companies in this country, as well as to higher overall innovation activity. On the other hand, our results indicate that the choice of a specific tax incentive might influence the structure of innovation schemes. For instance, while R&D tax incentives stimulate the enterprise R&D activity, they affect negatively the non-R&D activity in the companies.

From the perspective of the CCTB proposal, the effects of tax incentives having impact on tax base (enhanced allowances and accelerated depreciation) can be compared to those having impact on tax rate (tax credits and patent boxes). There is an evidence towards the preferable use of base-affecting tax incentives in the form of enhanced allowances which seem to be the only tool to affect positively both the R&D and non-R&D business activity. On the other hand, if the form of accelerated depreciation is applied in the new CCTB proposal in order to stimulate enterprise innovation activity in European companies, this can produce a strong negative effect in companies' non-R&D activity. Thus, our results can lead to suggestion to implement the best practices of R&D tax incentives of the EU member states considered as having the best scores in tax incentives ranking (especially Denmark) into a new CCTB proposal. However, the results we obtained should be treated with attention because only a static approach was applied in this part of our research for the reason of a limited access to data about development of tax incentives in the EU member states. More detailed data in this field would lead to adoption of more appropriate methods (like panel data regression analysis) in empirical research of tax incentives efficiency. From this point of view, the further research in this area is needed.

References

Cappelen Å, Raknerud A, Rybalka M (2012) The effects of R&D tax credits on patenting and innovations. Res Policy 41(2):334–345

Czarnitzki D, Hanel P, Rosa JM (2011) Evaluating the impact of R&D tax credits on innovation: a microeconometric study on Canadian firms. Res Policy 40(2):217–229

European Union (2008–2015) Innovation Union Scoreboard/European Innovation Scoreboard. European Commission, Brussels

European Union (2014a) Study on R&D tax incentives. http://ec.europa.eu/taxation_customs/resources/documents/taxation/gen_info/economic_analysis/tax_papers/taxation_paper_52.pdf

European Union (2014b) Taxation trends in European Union. http://ec.europa.eu/taxation_customs/taxation/gen_info/economic_analysis/tax_structures/index_en.htm

European Union (2015) Tax reforms in EU member states 2015. http://ec.europa.eu/economy_finance/publications/eeip/ip008_en.htm

Falk R (2005) What drives business R&D intensity across OECD countries. WIFO working paper No. 2004

Hunady J, Orviska M, Sarkanova Z (2014) Determinants of European firm's innovation and the role of public financial support. Eur Financ Account J 9(1):62–84

Ientile D, Mairesse D (2009) A policy to boost R&D: does the R&D tax credit work? EIB papers 14(1):144–169

Mulgan G, Albury D (2003) Innovation in the public sector. ver.1.9, October 2003 [cited 9.11.2014]. www.childrencount.org/documents/Mulgan%20on%20Innovation.pdf

Tassey G (2007) Tax incentives for innovation: time to restructure the R&E tax credit. J Technol Transf 32(6):605–615

Uramová M, Lacová Z, Suplata M, Hasulova K (2016) Quelles consequences de l'harmonisation des assiettes pour l'impot sur les sociétés en Europe ? In: PGV network "For stronger Europe: renovation of the economic, entrepreneurial, political, territorial and educational structures". Wolters Kluwer, Prague, pp 251–263

Zemplinerová A, Hromádková E (2012) Determinants of firm's innovation. Prague Econ Pap 2012 (4):487–503

Žaneta Lacová is an assistant professor at the Faculty of Economics, Matej Bel University, Banska Bystrica Slovakia. She received her PhD. in Economics both at the Faculty of Economic Sciences, Université Nancy 2 (France) and at the Faculty of Economics, Matej Bel University in Banska Bystrica (Slovakia) in 2007. Since 2007, she has been teaching Microeconomics and Macroeconomics courses at basic and intermediate level, courses of European Integration and European Monetary Integration Issues in Slovak, English and French language. Zaneta Lacova has been doing research in areas of international economic integration, consequences of economic policy on companies and households. She is a member of various national and international research teams, a member of International Network of Vysegrad Group (PGV Network) as well as of International Network for Economic Research (INFER).

Ján Huňady is an assistant professor at the Faculty of Economics, Matej Bel University, Banska Bystrica Slovakia. He received PhD. in Public Economic and Services at the Faculty of Economics, Matej Bel University. His research interests range from public finance, taxation and innovation. He also participates on several research projects in the area of research policy, innovation and public finance. He is the author or the co-author of several research papers published in renown scientific journals and he has given conference presentations in several European countries.

Absorptive Capacity of R&D in Space: A Conceptual Approach to the Productivity Paradox

Mário A.P.M. Da Silva and Peter Nijkamp

Abstract This paper addresses R&D acceptance in a spatial-economic system. We assume that the absorptive capacity of firms in a region is positively influenced by territorially-dependent factors. We analyze the effects of various spatial factors that explain the differences between territories to access and absorb external knowledge on the innovative performance of regions as well as the possibility of rising local increasing returns. Our pedagogical model shows that cognitive determinants associated with the spatial organization and culture of a territory—when combined with cognitive factors related to the organization of R&D and the culture of firms—help to understand the productivity paradox observed in regions with a considerable level of innovative effort.

Keywords Absorptive capacity • Knowledge spillovers • Complementarities • Proximity • Innovation • R&D

JEL Classification O33—Technological change: choices and consequences • diffusion processes • R11—Regional economic activity: growth, development, environmental issues, and changes

1 Introduction

The concept of absorptive capacity is key in understanding knowledge spillovers; it has already a long history and was introduced by Cohen and Levinthal (1989, 1990). Absorptive capacity comprises the organizational capabilities of acquisition and assimilation of external knowledge as well as its transformation and

M.A.P.M. Da Silva (✉)
Universidade do Porto, Porto, Portugal
e-mail: msilva@fep.up.pt

P. Nijkamp
VU University, Amsterdam, The Netherlands

Adam Mickiewicz University, Poznan, Poland

A. Dias et al. (eds.), *Modeling Innovation Sustainability and Technologies*, Springer Proceedings in Business and Economics,
https://doi.org/10.1007/978-3-319-67101-7_4

exploitation capabilities. Absorptive capacity stems from knowledge accumulation, while its effective development requires familiarization with relevant prior knowledge. Learning performance is greatest when the object of learning is related to what is already known, i.e. the level of prior related knowledge.

A cognitive element which enhances absorptive capacity besides similarity is complementarity of different but related business innovation strategies and R&D approaches. The diversity versus similarity dimensions of knowledge are according to several researchers (see e.g. Abecassis-Moedas and Mahmoud-Jouini 2008) critical determinants of learning. It is necessary to balance similarity and differences between the source and recipient type of knowledge (Cohen and Levinthal 1990). In addition, there is also a trade-off between novelty and understandability (Nooteboom 2000). For exploitative learning one needs cognitive proximity, while for exploratory learning cognitive distance is more important. The important challenge for management is therefore, to balance exploitation and exploration.

In a dyadic perspective with a recipient and source of knowledge, the source-recipient knowledge complementarity plays an important role in the transformation and exploitation of source knowledge. The two types of knowledge are related in a sense that they address similar problems and use the same objects. Abecassis-Moedas and Mahmoud-Jouini's (2008) main result is that the complementarity between the recipient and the source knowledge is a critical aspect of the absorption process and therefore of the designing of new products.

The notion of complementarity between elements of a system can also be found in Hopner (2005). Compatibility between different but related elements can be caused by structural similarity between them. Economists use this concept to identify elements of company strategies that increase output, if they are combined (Milgrom and Roberts 1995). Complementary institutions and resources generate positive externalities and spillovers which are beneficial to the innovative activity (Feldman 1994). Diversity plays an important role in innovation networks; heterogeneity may be preferable to homogeneity for the effectiveness of innovation (Corsaro et al. 2012). The notion of absorptive capacity plays usually a critical role in this scientific debate on knowledge acquisition.

In this paper, we develop a comprehensive notion of absorptive capacity which, in addition to recognizing the importance of similarity of R&D approaches to access and assimilate external knowledge also considers the relevance of complementarity of different R&D designs adopted by firms to improve their absorptive capabilities. The adoption of different but compatible R&D approaches may be preferable for enhancing the effectiveness of innovation of firms in networks and regions.

We present here a pedagogical application of the absorptive capacity concept first introduced by Cohen and Levinthal (1989, 1990), in the context of a regional system. In this paper, absorptive capacity is dealt with in a theoretical-conceptual framework from an analytical perspective. An essential hypothesis here is that similarity and variety are important determinants of the absorptive capacity of a firm located in a region. Another basic hypothesis is that similarity and variety are explained both by the individual strategies of firms and by regional conditions and policies of the common space where firms are located. Quite some recent empirical work defines and employs this generalized notion of absorptive capacity (see e.g. Caragliu and Nijkamp 2012), but solid theoretical contributions on the spatial dimension of absorptive capacity are more rare.

The concept of absorptive capacity, whose foundations were originally designed in the context of firm theory, can be extended to more complex institutions or actors, such as countries and regions (Caragliu and Nijkamp 2012). Human capital and a proper knowledge base are important in enhancing the growth capabilities of regions or countries. The stock of accumulated knowledge plays a role in the capability of a region to identify and utilize new knowledge, both locally produced and originating from outside.

Cognitive elements and cognitive mechanisms explain the differences among territories to grasp innovation and translate it into growth, giving rise to increasing returns (Capello et al. 2009). The capabilities of economic agents within a region to access and absorb the knowledge being generated and ultimately utilize it are not invariant with respect to geographic space. Particularly, diversified areas, which are characterized by differences among people thereby causing them to look at and appraising a given information set differently, are expected to gain more from new knowledge.

Based on the notion of "related variety", Broekel and Boschma (2012) found that cognitive proximity may increase the likelihood of cross-fertilization among actors within a region. Technological cross-fertilization requires that people speak the same scientific language, while at the same time they are separated in terms of technologies. In turn this generates a mechanism of creative resonance based on the pool of ideas from relatively different technologies available within the same technological class which ultimately leads to innovation. Boschma et al. (2012) posit that industries and regions must be cognitively neither too close, nor too distant, in order to trigger mutual learning processes. A certain degree of cognitive proximity stimulates effective communication, while some degree of cognitive distance is required to avoid cognitive lock-in.

The creation of business incubators to nurture startups say, in creative industries, and the provision of room to accommodate foreigners originating from diverse cultures may be a way of promoting local development and turning a less developed region into an entrepreneurial region. Business incubators provide a supporting environment during the critical stages of creating a new business. Knowledge creation and acceptance is a critical factor here. The crucial question in a study on this topic is: what are the important determinants of absorptive capacity in a region? These include the cognitive elements of similarity and complementarity, and both elements are explained by firms' decisions and how space is organized and structured.

In this paper, we assume that the absorptive capacity of firms located in a given region is positively influenced by territorially-dependent factors of similarity and complementarity, and analyze the effects of the spatial elements that explain the differences between territories to access and absorb external knowledge on the innovative performance of regions, the occurrence of the productivity paradox observed in regions, and the possibility of rising local increasing returns.

We show that the spatial elements of absorptive capacity have an important impact on the innovative performance of firms located in a region and therefore on the rate of technical progressiveness of the region. If the territorial elements of absorptive capacity are not sufficiently high, there are manifestations of the productivity paradox at the regional level. If the territorial elements of connectivity and complementarity are significant enough, they may well give rise to local increasing returns in the rich region and to the increase of absorbable spillovers to surrounding areas. Under no territorial circumstance, however, the paradox of productivity at regional level and the increase of usable spillovers towards other, less developed regions take place at the same time.

The remaining part of the paper is organized as follows. In Sect. 2, the hypotheses of the theoretical model are set out and the structure of the model is developed. Section 3 presents the equilibrium results of the market game of firms, in particular, the optimal choices of the various components of innovation strategies of firms. In Sect. 4, the model of absorptive capacity and its equilibria are applied to our study of major regional innovation and growth issues. Finally, Sect. 5 sets out the main findings and proposes further research trajectories.

2 The Theoretical Model

In this section, we analyze a game of three stages with the following timeline. In the first stage, firms simultaneously select their R&D designs, in the second stage, their R&D expenditure levels are chosen, and finally output quantities are chosen. Business decisions concerning R&D orientations and R&D expenditures can be taken non-cooperatively or cooperatively. We shall consider in turn each of these cases. To conclude our description of the setup of the model, we assume that firms are maximizing profits when they choose their innovation strategy, while the R&D technology they employ exhibits diminishing returns in R&D expenditures. Our conceptual exposition will be illustrated by means of a numerical example in Sect. 3.

Firms serve only a given regional market. Firms compete with each other in the final product market because they are located or agglomerated in the same region. There is no trade, no interdependence between regions, and no interregional or international competition in the given final market.

We define next the measure of innovative performance of firms agglomerated in a region, and so the rate of technical progress of the region. The effective R&D level of a firm as assumed in this paper is

$$X_i = x_i + b(1-d_i)(1-d_j)x_j + p\ d_j d_i x_i, \text{ with } \quad i,j = 1,2 \text{ and } \quad i \neq j.$$

By committing themselves to R&D investments, firms have an opportunity to reduce their unit costs of production. The effective cost reduction obtained by firm i, X_i, amounts to its own R&D effort, x_i, a fraction of its rival's R&D efforts, x_j, and an increase of firm i's technical competence due to the complementarity between the R&D approaches of the two firms, d_i and d_i. Parameters b and p, with $0 \leq b$, $p \leq 1$, denote respectively the productivity or effectiveness of knowledge similarity and complementarity exogenously determined by the structure, organization and culture of the region or territory where firms are located. This space-based interpretation of similarity and complementarity in absorptive capacity is explained and justified on the basis on the brief literature review provided in the introduction of this paper.

In this representation of a firm's effective level X_i, firm i's absorptive capacity is given by two elements, similarity of knowledge, $b(1 - d_i)(1 - d_j)$, and complementarity of knowledge, $p\, d_j\, d_i$. Effective level of knowledge comprises not only firms' choices of R&D approaches, but also absorptive capacity aspects dependent upon the connectivity and complementarity of the territory. The complementary element of knowledge is new to the game-theoretical literature and makes our model an extended version of the absorptive capacity's model by Wiethaus (2005). The territorial elements of absorptive capacity, or the cognitive elements of space, as represented by parameters b e and p are new to the theoretical literature too.

The territorial element of connectivity b is enhanced by investments made over time by regional authorities in communication infrastructure and inter-firm networks. As an illustrative example of the effects of connectivity on the usefulness of information which is spilled over between firms, $b = 0.6$ means that 40% of knowledge that spills over towards rival firms is lost in the diffusion process due to inefficiencies in communication networks within the region, and so the extent of usable knowledge that is actually available to rival firms is only 60% of spillovers from the industry. In turn, the 60% potential of shared knowledge is only partly assimilated and absorbed by rival firms and that is a function of firms' absorptive capacity.

The territorial element of complementarity p magnifies technological complementarities of firms, giving rise to local increasing returns in R&D activities. The specificity of a firm's R&D orientation may have its source in the specific culture of most of its employees. Denote the proportion of migrant workers originating from a different culture employed by firm i as f_i, with $0 \leq f_i \leq 1$. Then, we simply have $d_i = f_i$. Cultural diversity would then be the explanation for the existence of different approaches in conducting R&D and the emergence of technological complementarities and local increasing returns in the activity of knowledge production and absorption.

From the point of view of firms, similarity and complementarity components seem to be in conflict to each other. There is a trade-off between these two components and both firms need to face them at the same time. Firm i faces a trade-off between similarity and complementarity in its choice of R&D orientation, d_i. Increases in d_i increase the third additive term of X_i at the expense of the second term, while increases in the complement $1 - d_i$ increase the second term at the expense of the third. The strategic interactions at the initial stage of the game when firms choose simultaneously their respective d's are interesting and, and drive to a great extent the equilibrium results of this paper.

We assume a tradeoff between similarity and complementarity that is ex ante symmetric, i.e. symmetrical before the firms' choices of their technological distances d_i and d_i. We impose by hypothesis that p does not exceed 1, besides that we logically set that b does not exceed 1. Hence, the maximum that one gets in terms of effective cost reduction with territorial connectivity equals the maximum that one can get with synergies and local increasing returns arising from the different R&D approaches adopted by firms. However, we should note that there is no particular reason for restricting p to values not exceeding 1. The phenomenon of local increasing returns would be most likely more pronounced for values of p greater than 1.

This conflict between similarity and complementarity as perceived by firm i, however, ceases to exist as a result of a particular choice of the other firm j, that is, a soon as it either picks $d_j = 0$ or $d_j = 1$. Thus, the connectivity between firms ceases to exist at the moment one of them chooses an idiosyncratic, firm-specific R&D approach $d = 1$, and the potential of complementarities ceases to exist the moment that one of the firms chooses a purely broad, identical R&D approach $d = 0$.

The business decision to choose d therefore involves a second trade-off in addition to the one analyzed by Wiehthaus (2005), the last one being the trade-off between high appropriability of a firm's own R&D results and connectivity to external sources of technological knowledge. In this paper, however, we implicitly assume an exogenous level of knowledge spillovers of 1, as we are not interested in the analysis of the effects of appropriability of innovation benefits determined by intellectual property rights considerations on the strategies of firms.

3 Equilibrium Results

The R&D model from Sect. 2 is critical in our experiments. Rather than using complicated mathematical expressions, we will use illustrative simulation exercises. We employ the Mathematic software to perform numerical simulations and provide in Table 1 in the appendix the equilibrium results of the model in the non-cooperative case. We skip here the presentation of the technical treatment of the analysis, involving the derivation of first-derivatives of profit functions and assessments of the signs of each firm's incentives at the margin, and present the main results of it.[1]

The presentation of results when firms compete in R&D begins with the characterization of existence and multiplicity of non-cooperative equilibria at the initial stage of R&D design choices. We limit our attention to symmetric equilibria regarding the choice of R&D orientations d by each of the two firms located in an innovative region.

[1]The mathematical derivations are available from the corresponding author on request.

Management of each competing firm seeks to establish the optimal balance between similarity and complementarity with the choice of its own technological distance d in order to better control spillover flows between firms and enjoy potential synergies between firm-specific orientations of R&D at the same time. Firms interact strategically by competing in R&D while striking this balance, and this makes the equilibrium analysis rather demanding.

Proposition 1 Suppose that firms compete in R&D. For $b \leq 0.977745$ and $p = 0$, there are two corner solutions, the general and firm-specific R&D approaches $d = 0$ and $d = 1$, and for $b \leq 0.977745$ and $p > 0$, a unique solution $d = 0$. For $b > 0.977745$ and $p > 0$, an interior solution emerges, where d increases in both b and p, continuing to exist equilibrium $d = 1$ for $b > 0.977745$ and $p = 0$.

According to Proposition 1 and Table 1 in the appendix, the way competing firms strike such a balance at the beginning of the game leads to the extreme choices of $d = 0$ or $d = 1$ for $b \leq 0.977745$, while an interior solution of d is emerging for $b > 0.977745$. The absolute magnitudes of b and p determine whether the optimal solution of d is a corner solution $d = 1$, or an interior solution where a symmetric equilibrium d is increasing in both b and p. In equilibrium, for sufficiently high values of b and p, distance d increases in response to an increase of b, and this increase is greater the higher complementarity p is. Specifically, each firm's managerial response to increases of b is an increase d in order to better control the flow of information to its rival and benefit from complementarity or synergy gains if $p > 0$.

R&D competition between two firms means essentially similarity in absorptive capacity in the following sense. Broadly speaking, there is similarity in interior solutions of R&D designs, as the maximum of d is much smaller than 1. For $b = 1$ and $p = 1$, the symmetric equilibrium is $d = 0.0227415$ (the maximum value is rather close to zero). As we shall see below however, we can have cooperation in R&D with similarity or variety.

To the extent that d and b are the outcome of choices made by different economic agents or entities, we can say that business similarity $1 - d$ and territorial similarity b are strategic substitutes (Bulow et al. 1985). Likewise, we can say that d and p are strategic complements. We can speak about geographical proximity and cognitive proximity, and knowledge spillovers vary according to the definition of proximity (Caragliu and Nijkamp 2013).

The characterization of the innovation strategy of a competing firm in a cluster or region is not complete before we determine each firm's own R&D investment in symmetric equilibrium and describe how the optimal level of R&D effort varies with the elements of absorptive capacity that are influenced by or dependent on the region or territory in question.

Proposition 2 Suppose that firms compete in R&D. For $b \leq 0.977745$, in the case of a corner solution $d = 0$, the level x of R&D investment decreases in b, but doesn't change with changes in p, and in the case of a corner solution $d = 1$, x doesn't change with changes in b. For $b > 0.977745$ and $p > 0$, x increases in b and p, but for $b > 0.977745$ and $p = 0$, x doesn't change with changes in b in either case $d = 0$ or $d = 1$.

This outcome stresses the impact that spatial elements may have on firms' behavior. We can see that a tiny increase in p is all that is needed to change x's evolution. For $b \leq 0.977745$, x decreases in b, but once b is greater than 0.977745, x doesn't in b if $p = 0$, but increases in b if $p > 0$. This second equilibrium result is obtained from Table 1. The explanation of this result is based on the incentives at the margin for a given firm i to undertake R&D, $\partial(2X_i - X_j)/\partial x_i = 2(1 + pd^2) - b(1 - d)^2$. The amounts of incentives at the margin to undertake R&D are consistently aligned with the R&D investment levels to be chosen in equilibrium, and are shown in a separate table available from the authors upon request.

Next, we present a result on the effective level of R&D X which is dependent not only on the individual R&D effort of both firms, but also on cognitive elements of absorptive capacity influenced by the connectivity and complementarity that exists in the region or territory. The effective level of technological knowledge X is a good indicator of the innovative performance of a firm in our model. Firm i can lower its unit production cost by $2X_i - X_j$ as a result of R&D efforts of both firms. In symmetric equilibrium, this difference equals the common value X. Table 1 allows one to establish the following proposition.

Proposition 3 Suppose that firms compete in R&D. For $b \leq 0.977745$, in the case of a corner solution $d = 0$, the effective level X of R&D effort increases in b, until it reaches a global maximum at $b = 0.5$ and then decreases in b to a local minimum at $b = 0.977745$, but doesn't change with changes in p, and in the case of corner solution $d = 1$, X doesn't change with changes in b. For $b > 0.977745$ and $p > 0$, X increases in b and p, but for $b > 0.977745$ and $p = 0$, X doesn't change with changes in b in either case $d = 0$ or $d = 1$.

The explanation of the non-monotonous evolution of the symmetric level X as a function of b and p requires its decomposition in two terms. In fact, X can be represented as the product of the common level of individual effort x by the absorptive capacity per common x (plus one unit), X/x, or equivalently the marginal incentives associated with the similarity and complementarity elements of absorptive capacity plus one unit, $\partial X/\partial x$: $1 + b(1 - d)^2 + pd^2$. Specifically, the non-monotonic evolution of X with respect to b is explained by the evolution in opposite direction of x and $X/x = \partial X/\partial x$. In a table available from the authors upon request, the latter term is represented.

Now we discuss another way firms interact strategically in the selection of their innovation strategies. Firms cooperate in succession in the stages of choice of R&D orientation d and choice of individual R&D effort levels x. We exclusively focus our attention here on the choices of d's in the case of R&D cooperation. As regards the choice of d, we establish the following symmetric equilibrium.

Proposition 4 Suppose firms cooperate in R&D. In the cooperative solution, the optimum R&D design is the corner solution $d = 0$, if $b > p$ and the corner solution $d = 1$ if $b < p$.

Cooperating firms deal with the trade-off between similarity and complementarity by picking extreme solutions only. There can be R&D cooperation with either strict similarity $d = 0$ or strict diversity $d = 1$ in absorptive capacity, depending on the relative magnitude of territorial parameters b and p. The relative magnitude of b and p determines whether the optimal solution is $d = 0$ or $d = 1$ because second-order conditions for maximizing industry profits set when each firm chooses its R&D design d rule out critical levels of d satisfying the first-order condition $\partial(2X_i - X_j)/\partial d_i = -b(1 - d_j) + pd_j$. This is why the cooperative solution is not an interior solution.

We can say that similarity of technology $1 - d$ and territorial similarity b are now strategic complements in a sense. To the extent that a further increase of b changes the relative magnitude of b and p, causing $b > p$, optimal d goes from 1 to 0, which increases the difference $1 - d$ from 0 to 1. Likewise, we can say that d and p are also strategic complements. An increase of p, such that the relationship between b and p changes to $b < p$, leads to a change of d from 0 to 1. Therefore, we can say that d weakly increases with increases in p.

In the cooperative case, can firms achieve a given reduction in the unit cost of production, X, with the minimum total cost of R&D? It can be shown that firms allocate their R&D resources efficiently when they act cooperatively in the following sense. Choices of R&D orientations are said to be efficient if they maximize the effective level of knowledge per unit of individual effort, X/x.

Proposition 5 The R&D cooperative solution is efficient in the previous sense.

In Sect. 4 we will offer an illustrative application of this model and its equilibrium results to important regional issues.

4 Application of Absorptive Capacity to Regional Issues

Empirical evidence shows consistent discrepancies between knowledge inputs and economic performance. Sicily is one such illustrative real-world case. Ranking Sicily among European regions for human resources employed in science and technology tells us a different story from ranking Sicily according to productivity data. Caragliu and Nijkamp (2012) ask where does knowledge produced in Sicily go, and why does it not show up in Sicily's statistics on productivity? The authors argue that these questions are somehow linked to Solow's paradox about the new economy.

Does our model predict such a regional productivity paradox under any circumstance? We address the particular question of whether the paradox of productivity at regional level can be observed for high values of parameters of territorial connectivity and complementarity b and p, respectively. In this model, the innovative performance of a region is best represented by the effective R&D level X of each firm clustered in the region. Will increases in regional absorptive capacity, i.e. increases in either b or p lead to non-increases in X, or even actual decreases in X? We are assuming here that this paradox has a stronger and a weaker version.

When firms compete in the formulation of their innovation strategies, in particular in their choices of R&D designs, there are reasons to believe that increases in b and/or p lead in the end to reductions of X or at least leave X unchanged. In the case of R&D competition, it can be shown that managing R&D approaches in order to strike the optimal balance between similarity and complementarity seems to counteract any positive effect on the innovative performance of firms that may arise from exogenous increases in cognitive similarity and territorial connectivity and/or complementarity and local increasing returns.

We can draw the following results on the conditions under which we can observe the regional productivity paradox. There are manifestations of this productivity paradox in our model, both in its strong formulation (Result 1) and in its weaker formulation (Result 2), when the territorial parameters b and p are not sufficiently high. The levels of effective R&D for different values of parameters (b, p) are described in Table 1.

Corollary 1 Suppose firms compete in R&D. Increases in territorial similarity b decrease the effective R&D effort X for $0.5 < b \leq 0.977745$ (and $d = 0$).

Corollary 2 Suppose firms compete in R&D. Increases in territorial similarity b leave X unchanged for $b > 0.977745$ and $p = 0$. Increases in territorial complementarity p don't change X for $b \leq 0.977745$ (and $d = 0$). Suppose that firms cooperate in R&D. Increases in b don't increase X for $b < p$, while increases in p don't increase X for $b > p$.

We infer that increases in the similarity and/or complementarity elements of absorptive capacity associated with a territory do not lead to an increase in the rate of technical progress of the region whenever initial values of either b or p are not considered to be high enough.

When firms compete in R&D, for $b > 0.5$, X decreases in b until reaching a local minimum at $b = 0.977745$, slightly increasing thereafter with additional increases in b if $p > 0$. The effective level of effort X does not increase—or hardly increases—in b or p despite the fact that individual level of effort x may increase when $b > 0.977745$. Competition in R&D at the stage of choosing R&D orientations has an important role in the explanation of this result. Competition between firms in the choice of d's causes increases in b and/or p to result even in reductions in the level of absorptive capacity per common x (plus one unit), X/x.

In the alternative case of R&D cooperation, both multiplicative components of the effective effort level X, x and X/x, actually increase in b, if $b > p$ and in p if $b < p$, because incentives at the margin to undertake individual R&D effort x and the level of absorptive capacity per unit of x increase when managers of firms respectively choose optimal R&D designs $d = 0$ and $d = 1$. Incentives at the margin to realize x and the level of absorptive capacity per unit of x are described in separate tables available from the authors upon request.

We draw now some policy implications from these results of our model. What must public authorities do in order to alleviate or solve that productivity problem at the regional level? Significant innovative performance at the regional level, with

X increasing after increases in b and p, can only be obtained if either b or p exceed some critical levels.

In case firms compete in R&D, unless public authorities with regional concerns invest massively in the territorial element of connectivity, b, so that its value exceeds the threshold of 0.977745, even if it is already greater than 0.5, any increase in territorial complementarities alone is simply fruitless and does not increase X. After b exceeding this critical level, unless public authorities also invest in the territorial element of complementarity, p, so that it is at least positive, any increase of b does not produce the desired effect of increasing X. When $p = 0$, competition in R&D explains the persistence of a regional productivity paradox. Apparently, only significant territorial complementarities and associated strong local increasing returns as a result of public investment can counter and overcome the dominance of the competitive force and thus eliminate this unwanted innovative performance of innovative regions.

Cognitive elements as part of territorial capital magnify the contribution of knowledge by determining the formation of increasing returns to knowledge exploitation (Capello et al. 2009). Innovation effort does not necessarily lead to innovation. These authors showed that territorial capital is positively correlated with economic performance. Within a rich definition of territorial capital, a particular subset of variables (defined as cognitive elements) is found to be responsible for fostering socio-economic interactions and mutual understanding in a spatial context. Collective learning and mutual understanding play a major role in determining long run economic performance.

Policy interventions are often associated with trade-offs when policymakers allocate economic resources, but sometimes there is a possibility of public authorities alleviating such trade-offs and even taking advantage of potential synergies (Swart 2008). Trade-offs and synergies are opposing concepts in this regard.

Regional authorities, once the above-mentioned thresholds for b and p in our model are exceeded, must face the relationship between b and p as being synergetic. When allocating scarce resources in promoting and developing their region, public authorities must finance activities that promote b and p at the same time, or finance specific activities to promote b and p that mutually support each other.

In case firms cooperate in R&D, as long as public authorities with regional concerns invest in the most important territorial element of absorptive capacity, that is, b if $b > p$ or p if $b < p$, increases in b or p always induce increases in X. A rational policy of persistent investment only in the most efficient territorial element of absorptive capacity is all that is necessary for regional development. The most effective territorial element in a given circumstance (be it either $b > p$ or $b < p$) doesn't even need to have a high value in absolute terms. Even increases in say b starting at low values of b are able to increase X. All that is required then is that b takes the relatively higher value, that is, $b > p$.

When deciding how to allocate resources to promote a region, public authorities must now face the relationship between b and p conceived of as being in conflict, leading them to face a trade-off between regional connectivity and complementarity. The economic activities and areas to be funded are those that must enhance and promote either b or p.

It should be noted that polycentric metropolitan areas have a potential which allows reaping the benefits of agglomeration in a network of cities (Meijers 2007). The concept of complementarity is central in regard to situations in which the functionality of an institutional form is conditioned by other institutions (Hopner 2005).

Will space structure and the culture of the territory decisively determine the values of b and p in our model and change the focus of regional and urban policies? Will the way firms are located in space and cities are networked be crucial to explain the values of b and p, and consequently the level of regional development? Or, instead, will the existence of institutional complementarities between the spatial organization and industrial relations determine industrial and regional success?

Empirical results show that a lower regional absorptive capacity increases knowledge spillovers towards neighborhood areas, lowering the regions' ability to exploit new knowledge. The relatively low endowment of territorial capital is found to be associated with higher outward knowledge spillovers (Caragliu and Nijkamp 2012). Territorial characteristics including cognitive proximity play a strong role in exploiting knowledge produced locally and originating from outside.

We will now address a new question assuming that the regional absorptive capacity decreases. In our application of the model to regional issues this actually is the only occasion in which we assume a decrease in absorptive capacity. Will knowledge spillovers towards surrounding areas increase when the absorptive capacity of firms in the developed region decreases? Does a decrease in either b or p increase spillovers towards other surrounding regions, or at least maintain them constant? We define knowledge spillovers to neighboring areas here as an exogenously determined fraction of x. The next results establish the territorial conditions under which this could happen. Looking at Table 1 in the appendix allows us to immediately obtain these conclusions.

Corollary 3 When firms compete in R&D, reductions in b lead to an increase of knowledge spillovers towards neighboring areas for $b \leq 0.5$ (and $d = 0$).

Corollary 4 Suppose that firms compete in R&D. Reductions in b don't cause knowledge spillovers to neighboring areas to change for $b > 0.977745$ and $p = 0$. Reductions in p don't lead to a reduction of knowledge spillovers for $b \leq 0.977745$ (and $d = 0$). Suppose that firms cooperate in R&D. Reductions in b don't lead to decreases of knowledge spillovers for $b < p$, and reductions in p don't lead to decreases of knowledge spillovers for $b > p$.

Developed regions don't seem to have much reason for fearing the appropriability problem of R&D results following reductions in their regional absorptive capacity. The phenomenon of increases in knowledge spillovers towards other less developed regions as a result of reductions in the innovative region's absorptive capacity can only take place for initial values of b and p considered to be not high enough.

When firms compete in R&D, spillovers increase with reductions in b for $b \leq 0.5$, but in this case regional absorptive capacity may be not regarded as that high. Looking at all the possible values of parameters b and p, for the most part spillovers remain constant with reductions in either b or p. When firms cooperate in R&D, reductions in b when the initial value of b is considered as insufficiently high, or reductions in p when p is considered not high enough don't cause knowledge spillovers towards surrounding areas to increase.

Caragliu and Nijkamp (2013) found that local absorptive capacity does not suffice to prevent outward spillovers to other surrounding regions. Knowledge spillovers are calculated in their empirical study on the basis of different definitions of proximity, namely geographical proximity and cognitive proximity. Local absorptive capacity is found to have a negative impact on outward spillovers only when a region is surrounded by regions with similarly high levels of absorptive capacity.

We address now a question concerning increases in the absorptive capacity as a result of connectivity and/or complementarity investments made in the territory. Regions with high innovative performance should have a high territorial similarity and complementarity simultaneously translated into high values of b and p, respectively. Is it possible that additional increases in the value of any of these parameters increase x and reduce the flows of absorbable spillovers towards neighboring areas, $x(1 - d)$ at the same time?

The next corollary establishes under what territorial circumstances such a reduction of $x(1 - d)$ takes place. Table 1 describes individual absorbable spillovers to neighboring regions.

Corollary 5 When firms compete in R&D, the individual R&D investment, x increases and absorbable spillovers towards other areas, $x(1 - d)$ decrease with increases in b for $b > 0.977745$ and $0 < p < 1$.

In regions with important territorial capabilities, it is possible that increases in the similarity element b of absorptive capacity increase each competing firm's investment in R&D at the same time that they reduce useful outflows of knowledge benefiting the neighboring regions. Increases in the territorial complementary p do not contribute to simultaneously increase x and reduce $x(1 - d)$, whatever the nature of the strategic interaction of firms in R&D is. When firms cooperate in R&D, there is no simultaneous increase in x and reduction in $x(1 - d)$ for any increase in b. If p increases when $b < p$, which leads cooperating firms to choose $d = 1$, then we have that $x(1 - d) = 0$, i.e. no increase. For the most part of the parameter range, useful spillovers don't change with increases in either b or p, whether firms compete or cooperate in R&D.

A new question of regional nature related to or complementing this is one is raised next. Does an increase in regional absorptive capacity increase useful spillovers towards other regions? This seems to be of relevance, as after all it is conceivable that very innovative regions be concerned with increases in useful spillovers towards other surrounding areas following their public investments in territorial capability.

Corollary 6 Suppose firms compete in R&D. Increases in territorial similarity b increase individual flows of absorbable spillover to surrounding areas, $x(1-d)$ for $b > 0.977745$ and $p = 1$. Increases in territorial complementarity p increase $x(1-d)$ for $b > 0.977745$. Suppose that firms cooperate in R&D. Increases in b increase $x(1-d)$ for $b > p$.

A region can be considered as having a very high innovative performance, if increases of its territorial components b and p increase individual knowledge flows $x(1-d)$ towards neighboring regions.

When firms compete in R&D, increases in $x(1-d)$ are an indicator of high territorial complementarity and existence of local increasing returns when b increases, and an indicator of high territorial connectivity when p increases. Increases in b decrease knowledge flows to other regions except when the complementarity component of absorptive capacity is strong enough, i.e. when $p = 1$ or p takes values in the close neighborhood of 1. Increases in $x(1-d)$ following increases in p is a good sign and an indicator that b is high enough.

Suppose now that firms cooperate in R&D. Increases in $x(1-d)$ following increases in b is an indicator that b is higher than p. In this sense, increases in $x(1-d)$ when b increases is an indicator of high territorial connectivity. Note that if b increases when $b > p$, which leads firms to choose $d = 0$, then $x(1-d) = x$ also increases.

Traditional regional policies apparently cannot achieve the objectives of higher aggregate growth and regional equity (Martin 1999). A public policy that improves infrastructures that reduce transaction costs will have an effect on the geography of economic activities, the growth rate of the whole economy, and the spatial income distribution. There will be a policy trade-off between growth and spatial distribution of economic activities.

However, there may be no such downside for regional policies that reduce the cost of innovation and diffusion of technology, as public authorities by implementing these policies eventually face a non-conflicting relationship between regional equity and efficiency. Firms innovate more in a rich region because they are agglomerated in the region concerned and benefit from dynamic advantages of agglomeration. The agglomeration of firms brings growth and dynamic efficiency, but also leads to declining profits among rival firms. On the other hand, the increase of knowledge diffusion accompanied by innovation in the rich region and useful spillovers to other, less developed regions will increase both the profits of imitative firms located there and the welfare of the less developed regions.

In our model, due to the past intervention of public authorities which have implemented regional policies promoting regional connectivity and complementarity, we may no longer observe the existence of such a conflicting relationship between regional growth and inequality. As soon as b and p reach sufficiently high values in a rich region, it appears to be possible to achieve both a higher long-term growth and smaller territorial inequalities through further political action. When firms compete in R&D, once a region has reached as a result of political action a value of $b > 0.977745$ and a value of p around 1, our model predicts that further increases in b and/or p are translated into simultaneous increases in x, X and $x(1 - d)$.

Finally, we raise another question of a regional nature that is related to the previous ones. One might think that innovative regions, with rising values of territorial similarity and complementarity b and p, respectively, while enduring the paradox of regional productivity, will show a high and growing amount of their absorbable knowledge being inevitably spilled over to other regions. The second phenomenon would be causally related to the first, one might argue, thus contributing to its explanation. However, that is not what happens in our model. The following corollary is an immediate implication of our previous results.

Corollary 7 The paradox of productivity at regional level and the increase of absorbable spillovers to surrounding areas don't take place simultaneously following increases in territorial similarity and/or complementarity.

There is no reason for developed regions to fear the occurrence of these two setbacks at the same time, whether their firms compete or cooperate in R&D. We do not observe at the same time a reduction or even a constancy of effective R&D effort, X and an increase of usable spillovers, $x(1 - d)$ following increases in territorial connectivity, b. This negative answer to our last question needs explanation. Suppose firms compete in R&D. When b increases, for $b > 0.977745$ and $p = 1$, x grows significantly more than d, thus increasing $x(1 - d)$. The equilibrium d increases relatively more than the equilibrium x with b if the local increasing returns incentive effects on x and d outweigh the competitive disincentive effects. When firms cooperate in R&D, the productivity paradox in its strong form doesn't even take place. As to the weak version of this productivity problem, this is possible to occur as either increases in b for $b < p$ or increases in p for $b > p$ do not change X, but then x and $x(1 - d)$ do not change as well. Results 1, 2 and 6 above allow us to draw these conclusions.

5 Conclusions and Further Developments

In this paper, we have shown that cognitive considerations associated with the spatial organization and culture of territory when combined with cognitive elements related to the organization of R&D and culture of firms contribute to justify and explain the productivity paradox observed in regions with a considerable level of innovative effort. Intra-regional competition between firms clustered in a given developed territory is shown to have the effect of counteracting any benefit in terms of firms' innovative performance from increases in technological distance d induced by increases in territorial complementarity promoted by regional policies.

We have also drawn policy implications from the model concerning the solution for the regional productivity problem, which are dependent upon the nature of the strategic interaction of firms in the region. Public authorities should invest massively and comprehensively in all territorial elements of absorptive capacity when firms compete in R&D, but invest persistently and exclusively in the most efficient territorial element when firms cooperate in R&D.

A complete analysis of local increasing returns and their effects on the technical progressiveness of a region and the amount of absorbable knowledge outflows towards surrounding areas requires the relaxation of the assumption that territorial complementarity p cannot exceed one. This is a first theoretical development worth pursuing. There is a second line of research also linked to firm competition and strategic interaction. Interregional competition can be included and treated in an enlarged version of our model to explain how interregional flows of information between competitors can influence business management at the early stage of R&D orientation and what impact the regional policies for promotion of intra-regional connectivity and complementarity might have on these business choices and information flows.

As a last and most promising research challenge, it would be interesting to study institutional interactions between the goals and behaviors of public authorities with regional interests and private for-profit firms. There may be an interesting situation of institutional complementarity between industrial relations and territorial organization, which is worth to be treated and studied using regional economic models in conjunction with the analysis carried out in this paper. Thus, an important development is to propose a regional model complementary to the market model in this paper, in which b and p are now choice variables under the control of local public entities.

Acknowledgements The authors are grateful to Roberta Capello and Andrea Caragliu for valuable suggestions and comments.

Appendix

Table 1 Equilibrium R&D design, R&D effort level, effective R&D level, and individual absorbable spillovers to surrounding areas in the non-cooperative case

b\p	0	0.25	0.5	0.75	1
0					
d	0; 1	0	0	0	0
x	1.42653; 1.42653	1.42653	1.42653	1.42653	1.42653
X	1.42653; 1.42653	1.42653	1.42653	1.42653	1.42653
x(1 − d)	1.42653; 0	1.42653	1.42653	1.42653	1.42653
0.25					
d	0; 1	0	0	0	0
x	1.26514; 1.42653	1.26514	1.26514	1.26514	1.26514
X	1.58142; 1.42653	1.58142	1.58142	1.58142	1.58142
x(1 − d)	1.26514; 0	1.26514	1.26514	1.26514	1.26514
0.5					
d	0; 1	0	0	0	0
x	1.08932; 1.42653	1.08932	1.08932	1.08932	1.08932
X	1.63399; 1.42653	1.63399	1.63399	1.63399	1.63399
x(1 − d)	1.08932; 0	1.08932	1.08932	1.08932	1.08932
0.75					
d	0; 1	0	0	0	0
x	0.903669; 1.42653	0.903669	0.903669	0.903669	0.903669
X	1.58142; 1.42653	1.58142	1.58142	1.58142	1.58142
x(1 − d)	0.903669; 0	0.903669	0.903669	0.903669	0.903669
0.977745					
d	0; 1	0	0	0	0
x	0.730274; 1.42653	0.730274	0.730274	0.730274	0.730274
X	1.4443; 1.42653	1.4443	1.4443	1.4443	1.4443
x(1 − d)	0.730274; 0	0.730274	0.730274	0.730274	0.730274
0.98					
d	0.00113624; 1	0.00131082	0.00154858	0.00189124	0.00242718
x	0.730252; 1.42653	0.730513	0.73087	0.731385	0.732192
X	1.44427; 1.42653	1.44454	1.44491	1.44544	1.44627
x(1 − d)	0.729422; 0	0.729555	0.729738	0.730002	0.730415
0.985					
d	0.00367465; 1	0.00423411	0.00499253	0.00607712	0.0077495
x	0.730252; 1.42653	0.731098	0.732248	0.733898	0.736453
X	1.44427; 1.42653	1.44515	1.446341	1.44805	1.4507
x(1 − d)	0.727569; 0	0.728002	0.728592	0.729438	0.730746
0.99					
d	0.00619381; 1	0.00712827	0.0083893	0.0101792	0.0129019
x	0.730252; 1.42653	0.731678	0.733609	0.736364	0.740587

(continued)

Table 1 (continued)

b\p	0	0.25	0.5	0.75	1
X	1.44427; 1.42653	1.44576	1.44777	1.45066	1.45509
x(1 − d)	0.725729; 0	0.726462	0.727455	0.728868	0.731032
0.995					
d	0.00869396; 1	0.00999382	0.0117401	0.0142007	0.0178959
x	0.730252; 1.42653	0.732252	0.734953	0.738786	0.744602
X	1.44427; 1.42653	1.44637	1.44921	1.45326	1.45944
x(1 − d)	0.723903; 0	0.724934	0.726325	0.728295	0.731277
1					
d	0.0111753; 1	0.0128312	0.0150461	0.0181449	0.0227415
x	0.730252; 1.42653	0.732822	0.736282	0.741166	0.748507
X	1.44427; 1.42653	1.44699	1.45066	1.45586	1.46374
x(1 − d)	0.722091; 0	0.723419	0.725204	0.727718	0.731485

References

Abecassis-Moedas C, Mahmoud-Jouini B (2008) Absorptive capacity and source-recipient complementarity in designing new products: an empirically derived framework. J Prod Innovation Manag 25:473–490

Boschma R, Minondo A, Navarro M (2012) Related variety and regional growth in Spain. Papers Reg Sci 91:241–256

Broekel T, Boschma R (2012) Knowledge networks in the Dutch aviation industry: the proximity paradox. J Econ Geography 12:409–433

Bulow J, Geneakoplos J, Klemperer P (1985) Multimarket oligopoly, strategic substitutes and complements. J Political Econ 93:488–511

Capello R, Caragliu A, Nijkamp P (2009) Territorial capital and regional growth: increasing returns in knowledge use. Tinbergen Institute Discussion Paper 059/3, Tinbergen Institute, The Netherlands. http://papers.tinbergen.nl/09059.pdf. Accessed 11 Aug 2016

Caragliu A, Nijkamp P (2012) The impact of regional absorptive capacity on spatial knowledge spillovers: the Cohen and Levinthal model revisited. Appl Econ 44:1363–1374

Caragliu A, Nijkamp P (2013) Space and knowledge spillovers in European regions – the impact of different forms of proximity on spatial knowledge diffusion. Tinbergen Institute Discussion Paper 148/VIII, Tinbergen Institute, The Netherlands. http://papers.tinbergen.nl/13148.pdf. Accessed 11 Aug 2016

Cohen WM, Levinthal DA (1989) Innovation and learning: the faces of R&D. Econ J 94:569–596

Cohen WM, Levinthal DA (1990) Absorptive capacity: a new perspective on learning and innovation. Admin Sci Q 35:128–152

Corsaro D, Chiara C, Tunisini A (2012) Actors' heterogeneity in innovation networks. Indus Market Manag 41:780–789

Feldman MP (1994) Knowledge complementarity and innovation. Small Bus Econ 6:363–372

Hopner M (2005) What connects industrial relations and corporate governance? Explaining institutional complementarity. Socio-Econ Rev 3:331–358

Martin P (1999) Public policies, regional inequalities and growth. J Public Econ 73:85–105

Meijers EJ (2007) Synergy in polycentric, interurban relationships, urban networks, complementarity, organizing capacity, critical mass. Delft University Press, Amsterdam

Milgrom P, Roberts J (1995) Complementarities and fit: strategy, structure, and organizational change in manufacturing. J Account Econ 19:179–208

Nooteboom B (2000) Learning by interaction: absorptive capacity, cognitive distance and governance. J Manag Gov 4:69–92

Swart R (2008) Climate change versus development: trade-offs and synergies. Progressive Governance, London

Wiethaus L (2005) Absorptive capacity and connectedness: why competing firms also adopt identical R&D approaches. Int J Indus Organ 23:467–481

Mário A.P.M. Da Silva is Professor of Industrial Economics and Competition Policy at the University of Porto, Portugal. He is currently doing research on Spatial Science.

Peter Nijkamp *Address Office*: VU University Amsterdam, Faculty of Economics, Dept. of Spatial Economics, De Boelelaan 1105, 1081 HV Amsterdam, The Netherlands, Tel. +31-20-598 6090, Fax +31-20-598 6004, E-mail: p.nijkamp@vu.nl

Personal: Born February 26, 1946, Dalfsen, The Netherlands

Education: Econometrics and Regional Economics, Erasmus University, Rotterdam (1964–1970); Ph.D. in Regional Economics at the Erasmus University, Rotterdam (1972), on *"Planning of Industrial Complexes by Means of Geometric Programming"* (Rotterdam University Press, Rotterdam, 1972) (cum laude)

Previous Appointments: Research Assistant, Econometric Institute, Erasmus University Rotterdam (1969–1970); Assistant Professor, Faculty of Economics, Erasmus University, Rotterdam (1970–1973); Lecturer, Faculty of Economics, VU University, Amsterdam (1973–1975), and Faculty of Economics, Erasmus University, Rotterdam (1973–1978); Professor and Chairman Department of Regional Economics, VU University, Amsterdam (1975–2002); Member of the Board of the Humanities Division of the Royal Netherlands Academy of Science (KNAW) (1993–2002); Vice-president Royal Netherlands Academy of Science (KNAW) (1999–2002); President Governing Board Netherlands Research Council (NWO) (2002–2009); President of the European Heads of Research Councils (EUROHORCs) (2005–2007)

Professional Position: Full Professor in Regional Economics and in Economic Geography, Faculty of Economics, VU University, Amsterdam (1975–2015); Executive Director, Joint Programming Initiative Urban Europe (2010–2012); Honorary University Professor, Faculty of Economics, VU University, Amsterdam (2009–2015)

Official Signs of Scientific Recognition: Honorary medal University of Poitiers, France (1981); Honorary medal Panthios University, Athens, Greece (1987); Honorary medal University of Catania, Italy (1994); Honorary medal State University of Groningen, Groningen (2009); Doctor honoris causa, Free University, Brussels (1989); Doctor honoris causa, National Technical University, Athens (2003); Doctor honoris causa, Academy of Economic Studies, Bucharest (2005); Doctor honoris causa, University of the Algarve, Faro (2012); Honorary award European Regional Science Association (1995); Walter Isard senior scholarship award of the North American Regional Science Association (1997); Honorary Scholar International Institute for Applied Systems Analysis (IIASA), Vienna (2009); Honorary Research Fellow, Economic & Social Research Institute, Dublin; Fellow Royal Netherlands Academy of Arts and Sciences (KNAW); Fellow World Academy of Arts and Sciences (WAAS); Fellow International Academy of Regional Development and Cooperation, Moscow; Fellow Holland Society of Sciences; Fellow Royal Belgian Academy of Science and Arts; Fellow Academia Europaea (and Chair of the Economics Section); Winner of the 1996 Dutch Spinoza Prize (highest scientific award in the Netherlands); Member Scientific Advisory Board, Department of Systems Ecology, Academia Sinica, China; Fellow Tinbergen Institute, Amsterdam/Rotterdam; Honorary member of Japan Section of Regional Science Association International; Founder's Medal of the Regional Science Association International (RSAI); Fellow Regional Science Association International

Detailed information can be found on http://personal.vu.nl/p.nijkamp.

A complete list of publications is available upon request. Send an e-mail to p.nijkamp@vu.nl.

Part II
Innovation Management and Policy Perspectives: Case Studies in the Central European Countries

The progressive enlargement of the European Union (EU) raised several questions such as the legal harmonization in the EU, regional policy and cross-border cooperation. As an internal market with over 500 million people, the EU integration process demanded for the EU institutions to become more effective at the regional and structural policy levels aiming at reducing disparity of development between regions and to support an innovation-driven economy.

Nevertheless, regional decision makers have to increase their collaboration and participation in such European innovation systems in order to answer the challenges of the global digital era. Designing and implementing effective mechanisms in European market-based economies requires the identification of relevant needs in each and every single country. However, building and implementing the all systems entails significant changes and requires not only targeting technological innovation aspects but also looking beyond and paying more attention to the cultural and business-related dimensions of each country and region.

This part focuses on a set of real case studies aiming to offer new insights and practical lessons on the issues of innovation management both in the Central European countries and in the context of the EU global internal market.

The Regional Innovation Policy: The Situation of Slovakia

Maria Horehajova and Jana Marasova

Abstract Main drivers of innovation in the economy are undoubtedly businesses. Entrepreneurs are expected to be inventive, to have initiative in discovering new products, methods, markets and be willing to take risks, which is also a prerequisite for successful business. Our research, while accepting the crucial role of enterprises in innovation, focuses on the role of regional authorities, which through innovation policies can greatly influence the development of innovation potential in the region. In our research, we rely primarily on existing studies of regional policy in general and we highlight the reasons for which the regional policy started to address the innovation potential. Next, we have analyzed the situation in Slovak Republic compared with the innovative potential of 28 EU countries and the changes that occurred in the years 2004–2013. In order to evaluate the innovation performance we have used six selected indicators that are part of a composite indicator—the Summary Innovation Index (Hollanders et al., Innovation Union Scoreboard: The European Commission Report, 2015. http://ec.europa.eu/enterprise/policies/. Accessed 16 June 2015). Finally, this chapter shows the importance and role of regional territorial authority in creating the innovative potential of the region, focusing on the Košice region. We present a specific case of cross-border innovation regional policy, common to the Kosice region in Slovakia and Northern Hungary region, and we summarize the results that this policy has made.

Keywords Innovation policy • Regional authority • Innovation potential • Innovation performance • Regional innovation strategy

1 The Theoretical Basis of Regional Policy

The concept of regional policy emerged in the 1930s of the twentieth century, in the time when we started to consider the application of economic and policy instruments at two levels—at the national but also at the regional level. The general

M. Horehajova (✉) • J. Marasova
Faculty of Economics, Matej Bel University, Banska Bystrica, Slovakia
e-mail: maria.horehajova@umb.sk

A. Dias et al. (eds.), *Modeling Innovation Sustainability and Technologies*,
Springer Proceedings in Business and Economics,
https://doi.org/10.1007/978-3-319-67101-7_5

definition of regional policy identifies it with the economic policy of a region and includes, according Klassen and Vanhove (1987), all public interventions that are leading to an improvement of territorial organization of economic activities. Indeed, regional policy supposes the correction of some regional consequences of the free market in terms of the following two objectives: economic growth and better social redistribution. For simplicity, we can characterize regional policy as a public policy aimed at dynamic development and harmonization of regions.

According Bachtler and Yuill (2001), paradigms of contemporary regional policy, results from the change from approaches based on the classical theories oriented to the reduction of cost and the acquisition of the workforce, to current neoclassical theories that emphasize a need to strengthen the competitiveness, quality of human resources and innovation. The growing importance of innovation in regional policy is linked to a change in its organization and management, which was originally centralized at national level. After quite a long period of the policy centralized at the national level, in recent decade, the regional authorities seek to participate both in determining the objectives of innovation policy, as well as on their achieving, emphasizing the principle of participation and partnership building.

The Regional Innovation Strategy (RIS) is one of the contemporary regional policies resulting from the development of the new economy. The RIS emphasizes endogenous growth factors of the region and considers corporate innovation capabilities a key element of corporate growth. Thus, it has become means of the search for a new paradigm of regional policy. The reason for the innovation oriented regional policy is, in particular, increasing mobility of capital. Especially peripheral regions of developed economies have been affected by the shift of production to developing countries, while the importance of new technologies and innovations in the competitive capacity of enterprises continued to increase (Kaufmann 2002). However, the business investments in innovation projects are associated with a high level of uncertainty and their profitability depends on the strength of their competitors. Especially in peripheral and lagging regions, research and development outputs are low, product innovation is weak, and the introduction of new practices is slow. These negative facts could be partly eliminated by an appropriate regional innovation policy that would stimulate the development of new technologies, technology clusters, and thus the competitiveness of businesses. Measures implemented within the regional innovation policy can be in the form of financial support for research and development activities, innovative projects, business advice, as well they may relate to transport and telecommunications infrastructure, support to educational institutions, which may increase the innovation potential of the region.

With the objective to increase the innovation capacity of regions and contribute to the optimization of regional innovation policy, since 1994 the RIS is supported by the European Fund for Regional Development. The original pilot innovation project has already been revised and improved several times, most recently in the form of RIS3 called Smart specialization. European Commission (EC) defines smart specialization as a strategy designed for a specific region, which will lead to its economic transformation. To achieve this goal, according to the EC (2012) it is necessary:

- To focus policy support and investment in key national and regional priorities;
- To build on the strengths of each region;
- To promote technological innovation as well as "practice-based" innovation;
- To involve all stakeholders in the promotion of innovation and experimentation;
- To identify monitoring and evaluation systems of innovations.

2 Support of Innovation in Slovakia

The support of research and innovation in Slovakia is related to a new perspective of regional policy and the overall Europe 2020 Strategy. This support is based on two main documents: *Innovation policy of the Slovak Republic* and *Innovative strategy of the Slovak Republic* developed for the period 2011–2013. These documents reflect the objectives of the European strategy and aim to reduce the lag of Slovak research in relation to the EU average (EC 2014a). According to the Innovation Union Scoreboard (Hollanders et al. 2015), in which the Summary Innovation Index (SII) is used as the evaluation tool of the member countries, Slovakia found itself between 2004 and 2010 in the group of five countries whose index value is the lowest. Both mentioned documents were intended to create a potential for innovation in the regions and thus enhance the overall performance of the Slovak economy (EC 2007).

In our analysis, we rely on a few indicators that comprise the index SII and explain best the problems with innovation in Slovakia. These indicators also show how cooperation between public institutions and private agents is important.

The Summary Innovation Index summarizes the performance of a range of 3 main types of indicators—Enablers, Firm activities and Outputs.[1] Among these three main groups of indicators, we have chosen two of each group. In the first group Enablers we followed the total R & D expenditure as a % of GDP and percentage of population aged 30–34 having completed tertiary education. In the second group Firm activities it was R & D expenditure in the business sector as a % of GDP and number of patents granted by the United States Patent and Trademark Office (USPTO) per million inhabitants. From the third group of Outputs we have selected employment in knowledge-intensive activities (manufacturing and services) as a % of total employment, and the high-tech product exports as percentage of total exports.

We compared the values of selected indicators in Slovakia for the years 2004 and 2013 (EC 2014b). Figure 1 shows that the situation has improved there, except the employment indicator, which is the most serious problem of the Slovak economy.

[1] IUS distinguishes eight innovation dimensions for each of the three main groups and the set of 25 indicators that analyse the results of the EU innovation system.

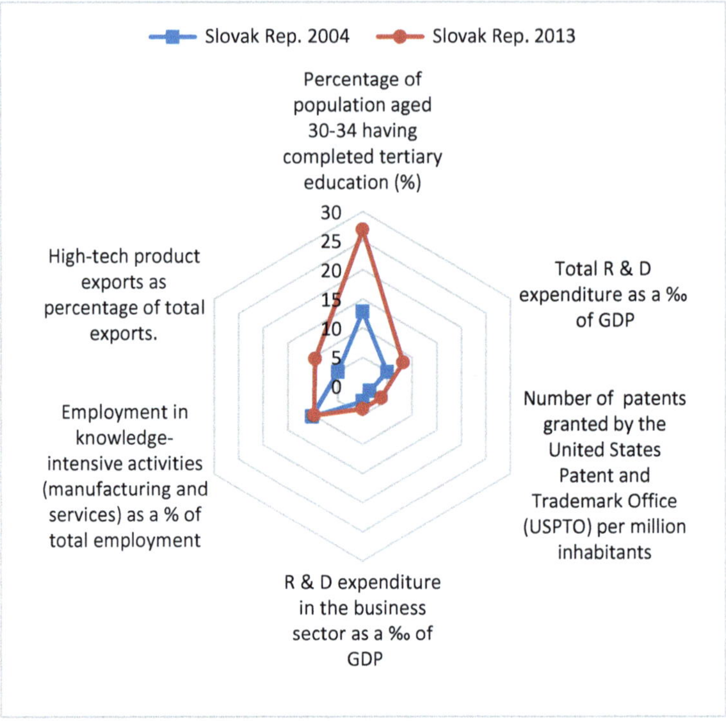

Fig. 1 Selected indicators in Slovakia for the years 2004 and 2013

The two indicators related to expenditures on R&D in Slovakia still belong to the lowest in the EU, and they are deeply below the 3% of GDP, both in the public and private sectors. One way to improve this situation would be to strengthen cooperation between the regional authorities, companies and universities or other research organizations. It represents a key base of accumulation of necessary resources to support regional research and development.

In order to show the current status of innovation in the country, we compare the results of 2013 in Slovakia with the average of EU 28 (Fig. 2). As can be seen, indicator of the number of granted patents in the country is very low, which is probably due to the low level of total R & D expenditure. Employment in knowledge-intensive activities achieved the best level of the selected indicators compared to the EU 28 and therefore we decided to analyze this indicator from a regional perspective.

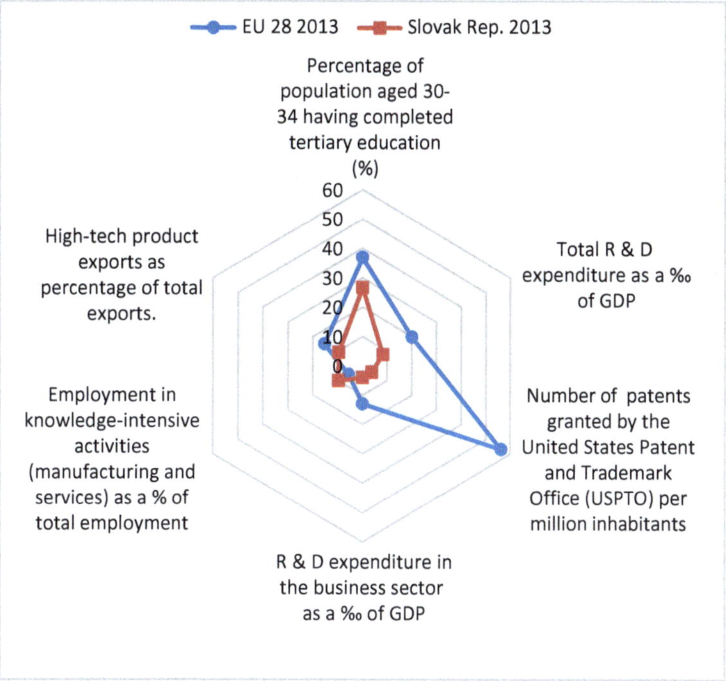

Fig. 2 Selected indicators in Slovakia and EU 28 for the year 2013

3 The Role of Regional Authorities in the Regional Innovation Policy

In 2002, as part of a decentralization process in Slovakia, there has been created 8 regions managed by the higher territorial unit (in Slovak language: Vyssie uzemne celky—VUC). One of the first responsibilities assigned to the VUC was the development of the regions with the aim to reduce regional disparities, especially by identification of natural, social and human capital in each region and by its mobilization.

Two legislative standards define the role and tasks of the VUC in the management and development of their regions. The first—NR SR Act No. 302 of 2001 assigns to the VUC precise activities related to the planning, investment, development and entrepreneurship in the region. The second—NR SR Act No. 539 of 2008 is aimed at supporting regional development.[2] Although this act highlights the role of VUC in regional development, it does not define specific competencies in the

[2]The regional authorities in fulfilling their duties rely on two documents: Program of economic and social development of VUC that results from the strategic planning of regions and Territorial plan VUC based on their territorial development.

promotion of innovation. An essential role in terms of competencies and investment financing by the VUC should play an Innovation Act. Government has been preparing Innovation Act for several years, but the final version has not yet been approved. Funding of science, research and innovation is thus not directly included in the competence of regional governments.[3] Consequently the budget of VUC does not include innovation expenditures even though the VUC plays a significant role in the formulation of Regional Innovation Strategy (RIS). Anyway, VUC can promote innovation, for example through funding, respectively co-financing of industrial parks, industrial zones, partnerships and clusters.

For a better understanding of the role and responsibility of the VUC, it must be emphasized that they do not have their own territories or "their" citizens. The territory of a region consists of cadastral areas of municipalities whose citizens participate in the management of the region through their elected representatives (deputies and chairman of the VUC) on the one hand and NGOs or some private sector institutions on the other hand.

The role of VUC in public administration is complex because of the many relationships in which they operate as regional governments with different partners and by inadequate instruments and insufficient space to perform regional development. The experiences and VUC analyzes show that some of their powers are very limited (Cernakova 2012). They demand strengthening of their coordination function towards the municipalities, increased funding and more opportunities for participation in decision making regarding the use of state (ministries) subsidies for the region. They should also be more involved in decisions concerning the redistribution of European resources. The fact that territorial authorities VUC were not directly involved in the preparation of the National Program of Reforms of the

---

[3]VUC are responsible for the following areas:

- Social, economic and cultural development of regions;
- Management of own budget, of investment and public contracts;
- International and trans-regional cooperation;
- Regional planning;
- Social welfare, including homes for children, social policy and coordination of all subjects related to this area;
- Healthcare, including the establishment of hospitals of second type, management of non-State healthcare as psychiatric hospitals and dental services);
- Education, including secondary, professional, art and vocational schools, construction and maintenance of buildings, payment of teacher on behalf of the state;
- Transport, including the construction and maintenance of regional roads, coordination of railway system on its territory;
- Culture, including regional theatres, libraries, museums, galleries and cultural centres;
- Tourism, including the planning of regional tourism and regional tourism development;
- Sport;
- Youth;
- Human pharmaceutics, including issuing licences for public pharmacies, executing control of public pharmacies, managing stand-by pharmaceutical services or decision-making on pharmaceutical reserves;
- Civil defence (in cooperation with state bodies).

Slovak Republic is a proof that the subsidiarity principle is not sufficiently applied in the creation of strategic documents. The Program aims at economic and social cohesion, but the territorial cohesion, in its sectoral logic, is absent.

Despite the above mentioned negative aspects, the VUC is a key element in the process of creation and implementation of regional innovation policy. They have a direct responsibility in the management of regions, initiative, and coordination and in the use of the potential of many agents acting on their territories. Slovakia has participated in the Regional Innovation Strategy Program in Newly Associated Countries in the years 2001–2004, when the first two strategies have been developed for the regions of Bratislava and Nitra. Between 2006 and 2008, innovation strategies were created for six other regions, including the strategy of the Kosice region which is special because it represents a first cross-border RIS within the EU. Common strategy of Kosice region and the region Northern Hungary is elaborated up to 2015.

It was developed as part of the bilateral project NORRIS (North Hungary and Košice Bilateral Regional Innovation Strategy) which was financed by the European funds (The Sixth Framework Program) for Research and Technological Development. Beside the main partners—VUC Kosice and regional authority of Borsod-Abauj-Zemplen, on development of this strategy also participated universities (Technical University of Kosice and Miskolci Egyetem), several foundations and development agencies (two Slovak agencies and three Hungarian) and an Austrian Institute for Technology and Regional Policy (Joanneum Research) in Styria.

This joint innovation strategy set for the period 2007–2013 priorities and measures to be taken by the regions in order to strengthen the knowledge economy of these territories in the selected areas. The RIS includes analytical part, strategic part and action plan. The analysis concerns the innovation potential of both regions and emphasizes the perspectives industries such as information technology, biomedicine, renewable energy resources and others, both from the point of view of demand as well as of supply. The strategic part formulates priorities and measures for their implementation, and identifies possible resources. The strategy defines six priorities:

1. Improving public management and professional capacities for innovation and knowledge policy
2. Creating a favourable business environment
3. The effective diffusion of knowledge and technology transfer in enterprises
4. Innovation poles and clusters
5. Supporting the creation and growth of innovative enterprises
6. Strengthening of applied research and development of new products and services.

At the end of the specified period, we can see that common RIS has been beneficial to the cooperation of several cross-border partnerships. We assume that the fulfilment of these priorities have also contributed to an improvement in several areas in the Košice region. We decided to check this assumption in the sphere of science and research using one of the Innovation Union Scoreboard indicators—employment in knowledge—intensive activities. In order to verify the presumption that the situation in the field of science and research has improved, we used the method of calculating the localization quotient for 2009 and 2014. The quotient measures the concentration of given economic activity in the region compared to the higher territorial unit, in this case to the whole Slovak Republic:

$$Q_k = \frac{\frac{E_k}{E}}{\frac{C_k}{C}}$$

E_k: number of employed in the k (knowledge—intensive activity) sector in the given territorial unit
E: number of employed in all sectors in the given territorial unit
C_k: number of employed in the k sector in higher territorial unit
C: number of employed in all sectors in the given territorial unit

Localization quotient for the Košice region with the reference territory of Slovakia for the year 2009 was $Q_k = 0.7822$ and for the year 2014 has the value of 0.8539. For comparison localization quotient for the most developed Slovakian region Bratislava in 2014 reached the value of 2.419. In both cases, as in 2009 and also in 2014 the localization quotient for the Košice region is less than 1, which means that there is less concentration of employment in knowledge—intensive activities compared with the situation in the whole country and this indicator is not sufficient to meet local, regional needs. Nevertheless, on the basis of calculation, there was some improvement, which is undoubtedly the result of the implementation of RIS.

The first existing evaluation reports prove that several scientific-research institutions were largely involved in fulfilling strategic priorities, particularly TUKE (The Technical University of Košice), that in the years 2011–2012 carried out more than 40 international and about 200 national projects. Likewise, the research results of two other universities UPJŠ (Pavol Jozef Šafárik University) and UVLF (The University of Veterinary Medicine and Pharmacy) show their major contribution to development of education, science and culture in the Košice region.

There are diverse regional innovation structures. The important role, in cooperation between academia, industry and the public sector in the Košice region, has played competency centres (CC). In the period under review four CC were established, two of them at UPJŠ (Software CC CELUM and CC for industrial research and development of light metals and composites), one CC ZATIPS at TUKE (CC of knowledge technologies for innovations of production systems in industry and services) and CC PROBIOTECH at the Institute of Animal Physiology (CC for bio modulators and supplements). Furthermore, there were built up 14 centres of excellence that document also interregional cooperation in science and research.

Successful outcomes of RIS in the Kosice region unquestionably include the existence of clusters despite the unfavourable conditions for their operation in

Slovakia (insufficient involvement of SMEs in clusters because of several barriers). There are two major clusters in Košice region:

- Cluster automation and robotics founded in 2010, which brings together six innovative companies and two research institutes and has the full support of VUC Košice and VUC Prešov
- IT Cluster VALLEY has contributed, in recent years, to significant increase in new jobs in the IT sector. One of its aims is to find and prepare talent among students of secondary schools and universities for the IT sector.

Achieved results of RIS show that the strategy plays an important role in regional policy, despite the fact that both the competence and financing of innovation at regional level are not resolved.

4 Conclusion

The new paradigm of regional policy emphasizes the need for innovation and the associated development of a knowledge economy and competitiveness. The economic crisis and its negative impact on the economy of the various European regions strengthened the importance of regional innovation strategy as a form of regional policy, which is focused on endogenous factors of economic growth typical for the region. By defining not only national but also regional priorities, a special innovation strategy can be developed, which helps enlarge the innovative potential of the region and realize its overall economic transformation.

According to the SII, which measures the innovation performance of the economy, Slovakia is among the countries that have to increase their innovative potential in order to achieve at least the average of EU 28 member states. Although some data included in SII show an improvement for the last 10 years, Slovakia's lagging behind most European countries is obvious. One of the main reasons is certainly the low level of spending on science and research both from public and private sources. From a regional perspective, the situation is even more complex, as there are large regional differences in innovation performance. Although, our research paper focused just on one of the SII indicators that was evaluated using a localization quotient, the difference, for example, between Kosice and Bratislava regions is enormous.

One way to reduce regional disparities in innovation is to develop and consequently implement RIS, which gives the possibilities of such regional innovation structures that can significantly contribute to the creation and expansion of innovative potential in the region.

This paper has been written as a part of research activities of following projects: ITMS 26110230082 "Mobility - Support of science, research and education at UMB" and KEGA 032UMB-4/2013 "Teaching economics in foreign languages as a prerequisite for the quality university education and the development of research activities in the European region".

References

Bachtler J, Yuill D (2001) Policies and strategies for regional development: a shift in paradigm? http://www.paca-online.org/cop/docs/J_Bachtler_Policies_and_strategies_for_regional_development.pdf. Accessed 12 Oct 2015

Cernakova V (2012) Moznosti podpory inovacii na regionalnej urovni. http://ekonomikavs.fvs.upjs.sk/pdf/Cernakova.pdf. Accessed 21 Oct 2015

European Commission (2007) Rastúce regióny, rastúca Európa. Štvrtá správa o hospodárskej a sociálnej kohézii. http://ec.europa.eu/regional_policy/sources/docoffic/official/reports/cohesion4/pdf/4cr_sk.pdf Accessed 22 Sept 2015

European Commission (2012) Guide to Research and Innovation Strategies for Smart Specialisation (RIS 3). http://ec.europa.eu/regional_policy/sources/docgener/presenta/smart_specialisation/smart_ris3_2012.pdf Accessed 22 Oct 2015

European Commission (2014a) Regionalna politika EU. http://europa.eu/pol/pdf/flipbook/sk/regional_policy_sk.pdf. Accessed 31 May 2015

European Commission (2014b) Tableau de bord de l'Union de l'innovation. http://ec.europa.eu/enterprise/policies/innovation/files/. Accessed 16 June 2015

Hollanders H et al (2015) Innovation Union Scoreboard: The European Commission Report. http://ec.europa.eu/enterprise/policies/. Accessed 16 June 2015

Kaufmann D (2002) Transparency, incentives and prevention for corruption control and good governance, empirical findings, practical lessons, and strategies for action based on international experience. http://info.worldbank.org/etools/docs/library/18380/quinghua_paper.pdf. Accessed 20 June 2015

Klassen LH, Vanhove N (1987) Regional policy: a European approach, 2nd edn. Avebury, Aldershot

Maria Horehajova is university teacher and researcher at the Faculty of Economics, Matej Bel University in Banska Bystrica. She teaches economic theory and history of economic thought. Research field of Maria Horehajova includes government social policy and regional development, regional disparities, poverty and social exclusion. She is also interested in issues of the European social model, the principles and criteria on which it is built. She is the author of "Justice in the Context of the Principle of Social Policy".

Jana Marasova is university teacher and researcher at the Faculty of Economics, Matej Bel University in Banska Bystrica. She teaches economic theory and business management. She focuses her research on the economy and the management of human resources, but she also addresses educational issues of teaching economic theory in the current university education, its content and its forms. She is the author of textbook "Microeconomics. Theory of consumers' behaviour".

The Perceived Value of Public Services as a Prerequisite for a Comprehensive Analysis of the Effectiveness of Public Sector Organizations Using the Czech Library as an Example

Simona Pichova and Jan Stejskal

Abstract At present, the efficiency of allocating funds from public budgets is an issue that is being increasingly debated in the public sector. Mainly, this is due to increasing debt but also to changes in the way public services are provided. Good decisions regarding allocations, however, are prevented by the inability to measure output volume and the benefits for consumers provided by various services. Outcomes of public libraries are benefits of a system or service producer to its users (Vakkari and Serola, Library Inf Sci Res 34(1):37–44, 2012). Their value is more complex in the public sector than in the private sector and can therefore be harder to measure (Bloch and Bugge, Struct Chang Econ Dyn 27:133–145, 2013). This paper provides evidence that it is possible to analyse the effectiveness in the public sector—which provides library services—of both providers (libraries) and individual components of the services. This can be done through the application of a methodology that allows the consumers themselves to determine the perceived value of the services being used. The result of the analysis of these selected services' effectiveness at the biggest Czech library is a determination of its degree of effectiveness, which fluctuates around a value of one. The next step is to further divide individual standardized services into groups by whether they are effective or ineffective. A completely unique representative survey carried out in the Czech Republic in 2012 has been used throughout the analysis.

Keywords Efficiency • Cost-benefit analysis • Public services • Library • Perceived value

S. Pichova (✉) • J. Stejskal
Faculty of Economics and Administration, Institute of Economic Sciences, University of Pardubice, Studentska 95, Pardubice 53210, Czech Republic
e-mail: simona.pichova@upce.cz; jan.stejskal@upce.cz

A. Dias et al. (eds.), *Modeling Innovation Sustainability and Technologies*, Springer Proceedings in Business and Economics,
https://doi.org/10.1007/978-3-319-67101-7_6

1 Introduction

Securing public investment for the production of public services is one of the main tasks of the public sector in all modern countries. In the last 40 years, however, these systems have undergone significant reformative changes. These changes have been mainly focused on public management and the implementation of various elements of business management (new public management) and on individual principles of good government (Osborne and Plastrik 1997; Lynn 1996; Denhardt and Denhardt 2000; Fernandez and Rainey 2006). In the public sector, elements such as decentralization and deconcentration occur, the administration's uniformity is disrupted and the role of non-profit organizations is made stronger with an emphasis on effectiveness and quality. The tendency to hand over the provision of public production into the hands of the private sector is becoming much more frequent (the tensions between the emphases on decentralization promoted in the market model and the need for coordination in the public sector; Peters and Savoie 1996). Some theories describe these "neo-managerial" tendencies as "public entrepreneurship" (Fernandez and Rainey 2006).

In the last two decades of the twentieth century, scholars and practitioners in Europe and elsewhere have frequently attempted to analyse systems for providing public production and to make suggestions for increasing productivity and finding alternative service-delivery mechanisms based on public-choice assumptions and perspectives (Dunleavy and Hood 1994). In variously named managerial trends, a focus on accountability, effectiveness and high performance has also been recommended in these areas. Additionally, in order to attain these goals, it is necessary to redefine the organizational and production mission, to limit the influence of bureaucratic agencies and to allow for the privatization of certain public functions. This was a huge revolution in the existing understanding of the state's role in the economy (Brown and Osborne 2012). In the last decade of the twentieth century, Janet and Robert Denhardt published an approach called the new public service. This contains a range of elements from new public management and is considered a normative model, which differentiates it from other models. Precursors of the new public service are the theory of democratic citizenship, models of community and civil society as well as organization humanism and discourse theory.

This approach is focused on and is accountable to citizens; it is aimed at ensuring maximum prosperity by using new approaches for providing public production. Building coalitions of public, non-profit and private agencies is used for these purposes (Pestoff et al. 2013). Individual units are engaged in creating a suitable legal environment, they contribute to the creation of high-quality values in society and they further create standards for individual services according to the needs of the consumers—the citizens (Bao et al. 2013). At the same time, it draws these consumers into the production chain—both actively and passively (Bach and Kolins Givan 2011).

In their article, the creators of the new public service approach themselves state that it is necessary to create a suitable policy in order to apply this approach: one that will be a complex system featuring complex governance networks comprising a plurality of actors—public, private and non-government—each bringing their own special interest, resources, and set of expertise (Denhardt and Denhardt 2015). They emphasize the change in the public sector's role—from the service delivery role (a rowing role) to policy development (a steering role). However, it is necessarily to convincingly resolve one problem in order to accomplish this. Public administrators have long struggled with how to measure outcomes of public programs, because performance measurement tools have traditionally neglected them (Slater and Aiken 2014).

The aim of this paper is to present a method for determining the perceived value of public services used by the consumer and to use this to measure the effectiveness of these services with the help of the standard methods of cost-benefit analysis or return on public investment. This paper presents evidence that the analysis of the effectiveness is practical for the public sector and is not limited to analysis of the economy.

The structure of this paper is divided as follows: Sect. 1 consists of theoretical background on measuring the provision of public service. Section 2 focuses on measurement methods and approaches. Section 3 is dedicated to describing our methodology, data set and calculations. The last section recapitulates the main results and conclusions.

2 Theoretical Background

All the research on public services will therefore be used as a backdrop for discussion on how to evaluate or measure the utility of public service consumption. Many papers and studies revolve around both providing services cost effectively and creating societal wealth. However, value is more complex in the public sector than in the private sector and can therefore be harder to measure (Bloch and Bugge 2013). The business sector can measure output by existing indicators (sales, added value). However, these do not exist (or they do not have a corresponding value) in the public sector. Therefore, other indicators are used for measuring public sector performance (usually, the macroeconomic perspective). For individual services or organizations in the public sector, the microeconomic perspective is necessary, taking into account costs as well as benefits derived from the needs of the collective good. Evaluation potentially provides the key to improved effectiveness at both organizational and policy levels as defined in terms of the capacity to satisfy the needs and improve the quality of life of citizens (Sanderson 1996).

For evaluating public service outputs or outcomes, input-output economic methods are used. These analyse either one criterion by itself or more criteria, often just the costs as inputs or outputs in the form of benefits. There are primarily three methods for the economic evaluation of non-market goods or effects (e. g., externalities) based on consumer surplus: the travel cost method, the hedonic price method and the contingent valuation method (Marella and Raga 2014). The problem with these methods is again the difficulty of measuring outputs or results (Modell and Wiesel 2008; Hajek and Stejskal 2014, 2015) and the need for direct interaction with the consumer. Other methods replicate the procedures commonly used in the private sector, e. g., return-on-investment (Kaufman and Watstein 2008). The goal is to provide a clearer picture of the benefits and costs of the service producer. These methods can be used both for analysing the effectiveness of individual providers as well as for a region or the overall system a country uses for a selected type of service (McIntosh 2013).

All the methods for further analysis mentioned here use service values as perceived by customers. These methods are based on the principle of contingent valuation (Cummings and Taylor 1999), which originated as early as 1947. The essence of the methods established on this principle is the valuation of the willingness-to-pay of public service customers. The contingent valuation method is a non-market valuation method that is widely used, especially in the areas of environmental cost (Venkatachalam 2004), health care (Klose 1999) and public libraries (Stejskal and Hajek 2015). The CV principle is the basis of the method that is currently used in practice—the contingent valuation method (CVM). The CVM is a survey-based technique generally accepted as a meaningful tool used to estimate the value of various non-market goods (Lee and Chung 2012); it reflects altruistic motivation, a major component of non-use value in contingent valuation. This method gained popularity after the two major non-use values namely, option and existence values were recognized as important components of the total economic value (Venkatachalam 2004). For a methodology of contingent valuation, see (Russell et al. 1995; Wedgwood and Sansom 2003). In literature, critics of this approach have appeared. Variations on the contingent valuation method place value on goods or services that are far removed from any market pricing mechanism. Second, the valuation rests on subjective notions of value, rather than market values, with little regard for level of income or the trade-offs with other goods and services (IBRC 2007 in McIntosh 2013). Also contributing to the criticism is Matthews, who proposes using the CVM method results as "fairly low" estimates of value (Mathews 2011).

Experiments are often used for acquiring perceived values sometimes suitably formulated questionnaires or direct interviews are also used. Contingent valuation studies ask questions that help to reveal the monetary trade off each person would make concerning the value of goods or services (by means of a questionnaire, a hypothetical market is described where the service in question can be traded). The researcher must obtain a so-called "stated preference" from the customer. Such surveys are a practical alternative approach for eliciting the value of public goods, including those with passive use considerations. Results from contingent valuation studies are used for many purposes in benefit–cost studies (Marella and Raga 2014; Carson 2012; Merickova and Stejskal 2014).

The evaluation techniques discussed have also been used for library services to analyse the "return" on public investments. There are three main groups of approaches:

The first studies were generally "efficiency" or output-oriented studies and demonstrate the value of libraries in operating efficiently in managing human and material resources, being financially responsible and, therefore, in providing value as a service, per se. In these studies, we can include costing library activities (how processes and services can be made more effective) and benchmarking (Wilson and Pitman 1999a, b). Second another approach to measuring the value of library services is aimed at demonstrating the success of the library in providing a financial return to the organization or to the region. The problem of these research approaches lies in establishing a dollar figure for the contribution of the library (Marshall 1993; Griffiths and King 1993).

Third, a new methodology came along in the 1990s—taking a broader view of libraries' value and seeking to establish their value with stakeholders and clients. The balanced scorecard methodology was used, which enabled setting goals to split the hard numbers under consideration to determine which services should be changed and, also, to consider improvements in process (Kaplan and Norton 1996; Walsh and Greenshields 1998). The popularity of these approaches for the evaluation of public library services can be seen in current value studies using return-on-investment and contingent valuation. These methods are generally conducted to determine the economic benefit of public libraries for citizens and the economic benefit of particular services, such as national union catalogues and bibliographic services (Missingham 2005).

Regardless of the method used, it can be argued that analysis results can help to comment on both their effectiveness and their return.

3 Data and Research Methodology

The methodology of the completed research corresponds to the certified procedure for calculating CBA (OECD 1996). For analysing effectiveness, it is necessary to express the values in numbers for the benefits of individual services and the costs for their provision over a definite period.

The process of evaluating benefits for analysis requires empirical data. This paper builds on the results of a project focused on ROI analysis in municipal libraries in the Czech Republic. During 2012, an empirical survey was conducted. The data was obtained from the largest public library in the Czech Republic—from the Municipal Library of Prague (MLP).

The research was qualitative and representative, and it was conducted from July to August 2012 with the help of an on-line questionnaire (CAWI). A total of 11,397 randomly selected readers from the MLP were approached. These readers were over 15 years old, listed an email address in their application and used the library services within the last quarter previous to obtaining the questionnaire. The questionnaire's return rate was approximately 20% and comprised a basic sample of 2227 respondents after the data set was cleaned.

The questionnaire determining the perceived value of selected services provided by the library was first subjected to pilot testing so that individual questions were understandable for readers and the questions were able to be answered. At the same time, it was drawn up so that neither the way questions were phrased nor their order influenced the readers; this ensured a high degree of predicative ability and that the valuation obtained for the individual services would be realistic. The experiences published by (Venkatachalam 2004) were used here. Questions on the perceived value of library services from the questionnaire are for example:

1. When you consider borrowed material or other information that you received during your last visit, has the library saved you money?
2. How much money has a library like this saved you?
3. If you could not use the library and had to use an alternative, what would be the cost?
4. If you could lower your taxes because of an annual contribution paid to the library, how much would you be willing to contribute to the library?

The questionnaire in its final form was used for obtaining evaluations (EVU) of individual library services from respondents (they expressed the perceived value using the WTP system). The average perceived values of individual services are listed in Table 1. The values observed were given in CZK and, for the purposes of publication, have been translated into EUR at the reported rate.

Costs for providing the evaluated portfolio of public services were obtained from the MLP (listed in the necessary subdivisions in Tables 1 and 2). Financial resources in the form of MLP costs make up the public investment, because they come from the public budget at the given regional level. The result of the study presented in this paper is the ROI (or CBA) calculation of the allocation of public resources (sometimes also investments). In a cost-benefit analysis measuring secondary economic impacts, the library's impact on the rest of the economy can be calculated, e.g., its contribution towards employment, income, consumption expenditures and state or local government revenue in the form of taxes. Economic impact studies are an established methodology in economics (Aabø 2009); the methods were mentioned above.

4 Results and Analysis

Based on the empirical examination described, the perceived value library services (per unit) was determined by the WTP research questions. The quantification of all the services provided by the MLP in the year 2013 is also presented in Table 1. Data for the quantification was obtained from the MLP database.

The details about regional investment (from the local authority's budget) were obtained from the accounting department. The MLP accounting department is able

Table 1 WTP evaluation and quantification of library services (2013)

Standardized library services	WTP (Eur/unit) (EV_U)	Quantification (Q)
Circulating loans with assistance	2.40	33,887
Circulating loans without assistance	2.40	138,293
Non-circulating loans with assistance	.30	7304
Non-circulating loans without assistance	.20	30,976
Copying or printing documents	.11	5667
Digital services on-site	.60	7520
Digital services with off-site access	.10	350,533
Information and research	.29	1884
Cultural and educational events	1.91	4625
Technical services	.82	8230
Services related to the life of the community	2.97	603
Residence in the library	.21	6773

Source: Own research

Table 2 Total costs and perceived utility of individual library services in 2013 (EUR)

	Costs		Utility	
Standardized library services	Total cost (TC)	Unit costs (TC/Q)	Total utility (TU = EVU.Q)	ROI (TU/TC)
Circulating loans with assistance	79,172.50	2.34	81,328.80	1.03
Circulating loans without assistance	156,406.06	1.13	331,903.20	2.12
Non-circulating loans with assistance	19,758.24	2.71	2191.20	.11
Non-circulating loans without assistance	27,520.31	.89	6195.20	.23
Copying or printing documents	1954.48	.34	623.37	.32
Digital services on-site	5223.18	6.95	451.20	.09
Digital services with off-site access	9546.22	.03	35,053.30	3.67
Information and research	4620.44	2.45	546.36	.12
Cultural and educational events	14,949.18	3.23	8833.75	.59
Technical services	3625.10	4.40	674.86	.19
Services related to the life of the community	1228.71	2.04	1790.91	1.46
Residence in the library	11,403.65	1.68	1422.33	.12

Source: Own research

track the cost (expenses) of every service category. Table 2 also shows calculations that are necessary when using the CBA method and that determine the ROI for all services.

The data in Table 2 shows the overall ROI of services provided by the MLP. The total ROI of the MLP is 1.80467 (a 1 EUR investment will bring 1.80467 EUR of benefit to the customer and will increase societal wealth in Prague).

The value of overall efficiency should be used as a general reference that can serve as a benchmark for comparing different libraries or as information for library management or donors from the public sector.

On the other hand, data on the partial efficiency of the individual library services exhibits significant ambivalence. Some of the services are seen as highly effective, some are only marginally greater than the limit value, and one of them is seen as highly inefficient. It is to be expected that the main service constituting the essence of the libraries' existence (circulating loans) would exhibit a higher degree of efficiency. This is true to a similar extent in both libraries. Conversely, marginal services, which only complement the range of services provided by the libraries, are rated as inefficient. These include non-circulating loans, which are fairly labor intensive and, therefore, relatively expensive; it also include providing information or high-cost services such as residential stays in the library. When these results are interpreted by public authority management or those providing the investment, it is necessary to draw attention to the fact that inefficient services are often provided in the public interest and the resulting inefficiency is, therefore, just a consequence of entering the public sector.

A surprising finding is the high efficiency observed for digital services, both on-site and off-site. This is probably due to the current trend towards digitization and providing information through the Internet, e-books or various databases.

5 Discussion and Conclusion

The methodology of ROI calculation for public service systems is a very valuable tool for regional providers of public services and their investments. With this assistance, regional providers can better orient themselves when spending money and can make better decisions as to which services they will provide and to what extent they will provide these services. It will no longer be a question of making standard decisions under conditions of high uncertainty; applying this methodology will reduce the uncertainty.

The methodology also monitors the extent to which individual services are used and evaluated economically by the consumers themselves. This data can be used for the management of every library as well as for their owners and regional donors.

For the library analysed, it was found that its operation and provision of a selected range of library services is generally effective, but that the rate of effectiveness is relatively low, approaching a value of one. The library in question has a range of services that were defined but were not effective.

Wider practical use is hampered by a lack of high-quality data on the outputs of public service systems or, more precisely, their value as specified by consumers. This can be aided by the methodology presented here, which provides evidence that consumers are able to perceive the value of the consumed services as well as interpret it. This is the first research of its type for this branch of the public sector that has a high degree of representativeness and has been conducted in Central and Eastern Europe. The results and data cannot be transferred between countries, but it has been shown that such a survey can be conducted in other countries as well - with respect to the conditions for operation in that branch of the public sector in that particular country.

Acknowledgements The authors thank the anonymous reviewer whose comments and suggestions helped to significantly improve the quality of the resulting paper. This work was also supported by a grant provided by Student Grant Competition of the University of Pardubice in year 2015 (no. SGSFES_2015003).

References

Aabø S (2009) Libraries and return on investment (ROI): a meta-analysis. New Libr World 110 (7/8):311–324

Bach S, Kolins Givan R (2011) Varieties of new public management? The reform of public service employment relations in the UK and USA. Int J Hum Resour Manag 22(11):2349–2366

Bao G, Wang X, Larsen GL, Morgan DF (2013) Beyond new public governance: a value-based global framework for performance management, governance, and leadership. Adm Soc 45 (4):443–467

Bloch C, Bugge MM (2013) Public sector innovation – from theory to measurement. Struct Chang Econ Dyn 27:133–145

Brown K, Osborne SP (2012) Managing change and innovation in public service organizations. Routledge, London

Carson RT (2012) Contingent valuation: a practical alternative when prices aren't available. J Econ Perspect 26(4):27–42

Cummings RG, Taylor LO (1999) Unbiased value estimates for environmental goods: a cheap talk design for the contingent valuation method. Am Econ Rev 89(3):649–665

Denhardt RB, Denhardt JV (2000) The new public service: serving rather than steering. Public Adm Rev 60(6):549–559

Denhardt JV, Denhardt RB (2015) The new public service revisited. Public Adm Rev 75 (5):664–675

Dunleavy P, Hood C (1994) From old public administration to new public management. Public Money Manag 14(3):9–16

Fernandez S, Rainey HG (2006) Managing successful organizational change in the public sector. Public Adm Rev 66(2):168–176

Griffiths JM, King D (1993) Special libraries: increasing the information edge. Special Libraries Association, Washington, DC

Hajek P, Stejskal J (2014) Library user behavior analysis use in economics and management. WSEAS Trans Bus Econ 11(1):107–116

Hajek P, Stejskal J (2015) Modelling public library value using the contingent valuation method: the case of the municipal library of Prague. J Librar Inf Sci 47(1):43–55

Kaplan RS, Norton DP (1996) The balanced scorecard: translating strategy into action. Harvard Business School Press, Boston, MA

Kaufman P, Watstein BS (2008) Library value (return on investment, ROI) and the challenge of placing a value on public services. Ref Serv Rev 36(3):226–231

Klose T (1999) The contingent valuation method in health care. Health Policy 47(2):97–123

Lee SJ, Chung HK (2012) Analyzing altruistic motivations in public library valuation using contingent valuation method. Libr Inf Sci Res 34(1):72–78
Lynn LE (1996) Public management as art, science, and profession. Chatham House, Chatham, NJ
Marella G, Raga R (2014) Use of the contingent valuation method in the assessment of a landfill mining project. Waste Manag 34(7):1199–1205
Marshall J (1993) The impact of the special library on corporate decision-making. Special Libraries Association, Washington, DC
Mathews J (2011) What's the return on ROI? Libr Leadersh Manag 25:1–14
McIntosh CR (2013) Library return on investment: defending the contingent valuation method for public benefits estimation. Libr Inf Sci Res 35(2):117–126
Merickova B, Stejskal J (2014) The value of collective consumption goods. Politicka ekonomie 62 (2):216–231
Missingham R (2005) Libraries and economic value: a review of recent studies. Perform Meas Metrics 6(3):142–158
Modell S, Wiesel F (2008) Marketization and performance measurement in Swedish central government: a comparative institutionalist study. Abacus 44(3):251–283
OECD (1996) Performance auditing and the modernisation of government. OECD, Paris
Osborne D, Plastrik P (1997) Banishing bureaucracy. Addison Wesley, Reading, MA
Pestoff V, Brandsen T, Verschuere B (eds) (2013) New public governance, the third sector, and co-production. Routledge, London
Peters BG, Savoie D (1996) Managing incoherence: the coordination and empowerment conundrum. Public Adm Rev 56(3):281–289
Russell S, Fox-Rushby J, Arhin D (1995) Willingness and ability to pay for health care: a selection of methods and issues. Health Policy Plan 10:94–101
Sanderson I (1996) Evaluation, learning and the effectiveness of public services: towards a quality of public service model. Int J Public Sect Manag 9(5/6):90–108
Slater R, Aiken M (2014) Can't you count? Public service delivery and standardized measurement challenges–the case of community composting. Public Manag Rev 17(8):1085–1102
Stejskal J, Hajek P (2015) Evaluating the economic value of a public service—the case of the municipal library of Prague. Public Money Manag 35(2):145–152
Venkatachalam L (2004) The contingent valuation method: a review. Environ Impact Assess Rev 24(1):89–124
Walsh V, Greenshields S (1998) The value of libraries and library professionals to Australia's top 100 companies: draft report of the study conducted by the Australian library and information association. Aust Libr J 31(3):59–101
Wedgwood A, Sansom K (2003) Willingness-to-pay survey–a streamlined approach: guidance notes for small town water services. WEDC, Loughborough University, Leicestershire
Wilson A, Pitman L (1999a) Guidelines for the application of best practice in Australian university libraries: intranational and international benchmarks. Department of Education, Training and Youth Affairs, Canberra
Wilson A, Pitman L (1999b) Best practice handbook for Australian university libraries. Department of Education, Training and Youth Affairs, Canberra

Simona Pichova was born on 30th May 1988 in Šumperk, Czech Republic. She is currently a PhD student at the University of Pardubice, Czech Republic. She received Bachelor degree in "Public Economics and Administration" from the University of Pardubice, Pardubice, Czech Republic in 2011. She then finished master degree in "Economics of the Public Sector" at the same university in 2013. At present, she is studying PhD studies in "Regional and Public Economics" at the same university where she also teaches Microeconomics and Public Economics. Ing. Pichova has interests including innovation and development, knowledge economy and valuation of public services.

Jan Stejskal is Assoc. Prof. in public and regional economics, Ph.D. obtained at University of Pardubice, Faculty of Economics and Administration in Pardubice, Czech Republic. His domain is connection of the public economy in the regional scope and view. In the papers he describes and analyses specially the regional policy, tool of the local and regional economic development, public policies after care of the investors, systems of support in national economy in EU countries.

Determinants of Innovation Activities: Public Financing and Cooperation: Case Study of Czech Republic and Hungary

Viktor Prokop and Jan Stejskal

Abstract Innovation represents one of the key factors in achieving competitive advantage of companies, hence the whole economies. Therefore, managers aim to acquire knowledge. Likewise public policy makers understand an importance of creating innovations and thus promote the generation and spread of positive effects through knowledge diffusion. In the context of modern innovation, the science-industry collaboration comes into its importance. Many foreign studies pointing to the fact, that this cooperation cannot be successful in each sector and that not every kind of innovation depends on the same knowledge flows. Therefore, we can notice inefficient attempts to cooperate in a number of cases, which are frequently accompanied by excessive use of national and European funds. The article aims to compare situation of companies in manufacturing industry in the Czech Republic and Hungary to analyze how is their growth of total turnover affected by (i) implementation of innovation (product and process); (ii) university-industry and government-industry collaboration; (iii) provision of public subsidies (national and European). We show, by using the multiple linear regression models, that cooperation with universities and with other enterprises within enterprise groups positively influences innovation activities. The results also show that public funds are more effectively provided in Hungary, more specifically the European funds. We provide comparison between Czech and Hungarian manufacturing industries and proposals how to improve the efficiency of national funds provision, which is not sufficient in these countries.

Keywords Cooperation • Knowledge acquisition • Modern innovation • Public funding

---

V. Prokop (✉) • J. Stejskal
Faculty of Economics and Administration, Institute of Economic Sciences, University of Pardubice, Studentska 95, Pardubice 53210, Czech Republic
e-mail: viktor.prokop@upce.cz; jan.stejskal@upce.cz

A. Dias et al. (eds.), *Modeling Innovation Sustainability and Technologies*, Springer Proceedings in Business and Economics,
https://doi.org/10.1007/978-3-319-67101-7_7

1 Introduction

Since the 1970s of the twentieth century there is a gradual change in the economies of countries. It spontaneously leads to decreased in industrial production (in - manufacturing-based economy) and shifted attention to the services production (service-driven economy). It is due to many factors, particularly to developments in countries, but also to technology boom, and its speed.

It fundamentally changed the producers' requirements for production factors, which they are demanding. There is a shift from material and capital inputs to the input information, respectively knowledge. Many authors and studies have begun to analyse the influence of knowledge, respectively their research was focused on the ability to produce knowledge and ability to gain competitive advantage in various markets. Studies suggest that the impact of knowledge is essential, but quite distinct from the type of industry where rise. Scholars suggest that for economic growth promotion it is necessary to take actions to support the creation and dissemination of knowledge, to support research and development activities, investment in appropriate infrastructure and communication technology. There has also been a significant shift in the use of intellectual capital, patents and licenses. An economy that based on knowledge-intensive activities is described as a knowledge-based economy (Powell and Snellman 2004). Already in the 1990s of the twentieth century studies have been published and they provide the evidences that the knowledge economy increases gross domestic product (Abramovitz and David 1996).

However, some authors point out the attention to divide knowledge flows (as determinants) from effects (result or outputs) that they cause (Peri 2005). The flows represent a situation where information and knowledge passes through "learning" to another entity (institution). The effects are the result of the application realized (mostly) in commercial sector. They lead to commercialization and thus to achieve business goals. The scholars still warn that it is necessary to analyse knowledge flows separately and then evaluate the effects of knowledge flows (processes).

The remainder of this paper has been divided into four sections. The paper first gives a brief overview of the evidence for cooperation and knowledge as the most important production factors nowadays. The next section lays out the theoretical foundations of structural equation modelling, and analysis of the research data. Section 4 provides the results of analysis, conclusion and discusses its implications.

2 Selected Determinants for Innovation Activities

Many studies have focused on analysing the determinants of the knowledge economy in order to better control the amount of the increase in GDP due to the knowledge economy. Absolutely fundamental determinants are as follows: (1) knowledge (mostly codified knowledge) and information; (2) infrastructure, particularly technical, to ensure the transfer of information and technology; (3) educated and skilled workers with responsibilities such as creativity and innovation; (4) the scientific and technical institutions and other helpful organizations willing to cooperate. It turns out that the mentioned determinants are the pre-conditions; respectively they help to create an environment for knowledge transfer and creation of innovations (milieu). However, this innovative milieu requires further factors which are the common denominator in all the previously mentioned "secondary" determinants. Especially in Europe, it is possible to "secondary" or supporting determinants include: (i) cooperation and (ii) public financial support, which aims to support some of the primary determinants, in particular to ensure knowledge acquisition and transfer service, respectively willingness to cooperate. Completely separate pre-condition form, for example, public policy and management strategy.

Cooperation is more a prerequisite for the realization of knowledge flows and processes. Prerequisite for cooperation is a group of companies and scientific research institutes, applying a model of open innovation. The role of universities and R&D organizations is far more important than it was 20 years ago (Etzkowitz and Leydesdorff 2000). The universities and public research institutes have emerged as key components of the national innovation system (Eom and Lee 2010), which were gradually transformed into regional innovation systems (Cooke et al. 1997; Matatkova and Stejskal 2013). Any form of system or network of relationships is beneficial to knowledge transfer, but not same weight (equally). It turns out that in some sectors (and some states); the cooperation between firms has higher efficiency than cooperation between firms and knowledge-based organizations (Prokop and Stejskal 2015). The innovation systems are operational structure of cooperating entities from the public and private sectors whose activities and interactions initiate, import, modify, and diffuse new technologies. And there are the relationships that lead to the production, diffusion and economic application of new ideas and knowledge (Lundvall 1992).

R&D and university—knowledge-based organizations—play the important role in knowledge and innovation systems. Eom and Lee (2010) suggest that there are two contrasting views on the role of universities in these chains. The first view regards the universities involvement to triple helix system, later also quadruple-helix system. Triple-helix model presents the connection among industry and university and government, where the university plays social and knowledge role. Individual interaction and knowledge transfer in this simple system makes innovations creation easier, faster and cheaper. This emphasizes the "third mission" of the university, that is, serving for economic development aside from teaching and research (Etzkowitz and Leydesdorff 1997). Moreover, this model emphasizes the cooperation between research and application sphere in order to use the most of new knowledge. It should be added that spill-over effects are formed in this form of collaboration (Stejskal and Hajek 2015).

The second view is associated with the definition of the New Economics of Science (Partha and David 1994). It is a narrow view of the university, which, according to these authors, especially educational institution, and research leading to the industrialization of knowledge is perceived as harmful. It should be noted that the assessment of the impact and contribution of universities to industry is to be done individually, in different regions separately (the same conclusion is also in Eun et al. 2006). However, there are a number of studies that deals precisely analysing the contribution of university-industry cooperation in various sectors of industry. Some scholars argued that learning from advancements in technology is crucial for science-based industries, such as electronics, biotechnology and chemicals, for which industry-university should be more important (findings are confirmed by the results of studies (Pavitt 1984; Blumenthal et al. 1986; Feng et al. 2015; Althoff Philippi et al. 2015).

The financing is a specific secondary determinant of innovation activity. This money is inserted by co-actors into the realization of interaction and application of knowledge. In this study, it is not private investment, but of *public financing* (sometimes government-sponsored R&D or cooperation). The theory of public economics shows that public funding may be used for market failure preventing to new knowledge provision. The question for future research is whether in this case there is indeed a market failure when demand for new knowledge is increasing now. The first justification of public funding can be this: knowledge spill-overs lead to incomplete appropriability of the R&D results, which gives rise to market failure (Griliches 1992). The second justification is that R&D involves three types of uncertainties with regard to technological success, commercial success, and competitor behaviour (Malmberg et al. 1996; Nishimura and Okamuro 2011). However, the research results show it is strongly required that public support for innovation activities should be targeted. And the achieving of the targets must be carefully checked (also ex ante checks). Acceptable objectives of public support can be for example these: R&D cooperation and generate learning effects to Increase the absorptive capacity of supported subjects; public money to enable the use experimental research activities with the high initiative costs; innovation creation for public or private sector, which in future will generate societal benefits. The public support for the cooperating chains is justifiable even if public money is to remove obstacles that hamper knowledge transfer and reduced ability to generate innovation.

In practice, however, the public support is provided to research and promote collaboration without proof of the market failure. The reason is that public support is provided (typically in Europe) very heavily in member states and not-providing this support would reduce the competitiveness of individual beneficent. When public support was provided, the crowding-out effects of private investments were recorded. If authorities select high-quality projects and low-risk objectives, they can avoid the inefficient allocation of a harmful crowding-out effect (Nishimura and Okamuro 2011). Likewise, there are studies that show a high degree of public support inefficiency (Cowling et al. 1999; Hospers et al. 2009). Scholars argue that decisions on public subsidies are made by public choice under information asymmetry conditions (politicians have less information than the managers of companies; Wolf 1993).

The aim of the paper is to make an initial comparison of firms situation in manufacturing industry in the Czech Republic and Hungary with an emphasis on the determinants of innovative activities: public financing and cooperation. We aim to analyze how is the growth of total turnover affected by (i) implementation of innovation (product and process); (ii) university-industry and government-industry collaboration; (iii) provision of public subsidies (national and European); in these countries. For the purpose of this study and with following previous arguments, we hypothesize that:

H_1 *Implementation of product and process innovation in manufacturing industries in the Czech Republic and Hungary positively influence firms' growth of performance.*

Prior studies stated that innovation represent an essential component of competitiveness and analyzed relationship between innovation and firms' performance (Yam et al. 2011) in manufacturing industry (Gunday et al. 2011; Hashi and Stojčić 2013), specifically in chemical industry e.g. García-Morales et al. (2012). There are no studies analyzing impact of innovation implementation on firms' performance in chemical industry in the Czech Republic and Hungary. Halpern and Muraközy (2012) showed positive relation between innovation and firm performance in Hungary. Therefore we want to fill the gap and make comparison between the Czech Republic and Hungary by using multiple linear models including the same variables for both countries.

Importance of this research is emphasized by the fact that we are following results provided by the World Economic Forum (WEF), which annually publishes The Global Competitiveness Report and assesses the competitiveness landscape of 144 world economies. Results show that both Czech Republic and Hungary have failed in the scale of the competitiveness. However, determinants of competitiveness are many and complex (WEF is currently monitors 12 pillars of competitiveness), therefore we compare the evolution of the pillar Innovations (see Table 1) and hypothesize that:

H_2 *Implementation of innovation in manufacturing industry in the Czech Republic lead to more significant results than the implementation of innovation in manufacturing industry in Hungary.*

Results in Table 1 show that both countries decreased in the overall competitive index. However, we can see that Czech Republic has improved in the individual index/pillar Innovation, while Hungary has decreased significantly.

Moreover, Srholec (2014) analyzed effects of cooperation on innovative performance in the Czech Republic however we see the lack of studies analyzing the impacts of cooperation with universities (and/or government or public research

Table 1 Evolution of competitiveness and innovation in the Czech Republic and Hungary

	Czech Republic		Hungary	
Year	Competitiveness	Innovation	Competitiveness	Innovation
2006–2007	29	28	41	31
2008–2009	33	25	62	45

Source: Own processing according to World Economic Forum (2015)

Table 2 Evolution of individual competitiveness pillar—institutions

Year	Czech Republic	Hungary
2006–2007	60	46
2008–2009	72	64

Source: Own processing according to World Economic Forum dataset

institutes) on innovative activities and firms' performance in the Czech Republic and also in Hungary, specifically in manufacturing industry. Again, we compare the evolution of individual pillars of competitiveness between countries—Higher education and training; between the years 2006–2007 and 2008–2009. Czech Republic has improved from 27 to 25 places; conversely, Hungary has decreased from 30 to 40 places. This assumption leads us to the next hypothesis:

H_3 *Cooperation with universities and public research centers influences the overall performance of companies in manufacturing industry in the Czech Republic more significantly than cooperation with universities and public research centers in manufacturing industry in Hungary.*

As we argued above, there is a growing importance of universities and R&D organizations within cooperation process. This kind of collaboration is also analysed and supported by number of foreign studies, e.g. Okamuro and Nishimura (2013), López et al. (2014).

Rodríguez-Pose and Di Cataldo (2014) show, that (i) there are a relationship between quality of government and innovative performance in the regions of Europe; (ii) ineffective and corrupt governments represent a fundamental barrier for the innovative capacity of the periphery of the EU. To support this study, we also analyzed another individual pillar that WEF provides—Institutions. Both countries have significantly fallen in the ratings of this pillar (See Table 2).

Moreover, we compared results of Corruption Perceptions Index (Transparency International 2016). The results showed that the Czech Republic has improved by one place (from 46 to 45), while Hungary fell by six places (from 41 to 47). Therefore, we follow these findings and hypothesize that:

H_4 *In the manufacturing industry in the Czech Republic and Hungary, there is the same inefficient spending of public funds (national and European) that are targeted to support innovative activities of companies.*

3 Data, Methodology, Results and Analysis

For data collection, Community Innovation Survey (CIS) were used. CIS are part of the EU science and technology statistics and provide a harmonised questionnaire of EU Member States. CIS (Eurostat 2015) is a survey of innovation activities of enterprises and is designed to provide information on the innovativeness of sectors by type of enterprises, on the different types of innovation and on various aspects of the development of an innovation (e.g. the objectives, the sources of information, the public funding, the innovation expenditures etc.).

Community Innovation Survey carried out in the Czech Republic and in the Hungary for the period 2006–2008 by combining sample (stratified random sampling) and exhaustive surveys was used for our analysis. The target population of the CIS 2008 is the total population of enterprises in NACE Rev. 2 sections A to M. Data on 6804 Czech and 5390 Hungarian companies with at least 10 employees was obtained in total. For the purpose of this study, we filtered 547 Czech and 417 Hungarian companies from the manufacturing industry into our data groups—specifically, companies covering countries NACE categories 19–23 (Manufacture of coke and refined petroleum products; Manufacture of chemicals and chemical products; Manufacture of basic pharmaceutical products and pharmaceutical preparations; Manufacture of rubber and plastic products; Manufacture of other non-metallic mineral products).

Multiple linear regression models were used to analyse relationships between variables. We analysed the relationship between the growths of total turnover (between the years 2006 and 2008), as a dependent variable representing companies' performance, and selected independent variables (all the variables are listed in the Table 3). Multiple linear regression models have the following general form (Jann 2008; Vlachogianni et al. 2011):

$$y = \beta_0 + \beta_1 x_1 + \beta_2 x_2 + \ldots + \beta_n x_n + \varepsilon \qquad (1)$$

where

y is dependent variable;
x_1, x_2 ... x_n are independent variables;
β_1, β_2, ..., β_n called the regression parameters or coefficients, are unknown constants to be determined (estimated) from the data;
ε ... the residual error = difference between observations and predicted values.

At the beginning, 20 identical variables have been selected (see Table 3). These variables were tested for each country by using regression models. Results allowed us to (i) carry out an initial comparison of situation in manufacturing industries in the Czech Republic and Hungary; (ii) answer defined hypotheses. Initially, we conducted Spearman test to verify that the data are not correlated.

Before we answer defined hypotheses, we analyzed which variables (listed in Table 3) directly influence the growth of firms' total turnover. Table 4 shows the results of two mutually independent regression models. The first was assembled to analyze the situation of companies in manufacturing industry in the Czech Republic and the correlation coefficient of this model reached the value of 0.545. The coefficient of determination reached 0.297. P-value of the model was measured at 0.001. There was thus a rejection of the null hypothesis. The model could be regarded as significant. In this model, seven independent variables were used. The second model analyzed situation of companies in manufacturing industry in

Table 3 Variables used in the model

Dependent variable	Independent variables—categorical	Independent variables—continuous
TURN	LARMAR	RRDIN/TURN
	GP	RRDEX/TURN
	PROD_IN PROC_IN FUNGMT FUNEU CO CO_GP CO_SUP CO_CUSTOM CO_COMPET CO_UNI CO_GOV CO_CONSULT	ROEK/TURN RMAC/TURN RTOT/TURN

TURN the growth of total turnover between the years 2006–2008, *LARMAR* Largest market in terms of turnover between 2006–2008 (1—local or national, 0—other), *GP* part of the group of enterprises, *PROD_IN* introduced onto the market a new or significantly improved product (good or service), *PROC_IN* introduced onto the market a new or significantly improved process (method of production; logistic, delivery or distribution system; supporting activities), *FUNGMT* public funding from central government, *FUNEU* public financial support from the EU, *CO* cooperation arrangements on innovation activities, *CO_GP* co-operation partner: other enterprises within enterprise group, *CO_SUP* co-operation partner: Suppliers of equipment, materials, components, or software, *CO_CUSTOM* co-operation partner: clients or customers, *CO_COMPET* co-operation partner: competitors or other enterprises in sector, *CO_UNI* co-operation partner: universities or other higher education institutions, *CO_GOV* co-operation partner: government or public research institutes, *CO_CONSULT* co-operation partner: consultants, commercial labs, or private R&D institutes, *TURN* total turnover, *RRDIN* Expenditure in intramural R&D, *RRDEX* Expenditure in extramural R&D, *RMAC* Expenditure in acquisition of machinery, *ROEK* Expenditure in other external knowledge, *RTOT* Total innovation expenditure

Hungary. The correlation coefficient of this model reached the value of 0.718. The coefficient of determination reached 0.515. P-value of the model was measured at 0.037. There was thus a rejection of the null hypothesis. The model could be regarded as significant. In this model, nine independent variables were used. Results show, that only two variables (PROD_IN, GP) were same for both countries and influence dependent variable. We can also see, that in the case of the Czech manufacturing firms, there were other variable LARMAR. On the other hand, in Hungary, there were more variables, especially PROC_IN, FUNEU, CO_UNI, CO_GOV.

Regarding to the results listed in Table 4, we can answer hypothesis H_1. We can see that implementation of product innovation (PROD_IN) influence dependent variable (more significantly in Hungary; $0.01013 < 0.03044$). Implementation of process innovation (PROC_IN) is significant and influences the growth of total turnover only in Hungary. According to these results, we can accept hypothesis H_1 and state that implementation of innovation in manufacturing industry in the Czech Republic and Hungary positively influence firms' performance.

Table 4 Variables used in model in the Czech Republic and Hungary

	Czech Republic		Hungary	
Variables	p-value	sd	p-value	sd
RTOT/TURN	–	–	0.16455	0.12411
RRDIN/TURN	0.20498	0.41436	–	–
PROD_IN	0.03044**	0.12396	0.01013**	0.20930
PROC_IN	–	–	0.00033***	0.71022
CO_COMPET	0.57222	0.05612	–	–
FUNGMT	0.61628	0.13439	0.34529	0.21542
FUNEU	–	–	0.00214***	0.55331
LARMAR	0.02090**	0.14392	–	–
GP	0.01083**	0.18598	0.01527**	0.54623
CO_UNI	0.36056	0.13386	0.00199***	0.39990
CO_GP	–	–	0.89682	0.31773
CO_GOV	–	–	0.00074***	0.46732

Source: Own research
sd standard deviation
**Significant at $P < 0.05$; ***Significant at $P < 0.01$

To answer next hypothesis H_2, we compared another interactions that were created during analysis. Results show that in Hungary were created more significant interactions between implementation of innovation (product, process) and other variables (see Tables 5 and 6). For example, implementation of innovation (only product innovation) in the manufacturing industry in the Czech Republic significantly influenced the growth of total turnover in combination with cooperation with universities (PROD_IN*CO_UNI = 0.03917). This implementation was more significant when companies participated in the group of companies (PROD_IN*GP = 0.00178). Implementation of innovation in group of companies also showed more significant results than the implementation of innovation within company by itself (we compare results in Tables 4 and 5).

On the other hand, combination of implementation of innovation (product and process) with other variable in the manufacturing industry in Hungary caused creation of other significant results (see Table 6). For example, firms implementation of product innovation in collaboration with universities (CO_UNI*PROD_IN = 0.00222) causes more significant effect on the growth of total turnover than the situation without collaboration (we compare result in Tables 4 and 6; 0.00222*** is more significant than 0.01013**). For these reasons, we reject the hypothesis H_2. In the manufacturing industry in Hungary, implementation of innovation leads to more significant results.

Results in Table 4 also allowed us to answer hypotheses H_3. We claimed that cooperation with universities and public research centers influences the overall performance of companies in manufacturing industry in the Czech Republic more significantly than cooperation with universities and public research centers in manufacturing industry in Hungary. However, as we can see, this kind of cooperation influences firms' growth of total turnover only in manufacturing industry in

Table 5 Cooperation with universities and influence of participation in the groups of companies on firms' performance in the Czech Republic

Variables	CO_UNI	GP
PROD_IN	0.03917 (0.10472)**	0.00178 (0.17168)***
CO_COMPET	0.17715 (0.07578)	0.04583 (0.05626)**
FUNGMT	0.00392 (0.11177)***	0.03125 (0.16959)**
LARMAR	0.00000 (0.11907)***	0.16172 (0.10054)
GP	0.01114 (0.08331)**	–
CO_UNI	–	0.01114 (0.08331)**
PROD_IN*FUNGMT	0.00181 (0.10630)***	0.00287 (0.15041)***
CO_COMPET*LARMAR	0.00240 (0.06392)***	0.00676 (0.05595)***
PROD_IN*LARMAR	0.00009 (0.09400)***	0.93162 (0.07595)

Source: Own research
Table shows p-values; values of standard deviations are shown in brackets
**Significant at $P < 0.05$; ***Significant at $P < 0.01$

Table 6 Efficiency of public funding and innovation implementations in Hungary

Variables	FUNGMT	FUNEU	PROD_IN	PROC_IN
CO_UNI	0.31551 (0.222)	0.00446 (0.348)***	0.00222 (0.416)***	0.04754 (0.458)**
CO_GP	0.76553 (0.217)	0.00219 (0.352)***	0.00079 (0.321)***	0.06521 (0.157)*
GP	0.06431 (0.411)*	0.00374 (0.289)***	0.34640 (0.105)	0.13905 (0.396)
CO_GOV	0.79881 (0.165)	0.60756 (0.167)	0.00056 (0.449)***	0.00003 (0.347)***
FUNGMT	–	0.00016 (0.322)***	0.00603 (0.286)***	0.00885 (0.327)***
FUNEU	0.00016 (0.322)***	–	0.43722 (0.171)	0.00347 (0.503)***
PROC_IN	0.00885 (0.327)***	0.00347 (0.503)***	0.00024 (0.193)***	–
PROD_IN	0.00603 (0.286)***	0.43722 (0.171)	–	0.00024 (0.193)***
CO_UNI*CO_GP	0.00448 (0.123)***	0.00040 (0.318)***	0.00075 (0.304)***	0.04151 (0.154)**

Source: Own research
Table shows p-values; values of standard deviations are shown in brackets
*Significant at $P < 0.1$; **Significant at $P < 0.05$; ***Significant at $P < 0.01$

Hungary. In the manufacturing industry in the Czech Republic, there are not efficient cooperation between firms and universities or public research institutions (or/and government) that will significantly influence dependent variable. Therefore we have to reject hypothesis H_3. On the other hand, results in Table 5 show, that proper collaboration with universities in the manufacturing industry in the Czech Republic significantly influences firms overall performance (e.g. PROD_IN* CO_UNI = 0.039; LARMAR*CO_UNI = 0.000).

To answer hypothesis H_4, firstly, we use results in the Table 4. We can see that financial support from national funds is totally inefficient in both cases. Public subsidies from European funds are insignificant in the manufacturing industry in the Czech Republic, not in Hungary (0.00214). Results also showed us, that proper targeting of public subsidies (national and European) leads to creation of further significant interactions that influence firms' growth of total turnover. For example, we can see that in Hungary there are created a number of other significant interactions with variables FUNGMT and FUNEU (see Table 6).

For example, we can see that collaboration with universities, that is supported from European funds, significantly influences growth of firms' performance (CO_UNI*FUNEU = 0.00446). On the other hand, the same collaboration supported from the national funds becomes insignificant (CO_UNI* FUNGMT = 0.31551). Conversely, in the manufacturing industry in the Czech Republic, there are not created the same number of further significant interactions.

We analyzed only a negligible amount of significant interactions (see Table 5). For example, FUNGMT*CO_UNI = 0.00392; PROD_IN*FUNGMT*CO_ UNI = 0.00181. Therefore, we reject hypothesis H_4 because in manufacturing industries in the Czech Republic and Hungary, there are not the same inefficient spending of public funds (that are targeted to support innovative activities of companies).

4 Conclusions

Results allowed us to confirm only hypothesis H_1 that showed the growing importance of implementation of innovation (product and process) in both countries. These results contribute to the literature (e.g. Ulku 2007; Brown et al. 2009) and highlight the importance of innovation. Rejection of hypothesis H_2 showed to us that implementation of innovation leads to more significant results in manufacturing industry in Hungary, even though that the Czech Republic has improved in the individual pillar Innovation and Hungary decreased. The hypothesis H_3 was aimed to confirm the importance of collaboration with universities and public research centers and compare situation between selected countries. We claimed that cooperation with universities and public research centers in manufacturing industry is more significant in the manufacturing industry in the Czech Republic. However direct impact of this cooperation on firms' performance was shown only in manufacturing industry in Hungary, therefore we reject hypothesis H_3. On the other hand, further results showed that cooperation with universities could lead to creation of significant effects on firms' growth of total turnover in manufacturing industry in the Czech Republic. These collaborations are often accompanied by firms' participating in the groups of companies that lead to creation of strong ties and positively affect university-industry collaborations because there are a prerequisite for more effective dissemination of knowledge (Capello and Faggian 2005; Laperche et al. 2011). Hypothesis H_4 showed that in manufacturing industries there is the same inefficient spending of public funds (national and European) in both countries. However, in Hungary there were found further significant links between financial support (especially from the European Union) and firms' overall performance through combinations with other variables. Therefore, hypothesis H_4 was rejected.

Our findings provide practical implications for policy makers. The innovation implementation and knowledge diffusion represent a complex process that involves number of factors. As results shown, both determinants of innovative activities (Public Financing and Cooperation) significantly influence the overall performance of firms in manufacturing industries in selected countries. However, there is a need for proper targeting of these determinants. In the manufacturing industry in the Czech Republic, companies should not cooperate with universities by themselves because this kind of cooperation (university-industry) brings with it certain disadvantages or drawbacks (Siegel et al. 2003; Bruneel et al. 2010). For example, each collaborating partner has different interests and different expectations of the results. Therefore, cooperation with universities should include other specific determinants that were shown in this analysis. Other implication that we suggest is implementation of dual system vocational education and training (VET) that could support practical skills that are necessary for researchers and that could support collaborating activities between firms and universities and public research centres in manufacturing industries in the Czech Republic and Hungary. VET system is for example deeply embedded and widely respected in Germany and offers qualifications in a broad spectrum of professions and flexibly adapts to the changing needs of the labour market. Next implication is aimed on the issue of financial support (national and European) on innovative activities in manufacturing industry in the Czech Republic and Hungary. Following the results, we recommend continuing with financial supporting innovation activities of firms in manufacturing industry in Hungary, especially from European Union. However, we showed that there is a need for proper targeting of public financial support on innovative activities in both countries. In Hungary, public funds are provided more effectively in the manufacturing industry than in the Czech Republic. Policy makers should also decide which kinds of cooperation they will financially support (particularly from the national funds) because we can see inefficiency in the Czech Republic and also in Hungary. In this case, there are not direct effects of national subsidies on the growth of total turnover. Therefore, we propose the use of non-financial types of support for innovative activities because it is clear that there is no clearly demonstrable relationship between growth of financial subsidies provided from national or European funds and increase of firms' performance. The issue of non-financial support was also analysed by number of researchers (e.g. Mole and Bramley 2006; Sonne 2012). Finding appropriate determinants of innovation activities is a complex process that is influenced by number of factors. Therefore, further research is aimed to follow significant results of this analysis and their conditions, and to analyse the situation of other industries in selected countries to make further comparison and bring appropriate practical implications for policy makers.

Acknowledgments This article was created as part of the resolution of the research task No. 17-11795S financially supported by the Grant Agency of the Czech Republic.

References

Abramovitz M, David PA (1996) Technological change and the rise of intangible investments. The U.S. economy's growth-path in the twentieth century. In: Employment and growth in the knowledge-based economy. OECD, Paris, pp 35–60

Althoff Philippi D, Maccari EA, Cirani S, Brito C (2015) Benefits of university-industry cooperation for innovations of sustainable biological control. J Technol Manag Innov 10(1):17–28

Blumenthal D, Gluck M, Louis KS, Stoto MA, Wise D (1986) University-industry research relationships in biotechnology: implications for the university. Science 232(4756):1361–1366

Brown JR, Fazzari SM, Petersen BC (2009) Financing innovation and growth: cash flow, external equity, and the 1990s R&D boom. J Financ 64(1):151–185

Bruneel J, d'Este P, Salter A (2010) Investigating the factors that diminish the barriers to university–industry collaboration. Res Policy 39(7):858–868

Capello R, Faggian A (2005) Collective learning and relational capital in local innovation processes. Reg Stud 39(1):75–87

Cooke P, Uranga MG, Etxebarria G (1997) Regional innovation systems: institutional and organisational dimensions. Res Policy 26(4):475–491

Cowling K, Oughton C, Sugden R (1999) A reorientation of industrial policy: horizontal policies and targeting. In: Cowling K (ed) Industrial policy in Europe: theoretical perspectives and practical proposals. Routledge, London, pp 17–31

Eom BY, Lee K (2010) Determinants of industry–academy linkages and, their impact on firm performance: the case of Korea as a latecomer in knowledge industrialization. Res Policy 39 (5):625–639

Etzkowitz H, Leydesdorff L (eds) (1997) Universities and the global knowledge economy: a triple Helix of university–industry–government relations. Continuum, London

Etzkowitz H, Leydesdorff L (2000) The dynamics of innovation: from national systems and "Mode 2" to a triple helix of university–industry–government relations. Res Policy 29 (2):109–123

Eun JH, Keun L, Wu G (2006) Explaining the "university-run enterprises" in China: a new theoretical framework for university–industry relationship in developing countries and its applications to China. Res Policy 35:1329–1346

Eurostat (2015) Community innovation survey (CIS). Available at: http://ec.europa.eu/eurostat/web/microdata/community-innovation-survey

Feng F, Zhang L, Du Y, Wang W (2015) Visualization and quantitative study in bibliographic databases: a case in the field of university–industry cooperation. J Informet 9(1):118–134

García-Morales VJ, Jiménez-Barrionuevo MM, Gutiérrez-Gutiérrez L (2012) Transformational leadership influence on organizational performance through organizational learning and innovation. J Bus Res 65(7):1040–1050

Griliches Z (1992) The search of R&D spillovers. Scand J Econ 94:29–47

Gunday G, Ulusoy G, Kilic K, Alpkan L (2011) Effects of innovation types on firm performance. Int J Prod Econ 133(2):662–676

Halpern L, Muraközy B (2012) Innovation, productivity and exports: the case of Hungary. Econ Innov New Technol 21(2):151–173

Hashi I, Stojčić N (2013) The impact of innovation activities on firm performance using a multi stage model: evidence from the community innovation survey 4. Res Policy 42(2):353–366

Hospers GJ, Desrochers P, Sautet F (2009) The next Silicon Valley? On the relationship between geographical clustering and public policy. Int Entrep Manag J 5(3):285–299

Jann B (2008) The Blinder-Oaxaca decomposition for linear regression models. Stata J 8 (4):453–479

Laperche B, Lefebvre G, Langlet D (2011) Innovation strategies of industrial groups in the global crisis: rationalization and new paths. Technol Forecast Soc Chang 78(8):1319–1331

López SF, Astray BP, Pazos DR, Calvo N (2014) Are firms interested in collaborating with universities? An open-innovation perspective in countries of the South West European space. Serv Bus 9:1–26
Lundvall BA (1992) National systems of innovation: towards a theory of innovation and interactive learning. Printer Publishers, London
Malmberg A, Solvell O, Zander I (1996) Spatial clustering, local accumulation of knowledge and firm competitiveness. Geografiska Annaler. Ser B, Hum Geogr 78:85–97
Matatkova K, Stejskal J (2013) Descriptive analysis of the regional innovation system – novel method for public administration authorities. Trans Rev Adm Sci 39:91–107
Mole KF, Bramley G (2006) Making policy choices in nonfinancial business support: an international comparison. Environ Plann C 24(6):885–908
Nishimura J, Okamuro H (2011) Subsidy and networking: the effects of direct and indirect support programs of the cluster policy. Res Policy 40(5):714–727
Okamuro H, Nishimura J (2013) Impact of university intellectual property policy on the performance of university-industry research collaboration. J Technol Transf 38(3):273–301
Partha D, David PA (1994) Toward a new economics of science. Res Policy 23(5):487–521
Pavitt K (1984) Sectoral patterns of technical change: towards taxonomy and a theory. Res Policy 13(6):343–373
Peri G (2005) Determinants of knowledge flows and their effect on innovation. Rev Econ Stat 87 (2):308–322
Powell WW, Snellman K (2004) The knowledge economy. Ann Rev Sociol 30:199–220
Prokop V, Stejskal J (2015) Impacts of local planning to competitiveness index change – using approximate initial analysis to the Czech regions. WSEAS Trans Bus Econ 12(1):279–288
Rodríguez-Pose A, Di Cataldo M (2014) Quality of government and innovative performance in the regions of Europe. J Econ Geogr 15(4):673–706
Siegel DS, Waldman DA, Atwater LE, Link AN (2003) Commercial knowledge transfers from universities to firms: improving the effectiveness of university–industry collaboration. J High Technol Managem Res 14(1):111–133
Sonne L (2012) Innovative initiatives supporting inclusive innovation in India: social business incubation and micro venture capital. Technol Forecast Soc Chang 79(4):638–647
Srholec M (2014) Cooperation and innovative performance of firms: panel data evidence from the Czech Republic, Norway and the UK. J Knowl Econ 5(1):133–155
Stejskal J, Hajek P (2015) Modelling knowledge spillover effects using moderated and mediation analysis–the case of Czech high-tech industries. Lect Notes Bus Inform Process 224:329–341
Transparency International (2016) Corruption perceptions index 2016. Available at: https://www.transparency.org/news/feature/corruption_perceptions_index_2016
Ulku H (2007) R&D, innovation, and growth: evidence from four manufacturing sectors in OECD countries. Oxf Econ Pap 59(3):513–535
Vlachogianni A, Kassomenos P, Karppinen A, Karakitsios S, Kukkonen J (2011) Evaluation of a multiple regression model for the forecasting of the concentrations of NO x and PM 10 in Athens and Helsinki. Sci Total Environ 409(8):1559–1571
Wolf C (1993) Markets or governments: choosing between imperfect alternatives, 2nd edn. MIT Press, Cambridge, MA
World Economic Forum (2015) World economic forum dataset – competitiveness rankings. Available at: http://reports.weforum.org/global-competitiveness-report-2014-2015/rankings/
Yam RC, Lo W, Tang EP, Lau AK (2011) Analysis of sources of innovation, technological innovation capabilities, and performance: an empirical study of Hong Kong manufacturing industries. Res Policy 40(3):391–402

Viktor Prokop, Ph.D. is an professor assistant and Ph.D. in regional and public economics at Institute of Economics, Faculty of Economics and Administration, University of Pardubice. The author is co-researcher of the grant project: odeling knowledge spill-over effects in the context of regional and local development; and explores the issue of measuring the knowledge economy in his dissertation.

Jan Stejskal, Ph.D. is an associate professor with the Institute of Economics, Faculty of Economics and Administration, University of Pardubice, Czech Republic. His domain is connection of the public economy in the regional scope and view. Especially, he analyses regional policy, tools of the local and regional economic development, and public services.

Innovation Through Treasury Centralization: The Potential of the Visegrad Countries for Establishment of Corporate Treasury Centres

Beata Šarkanová and Peter Krištofík

Abstract Corporate treasury management becomes much more complex and hence more difficult as globalisation increases, cross border activities grow, and companies choose to adopt a multinational status. Multinational companies (MNCs) need to organise their treasury management activities and their main aim is to choose the model that will maximise value added. The range of possibilities includes a centralized form of treasury i.e. a treasury centre operating in a certain geographical region as the centre of expertise, knowledge and innovation in financial management. MNCs often centralize their treasury management by setting up regional treasury centres in locations that offer optimal conditions for the effective management of treasury activities having regard to local needs. The recent trend of nearshoring has focused multinationals' attention on Central European countries. The paper concentrates on the factors and criteria considered in MNCs' location decisions for treasury centres, and on the ability of a selected group of Central European countries to become the appropriate location for such centres. Location decisions commonly depend on both the local tax environment and other factors. The main purpose of this paper is to evaluate and compare the suitability of the Visegrad countries as sites for corporate treasury centres.

Keywords Corporate treasury • Centralization • MNCs • Location criteria • Visegrad countries

1 Introduction

Corporate treasury management as a specialized function of financial management is a relatively new phenomenon, especially in the geographical area of post-communist Europe including the V4 countries. As stated by Polak et al. (2011), corporate treasury in multinational companies (MNCs) faces numerous challenges.

B. Šarkanová (✉) • P. Krištofík
Faculty of Economics, Matej Bel University, Banská Bystrica, Slovak Republic
e-mail: beata.sarkanova@umb.sk; peter.kristofik@umb.sk

A. Dias et al. (eds.), *Modeling Innovation Sustainability and Technologies*,
Springer Proceedings in Business and Economics,
https://doi.org/10.1007/978-3-319-67101-7_8

These can be divided in two categories: (1) traditional challenges representing the typical actions of corporate treasury—financial risk management, cash forecasting, optimization of access to credit and liquidity, assets and liabilities management, management of cost of capital, bank relationship management, and (2) emerging challenges—efficiency, globalization, centralization and outsourcing of the control and management of treasury including geographic organization of roles, standardization and integration of treasury activities, and adapting to new regulation. Furthermore, Blach et al. (2014) add that the emerging challenges can be identified as the key drivers of innovations within traditional challenges.

Since the core objective of treasury management is to add value, treasury centres concentrate financial specialists and experts deeply understanding the functioning of a particular MNC, creating new innovative ideas for proactive and forward-looking approach to processes and activities performed in financial management. This includes also the cooperation with other subjects (internal or external to the treasury centre) e.g. IT experts and providers, banks or other financial institutions operating in a certain region, and consequently, pressing them to create customized and innovative products and services complying to their requirements.

The organisation of a corporate treasury influences the treasury function in MNCs significantly. If an MNC has decided to centralize its treasury, then it needs to consider its optimal location, and if it wants to have regional centres. A variety of tax and non-tax factors feed into the decision on the degree of centralization. Tax factors include corporate tax, withholding tax, value-added tax, and double tax treaty agreements. Non-tax factors include costs, location of other operations, currency, language barriers, and time zone. Several authors have recently explored such factors. The main aim of this paper is to evaluate and compare Visegrad group countries, using a defined group of tax and non-tax location criteria, as potentially suitable locations for treasury centres nearshored by European multinational companies.

2 Literature Review

Giegerich et al. (2002) define a treasury centre (TC) as "a centralized treasury management function which is legally structured as a separated group entity or as a branch and is located in a tax efficient environment". A treasury centre provides financial management and transaction services for other group entities. Furthermore, Giegerich et al. list the most typical functions of a treasury centre:

- managing group liquidity and reducing interest rate payments through operating cash pooling,
- providing the group with payment services—both internal (intragroup clearing or netting) and external (payment factory),

- setting rules for re-invoicing within the group and externally, offering factoring services for the group companies' third party receivables, and structuring asset backed securities,
- negotiating group wide leasing arrangements,
- structuring and monitoring the capital requirements of group companies, financing, coordination and management of the group's external borrowings,
- vendor financing for other group companies,
- providing the group with asset management services,
- providing exchange risk and interest rate risk management to other group entities,
- financial markets research and risk management analysis,
- optimizing the group's global tax position and controlling the repatriation of funds through the group (e.g. via dividends or interest payments).

Potty and Sarraf (2004, cited in Polak and Roslan 2009) define four functions of a treasury centre: (1) asset and liability management, (2) sales and trading of currency, (3) dealing with credit and derivatives products in capital markets, and (4) financial management. According to Blair (2011), a typical regional treasury centre performs some or all of these functions: foreign exchange dealing and management, in-house banking, intercompany loans, trapped cash management, risk management, and asset and liabilities management.

Bragg (2010) recognizes four stages of treasury centralization: stage 1—complete decentralization, stage 2—centralized netting and hedging, stage 3—centralized investments, and stage 4—centralized working capital management. If a company operates in a multitude of countries (especially in developing countries where there may be currency controls), there may be a greater need for local treasury personnel in order to handle matters that may impact treasury operations e.g. maintaining local banking relationships or monitoring local regulations. According to Bragg (2010), the necessity to maintain regional treasury centres (RTCs) occurs even in the case of fully centralized treasury department. Such regional treasury centres usually operate in the major time zones to ensure the availability of some treasury personnel during the business hours of subsidiaries located in their regions. However, Anwar (1999) states that even if the theoretical concept of a regional treasury centre seems attractive, there is a number of considerations, e.g. logistical, cultural and market-specific, that need serious analysis before putting it into practice.

Murphy (2000a) notes that many multinationals have established regional treasury centres e.g. European Treasury Centres, Asia-Pacific Treasury Centres or International Treasury Centres, for all non-domestic treasury activities. Moreover, he adds that the effectiveness of control and the enhanced value delivered by a substantive treasury centre can be more satisfactory than simply "scattering" treasury personnel throughout group entities. Also in favour of regional centralization of treasury management activities, Fogarty (1992, cited in Mulligan 2001) states that "it can be more efficient to work on a regional basis, striking a good balance between central control and the recognition of local detail, opportunities and special situations". Barlow (2011) emphasizes that the rationale for regional centralization is straightforward since it can reduce processing times, simplify fund flows and avoid unnecessary duplication.

The value added that a treasury centre can create has been further specified by Geigerich et al. (2002) as: (1) qualitative added value: this is generated by focusing on core competencies within the group, improving reporting, pooling know-how, standardizing treasury and risk management procedures, and coordinating data through a central database, and (2) quantitative added value: generated by reducing payment and transaction costs through higher volumes and fewer transactions, reducing financing costs and financing requirements, and reducing hedging costs.

The main research on treasury centre location has been conducted by Polak (2010), Polak and Roslan (2009), Polak and Klusáček (2010); and by Polak et al. (2011). Under Polak's supervision Simkova (2005) developed location criteria for establishing treasury centres. Polak and co-authors primarily focused on locations in South East Asia. Dizkirici (2012) compared Istanbul as a potential location for a regional treasury centre to Hong Kong and Singapore.

On the decision-making process for finding the right regional treasury centre location, several authors have investigated the factors/criteria influencing the location of treasury centres. They include Anwar (1999), Murphy (2000a, b), Mulligan (2001), Polak (2010) and in a number of other papers, either by himself or in collaboration with others, and Dizkirici (2012). In addition, organizations such as Ernst and Young—Geigerich et al. (2002); Citigroup—Shah and Yung (2012) and van der Sar (2013), PricewaterhouseCoopers (2013a) and Deloitte (2014) have prepared guides to treasury location, especially in areas of Asia and the Pacific. Popular locations for treasury centres in Asia are primarily Hong Kong and Singapore, but also include Malaysia and Thailand. In Western Europe early centres include Luxembourg, the Netherlands, Switzerland and Ireland.

Recent literature generally divides factors influencing regional treasury centre location into two groups—tax and non-tax factors. Anwar (1999) notes a number of major criteria. The location should offer good geographical access to regional and international affiliates, ideally in a country with a stable political regime and without onerous regulations. An eligible location would have benign tax conditions, solid infrastructure, be in an appropriate time zone relative to the region the regional treasury centre serves, and have a competitive cost base.

Several authors point out the key role of the tax system and state that treasury centre locations are usually tax driven. As noted above, treasury centres are usually located in a "tax efficient environment" defined by Roslan and Polak (2009) as a location that offers beneficial tax regimes compared to other locations: for example, where the MNC pays a low tax rate. Polak (2010) emphasizes the importance of the company's total tax contribution to the host state. This comprises:

- withholding tax on interest, dividend and royalties,
- corporate tax on profits allocated to the treasury centre and its capital,
- thin capitalization rules,

- stamp tax (establishment of a new subsidiary may have stamp tax consequences),
- value-added tax (in most countries VAT is not payable on certain financial transactions),
- double taxation agreements.

According to Murphy (2000a) in the past treasury structures' locations were primarily tax driven. Treasury activity was located where tax on profits generated was relatively low, but now new treasury structures' locations depend primarily on business needs. The optimal location would combine both possibilities—a favourable tax environment that also offers some additional value.

Beside taxation, there is also a number of equally important non-tax factors influencing the location of treasury centres e.g. costs, location of other operations, currency, language barriers, and time zone. Shah and Yung (2012) identify financial costs (bank transaction fees, treasury staff salary, office rental), bank and other regulatory reporting requirements, access to financial markets, the degree of currency restrictions, the availability of skilled employees, time zone compatibility with respect to business operations under coverage. They also stress the quality of physical and IT infrastructure, political stability, a sophisticated banking system, and the reputation of the location as a financial hub. Murphy (2000b) considers the non-tax factors for establishing a treasury centre in Europe, and reaffirms costs as the key factor among non-tax factors, especially labour costs, premises, IT and telecommunications. He also emphasizes the variation in costs from one location or country to another. As additional influencing factors he points to the availability of outsourcing, the distance from other key business operations, a need for high quality of treasury expertise, CEOs and CFOs control over treasury activities, and the euro (an issue for the UK, Switzerland and most Scandinavian countries). He adds the need for a location to have a modern banking system and a strong regulatory framework, English as the predominant financial language, and the tendency of multinationals to choose already recognised locations (in Europe the Netherlands, Belgium, Ireland, Luxembourg, Switzerland, some Mediterranean islands, and the United Kingdom).

When focusing on the location of a treasury centre itself, it is important to recognise that Asia represents a very popular location for establishing a treasury centre. The Asia Pacific Treasury Management Barometer (Bank of America Merill Lynch and Sungard 2015), is based on a survey of around 1350 respondents who were treasury professionals from the Asia Pacific region, Europe and the United States. It reported that 63% of respondents replied that they had, or were implementing a centralized treasury organization policy through one or more regional treasury centres (RTC) and/or a global treasury centre.

Criteria for locating regional treasury centres are continuously changing. Anwar (1999) recognises Europe as a potentially suitable location for a regional treasury centre, since Europe offers a sophisticated banking and telecommunications infrastructure, a large pool of financial, treasury, and accounting professionals, and years of experience in regional treasury centre locations ranging from Brussels or Zurich to London. While locations in Western Europe remain popular, Polak (2010) states that many companies are taking a new approach known as "nearshoring"[1] and are looking for suitable locations in Central Europe e.g. in the Czech Republic or Hungary.

3 Data, Methodology and Results

Because companies have recently begun looking for locations to nearshore their treasury centres in Central Europe, the aim of this part of the paper is to compare the countries of the Visegrad group i.e. the Slovak Republic, the Czech Republic, Hungary and Poland as potential suitable locations for regional treasury centres. The research methodology is based on Simkova's location criteria used in 2005 for the Czech holding company CGS (Česká Gumárenská Společnost) to identify the best location for establishing an international treasury centre,[2] offering the best conditions for cash flow control and administration, and an advantageous tax system (Simkova 2005). The study involves the application of TOPSIS (Technique for Order Preference by Similarity to Ideal Solution)—a multi-criteria decision-making methodology.

Simkova's location criteria (LC_n), as noted by Roslan and Polak (2009), include eleven criteria. Seven of them require quantitative data (monthly banking fees, bank transaction fees, the cost of incoming foreign payments, the cost of outgoing foreign payments, the cost of foreign urgent payments, withholding tax, and corporate tax payments. The remaining four criteria (important treasury centres, reporting requirements, the currency environment, and ratings) require qualitative data. Depending on whether the criteria scores need to be minimized or maximized the criteria can also be divided into minimization criteria (LC_1–LC_7, LC_9) and maximization criteria (LC_8, LC_{10}, LC_{11}). Table 1 gives a detailed description of each criterion.

Information to measure performance on the criteria was collected mostly from online secondary sources. Data for quantitative criteria concerning fees and prices of bank services are calculated as the average minimum charge of the top three banks (by asset size). Charges and fees were averaged for individual banks, and then the criteria performances for LC_1–LC_5 were calculated as the average minimum charges of three banks. Quantitative data for tax criteria is based on tax rates (in the case of corporate tax both nominal and effective rates). Data for the remaining qualitative criteria came from online research. Table 2 summarizes the data sources for criteria across the analysed countries.

[1]Nearshoring is "moving the business process to a foreign country that is relatively close by, e.g. UK businesses nearshoring to Eastern Europe or American businesses nearshoring to Mexico" (Koning 2012).

[2]Here the terms international treasury centre and regional treasury centre are interchangeable.

Table 1 Description of Simkova's 11 location criteria

Location criterion (LC_n)		Description
Quantitative criteria		
LC_1	Monthly banking fees	Business account *minimum* monthly maintenance fees charged by banks.
LC_2	Bank transaction fees	*Minimum* fee per transaction charged by banks for business accounts.
LC_3	Price of incoming foreign payment	Inward remittance fee—*minimum* charge for funds transferred (buying foreign currency) by foreigners to their country of residence.
LC_4	Price of outgoing foreign payment	Outward remittance fee—*minimum* charge for funds transferred (selling foreign currency) by foreigners to their country of residence.
LC_5	Price of outgoing urgent payments	This service fee is similar for making outgoing payments but more expensive (*minimum* charge).
LC_6	Withholding tax	Percentage of payment (*minimum*) payers made to residents or non-residents that are withheld by the local tax authority.
LC_7	Corporate tax	Tax imposed on profits (*minimum*) made by companies by the local tax authority.
Qualitative criteria		
LC_8	Important treasury centres	The existence of treasury centres (*the availability of other RTCs shows that the region has the required treasury centre support*).
LC_9	Reporting requirements	The size of certain large amounts of funds transferred by RTCs or MNCs domestically or internationally that need to be reported to the central bank or monetary authority of the country *(minimum)*.
LC_{10}	Currency environment	The possibility of financial transactions in other currencies (especially EUR, USD, GBP), convertibility of domestic currency, foreign currency accounts and services. *The ability to provide instant conversion of currencies increases the suitability of a location for RTC establishment.*
LC_{11}	Ratings	Credit ratings by the rating company Coface (*as good as possible*; Rating A1—best, D—worst).

Source: Roslan (2008)

The data sources set out in Table 2 provided the basis of the data matrix used in the TOPSIS model (see Table 3). Data in foreign currency i.e. for the Czech Republic in Czech Koruna (CZK), for Hungary in Hungarian Forint (HUF), and for Poland in Polish Zloty were converted into EUR according to current exchange rates of the European Central Bank (1 EUR = 27.083 CZK, 1 EUR = 309.51 HUF, 1 EUR = 4.2367 PLN) to ensure data uniformity for criteria LC_1–LC_5. In the case of corporate tax, effective corporate tax rates are used in the TOPSIS model (nominal rates are provided, but are not used in the TOPSIS model) in order to assess the total tax contribution of companies to the state more precisely. As further described below, dummy variables replaced the values for the qualitative criteria

Table 2 Summary of data sources for each criterion

Location criterion (LC_n)	Data source			
	Slovak Republic	Czech Republic	Hungary	Poland
Quantitative criteria				
LC_1 monthly banking fees; LC_2 bank transaction fees; LC_3 price of incoming foreign payment; LC_4 price of outgoing foreign payment; LC_5 price of outgoing urgent payment	Own calculation as the average minimum charge of top 3 banks (by asset size): Slovenská sporiteľňa, VÚB banka, Tatra banka	Own calculation as the minimum charge of top 3 banks (by asset size): Česká spořitelna, ČSOB, Komerční banka	Own calculation as the average minimum charge of top 3 banks (by asset size): OTP Bank, K&H Bank, Erste Bank Hungary	Own calculation as the average minimum charge of top 3 banks (by asset size): PKO Bank Polski, Bank Pekao, Bank Zachodni WBK
LC_6 withholding tax	Deloitte withholding tax rates			
LC_7 corporate tax	(nominal rates) Deloitte corporate tax rates 2015 (effective rates) Effective average corporate tax—Eurostat			
Qualitative criteria				
LC_8 important treasury/financial shared service centres	Academic databases and world wide web			
LC_9 reporting requirements	PricewaterhouseCoopers Know your customer: quick reference guide, world wide web			
	AML control: National Bank of Slovakia, Ministry of Interior, Ministry of Finance	AML control: Czech National Bank, Ministry of Finance, Czech Trade Inspectorate	AML control: Hungarian National Bank, National Tax and Customs Administration	AML control: National Bank of Poland, National Savings and Credit Cooperative Union, Polish Financial Supervisory Authority, General Inspector of Financial Information
LC_{10} currency environment	Banks: Slovenská sporiteľňa, VÚB banka, Tatra banka	Banks: Česká spořitelna, ČSOB, Komerční banka	Banks: OTP Bank, K&H Bank, Erste Bank Hungary	Banks: PKO Bank Polski, Bank Pekao, Bank Zachodni WBK
LC_{11} ratings	Coface business climate and country risk assessment			

Table 3 Data matrix used in the TOPSIS model

Country	LC_1	LC_2	LC_3	LC_4	LC_5	LC_6	LC_7		LC_8	LC_9	LC_{10}
							Nominal	Effective			
Slovak Republic	4.17	0.11	0.11	4.59	35.00	0.19	0.22	0.194	1	1	1
Czech Republic	4.37	0.18	4.76	8.52	16.62	0.15	0.19	0.167	1	1	1
Hungary	20.08	0.71	8.97	14.36	17.00	0.00	0.19	0.193	1	1	1
Poland	14.16	0.59	3.54	7.25	11.21	0.20	0.19	0.175	1	1	1

LC_8–$LC_{10.}$ Since weights based on paired comparisons could be determined only for criteria LC_1–LC_{10}, only these were involved in the TOPSIS model.

Information gathered during our research indicates that many MNCs have established financial shared service centres (FSSCs) in our analysed countries to house their treasury function. According to the Shared Services Centres 2014 survey by PricewaterhouseCoopers (2015), 66% of shared service centres in the Czech Republic and in Slovakia are serving groups with headquarters in Europe and the UK, and a further 28% have headquarters in the United States of America. Central Europe is the destination number one for nearshoring from Western Europe, due to its skilled workforce, low labour costs and its good quality of life.

Examples of FSSCs in Slovakia include the presence of companies like Dell, Henkel, IBM, Johnson Controls International, Lenovo and Kraft Foods (Sario 2014). Among FSSCs in the Czech Republic are the centres of companies like TESCO Stores, SAP, InfoSys, Monster, Johnson & Johnson, Honeywell (CzechInvest 2009). Hungary hosts FSSCs of e.g. Celanese, KLM, UCMS Group, Computacenter, Roche (SSC Heroes 2015). Poland is also an attractive location for FSSCs of companies like Indesit, Oriflame and Rockwell Automation (KPMG 2015). The presence of an increasing number of FSSCs in certain locations over recent years could be a substantial prerequisite and basis for the future establishment of treasury centres. However, our search did not turn up any information or documents containing a summary of the presence of FSSCs for the Visegrad countries, covering comparable periods. Because of the consequences of our difficulties in determining and using the actual number of FSSCs established in each of the Visegrad countries as variables in the TOPSIS model, the values of criterion LC_8 were replaced by the dummy variable (1) if a country had FSSCs, and by the dummy variable (0) if it did not.

Monitored reporting requirements include country-specific information on anti-money laundering (AML), from PricewaterhouseCoopers Know Your Customer: Quick Reference Guide (2014). This analyses the regulatory environment, customer due diligence and reporting requirements, especially about suspicious activities. Table 2 provides the list of key regulators for AML control. In all four countries, the minimum due diligence transaction threshold is 15,000 EUR. AML reporting requirements seem to be similar across the four countries, so the dummy variable (1) replaced the LC_9 values in the TOPSIS model. Nevertheless, in future research additional criteria could reflect the reporting requirements on the observance of corporate governance principles and policies to counter tax evasion.

The currency environment can make a location more favourable for regional treasury centres. In all four countries, the currency environment is evaluated as suitable (indicated by dummy variable 1) due to the availability of accounts and transactions in important currencies (EUR, USD, GBP), but also in many other foreign currencies (e.g. CHF, JPY, CAD, AUD, DKK).

The weights representing the importance of each criterion are based on the paired comparison method using the Fuller's triangle (Simkova 2005) transformed into a matrix of preferences (see Table 4). Paired comparisons were only available for LC_1–LC_{10} so weights were calculated for only ten location criteria, and not for

Table 4 Matrix of preferences in paired comparisons

	LC_1	LC_2	LC_3	LC_4	LC_5	LC_6	LC_7	LC_8	LC_9	LC_{10}	p_{LC_n}	$p^*_{LC_n}$	w_{LC_n}
LC_1	NA	1	0	0	1	0	1	1	1	1	6	7	0.127
LC_2	0	NA	0	0	1	0	1	1	1	1	5	6	0.109
LC_3	1	1	NA	1	1	0	1	1	1	1	8	9	0.164
LC_4	1	1	0	NA	1	0	1	1	1	1	7	8	0.145
LC_5	0	0	0	0	NA	0	1	1	1	1	4	5	0.091
LC_6	1	1	1	1	1	NA	1	1	1	1	9	10	0.182
LC_7	0	0	0	0	0	0	NA	1	1	1	3	4	0.073
LC_8	0	0	0	0	0	0	0	NA	1	1	2	3	0.055
LC_9	0	0	0	0	0	0	0	0	NA	1	1	2	0.036
LC_{10}	0	0	0	0	0	0	0	0	0	NA	0	1	0.018
												$\sum_{n=1}^{10} w_{LC_n}$	1.000

p_{LC_n} sum of preferences of criterion LC_n
$p^*_{LC_n}$ sum of preferences of criterion LC_n after correction (+1)
w_{LC_n} calculated weight of criterion LC_n

Table 5 Results of TOPSIS and the corresponding rankings for evaluated countries

Country (variant)	c_i Relative index of distance of variants from basal variant	Rankings
Slovak Republic	0.5221383	1
Czech Republic	0.5188552	2
Hungary	0.4604049	3
Poland	0.4030688	4

the criterion of rating. The most important criteria identified by paired comparisons include withholding tax and the prices of foreign incoming and outgoing foreign payments. The less important criteria included the qualitative criteria i.e. the currency environment, reporting requirements and the existence of important treasury centres.

Based on the results of the TOPSIS method presented in Table 5, the country rankings depend on the declining relative indices (c_i) of the distance of variants (i.e. countries) from the basal variant. As the value of c_i declines, a country's suitability for the establishment of treasury centres based on Simkova's location criteria also decreases. It is important to note that the evaluation of countries depends on selected criteria defining specific aspects of their business environments.

The data matrix used in the TOPSIS model (see Table 3) indicates that two main groups of quantitative location criteria strongly influenced the overall results—the prices of banking services and taxation/tax rates. Qualitative criteria seem to have little or no influence on the overall results. While Slovakia offers less expensive banking services related to the functions of a regional treasury centre, the withholding tax rate is the second highest after Poland and corporate tax rates (both nominal and effective) are the highest in the Visegrad group. Nevertheless, overall Slovakia offers the best conditions for the establishment of regional treasuries for MNCs, followed by the Czech Republic. The Czech result is due to it having the second cheapest banking services, relatively low withholding tax rates and the lowest effective corporate tax rate. Hungary's ranking is caused by the most expensive banking services in the group and the second highest effective corporate tax rate. On the other hand, interest, dividends and royalties are not subject to a withholding tax. Poland's position of having the least attractive conditions for setting up treasury centres is due to it having the highest withholding tax, and relatively high corporate tax rates, plus relatively expensive bank services.

Because it was impossible to determine the weight and importance of rating in paired comparisons, LC_{11} was not involved in the TOPSIS model and so this location criterion needs a separate commentary. The ratings by Coface use a seven-level ranking system: in ascending order of risk A1, A2, A3, A4, B, C, and D, depending on a country's overall liquidity and solvency. Evaluations depend on macroeconomic, financial and political data. A comparison of the Visegrad countries gives all of them the same relatively low risk business climate rating A2. However, Hungary's country risk rating (B) is noticeably higher than the other three's A3 rank (Coface 2015). According to Coface the sources of Hungary's relative weakness lie in both economic and social factors. The economic factors include high levels of public and external debts, high exposure to exchange rate risk, a weak banking sector, and a high dependency on imported energy. The social factors are a large poorly integrated Roma minority, and an aging population. The business climate ratings would not influence the final TOPSIS results. However, the country risk assessments of Hungary and Poland could reverse their final rankings presented in Table 5.

4 Conclusion

Nowadays multinational companies often decide to set up a regional treasury centre to keep their treasury function centralized, but also to be able to recognize and manage local issues. The choice of the right location for establishing regional treasury centres is an important matter. Despite the fact that popular locations for regional treasury centres are mostly in the Asia Pacific region, the United States of America or in Western Europe, Central European countries could also provide a favourable location for a treasury centre. Many multinationals have already established financial shared service centres with incorporated treasury functions in Visegrad group countries. An MNC's decision about a suitable location for a treasury centre is influenced by a variety of factors. Although the tax environment is the most commonly mentioned key driver, other important factors include the prospects for reducing costs, a satisfactory regulatory and legal framework, the currency environment, the supply of skilled personnel and their language skills, a modern banking system, and the time zone of the location.

The main aim of this study was to evaluate and compare the countries of the Visegrad Group from the perspective of their suitability as locations for a regional corporate treasury centre, based on a selected group of tax and non-tax location criteria. The assessment used a multi-criteria decision-making approach—the TOPSIS method. This was based on both quantitative and qualitative criteria identified by Simkova (2005), using weights for the criteria determined by paired comparisons.

The results of the TOPSIS analysis showed that the Slovak Republic had the most favourable conditions for establishing regional treasury centres, followed by the Czech Republic and Hungary. Poland had the worst conditions for regional treasury centres. However, it is important to mention the specificity of evaluated aspects of business environment in the analysed countries, and there was only limited data available on the qualitative location criteria. This meant that except two key groups of criteria, the costs of banking services, and the costs of taxation, we could only use dummy variables in the absence of actual data. Better and more detailed information on the qualitative criteria could change the outcome of the analysis. Given this, it would also be appropriate to consider redefining some of the criteria, as well as adjusting the range of criteria used. Based on the literature review presented in this paper, there are additional criteria influencing MNCs' decisions on treasury centre locations. These include labour costs and the availability of skilled employees. Adding additional criteria in multi-criteria decision-making analyses is the object of further research.

Acknowledgements This work was supported by the Slovak Research and Development Agency under the Contract No. APVV-15-0322.

References

Anwar T (1999) Conducting a feasibility study for a regional treasury centre. In: Treasury management international special report. http://www.treasury-management.com/article/4/73/668/conducting-a-feasibility-study-for-a-regional-treasury-centre.html. Accessed 7 Aug 2015

Bank of America Merrill Lynch, SunGard (2015) Asia Pacific treasury management barometer. http://www.sungard.com/~/media/fs/corporate-liquidity/resources/reports/Asia-Pacific-Treasury-Management-Barometer-2015-MarketStudy-SunGard. Accessed 9 Sep 2015

Barlow J (2011) The rise of regional treasury centres. http://www.financeasia.com/News/268141,the-rise-of-regional-treasury-centres.aspx. Accessed 21 Aug 2015

Blach J, Wieczorek-Kosmala M, Gorczyńska M, Doś A (2014) Innovations in liquidity management – the potential of corporate treasury. J Econ Manag 2014(18):210–224

Blair D (2011) Choosing a regional treasury centre location in Asia. In: Treasury Today. http://treasurytoday.com/2011/05/choosing-a-regional-treasury-centre-location-in-asia. Accessed 15 Sep 2015

Bragg SW (2010) Treasury management: the practitioner's guide. Wiley, Hoboken, NJ

Coface (2015) Country risk assessment. http://www.coface.com/Economic-Studies-and-Country-Risks. Accessed 11 Oct 2015

CzechInvest Investment and Business Development Agency (2009) Business support services in the Czech Republic. http://czechinvest.org/data/files/bss-764-fr.pdf. Accessed 20 Sep 2015

Deloitte (2014) Ireland a corporate treasury location. Approaching treasury with certainty in uncertain times. http://www2.deloitte.com/content/dam/Deloitte/ie/Documents/Tax/ie_tax_treasury_brochure_corporate_treasury_2014.pdf. Accessed 8 Aug 2015

Deloitte (2015a) Corporate tax rates 2015. http://www2.deloitte.com/content/dam/Deloitte/global/Documents/Tax/dttl-tax-corporate-tax-rates-2015.pdf. Accessed 15 Aug 2015

Deloitte (2015b) Withholding tax rates 2015. http://www2.deloitte.com/content/dam/Deloitte/global/Documents/Tax/dttl-tax-withholding-tax-rates-2015.pdf. Accessed 15 Aug 2015

Dizkirici AS (2012) Comparison of Istanbul with Hong Kong and Singapore for regional treasury centers. J Bus Res 2012:31–44

Eurostat (2015) European Commission, Luxembourg. http://ec.europa.eu/taxation_customs/taxation/gen_info/economic_analysis/data_on_taxation/index_en.htm. Accessed 12 Oct 2015

Giegerich G et al (2002) Implementing a treasury center in Switzerland. https://www2.eycom.ch/publications/items/tax_fs_treasury/en.pdf. Accessed 10 Sep 2015

Koning L (2012) What does onshoring, offshoring, nearshoring mean? http://www.managementtrainee.co.uk/what-does-onshoring-offshoring-nearshoring-mean.html. Accessed 15 Sep 2015

KPMG (2015) Poland as the destination for business services centres. https://www.kpmg.com/PL/pl/IssuesAndInsights/ArticlesPublications/Documents/2015/Przewodnik-KPMG-Poland-as-the-destination-for-Business-Services-Centres.pdf. Accessed 12 Oct 2015

Mulligan E (2001) Treasury management organisation: an examination of centralised versus decentralised approaches. Ir J Manag 22(1):7–33

Murphy A (2000a) Solving the problems of globalisation. In: Treasury management international. http://www.ftitreasury.com/images/sampledata/pdf/Solving%20the%20problems%20of%20Globalisation.pdf>. Accessed 20 Aug 2015
Murphy A (2000b) Non-tax factors in treasury centre location decisions. In: The treasurer. http://www.treasurers.org/node /1988. Accessed 20 Sep 2015
Polak P (2010) Centralisation of treasury management in a globalized world. Int Res J Financ Econ 2010(56):88–95
Polak P, Klusáček I (2010) Centralization of treasury management. Business Perspectives, Sumy
Polak P, Roslan RR (2009) Regional treasury centres in South East Asia – the case of Brunei Darussalam. J Contemp Manag 14(1):77–101
Polak P, Robertson DC, Lind M (2011) The new role of the corporate treasurer: emerging trends in response to the financial crisis. Int Res J Financ Econ 2011(78):48–69
PricewaterhouseCoopers (2013) Luxembourg: the place to be for your treasury centre. http://www.pwc.lu/en/invest-in-luxembourg/docs/pwc-lux-treasury-centre.pdf. Accessed 7 Sep 2015
PricewaterhouseCoopers (2014) Know your customer: quick reference guide. http://www.pwc.com/gx/en/financial-services/publications/assets/pwc-anti-money-laundering-know-your-customer-quick-reference-guide.pdf>. Accessed 15 Sep 2015
PricewaterhouseCoopers (2015) Shared services centres 2014. Moving toward centres of excellence. Czech Republic & Slovakia. https://www.pwc.com/sk/en/shared-services-centres/assets/ssc-survey-2014.pdf. Accessed 15 Sep 2015
Roslan RR (2008) Brunei Darussalam: Asia's next leading location for regional treasury centres? Thesis, Swinburne University of Technology
Roslan RR, Polak P (2009) Brunei Darussalam: Asia's next leading location for regional treasury centres? Econ Altern 2009(1):84–101
SARIO (2014) Shared services centers, 2014. http://www.sario.sk/sites/default/files/content/files/shared_service_center_0.pdf. Accessed 20 Sep 2015
Shah G, Yung C (2012) Evolution of corporate treasury centers and location considerations for Asia Pacific. https://www.citibank.com/tts/about_us/articles/docs/2012/Citi_ Insights_ EvolutionofCTC.pdf. Accessed 20 Aug 2015
Simkova L (2005) Finding International Treasury Centre for the joint-stock CGS. VSB – Technical University Ostrava
SSC Heroes (2015) http://sscheroes.com/company. Accessed 20 Sep 2015
Van der Sar B (2013) Going Dutch: The Netherlands as a location for operation and treasury services. http://www.citibank.com/transactionservices/home/tts/corp/liquidity/docs/28januarygoingdutch.pdf. Accessed 20 Aug 2015

Beata Šarkanová is a young researcher at the Department of Finance and Accounting, Faculty of Economics, Matej Bel University in Banská Bystrica, Slovak Republic. Her main research area includes finance, corporate economics and financial management. She is the author of multiple papers, mostly in the field of corporate treasury management.

Peter Krištofík (44) is an associate professor at the Faculty of Economics, Matej Bel University in Banská Bystrica, Slovakia. The focus of his research and pedagogical activities is on corporate financial management and financial markets. He is author and co author of more than 100 scientific papers and monographs. Pedagogical activities include lectures and seminars at domestic university as well as at universities worldwide (Netherlands, Canada, Austria, Czech Republic, Poland etc.) and he is often invited to provide trainings for managers in commercial companies.

Modeling Open Innovation Effects in Czech Manufacturing Firms

Petr Hajek and Jan Stejskal

Abstract This chapter deals with the effect of open innovation on innovation activity in Czech manufacturing firms. We analyze the European CIS 2010 data using the methodology of Laursen and Salter (Strateg Manag J 27(2):131–150, 2006), Van de Vrande et al. (Technovation 29(6):423–437, 2009) and Ebersberger et al. (Res Policy 43(3):495–504, 2012) based on the depth (intensity) and breadth (variety) for open innovations. In our paper we focus on measuring the depth and breadth in various manufacturing industries (from low to high-tech firms) oriented on different markets (from local to non-EU regions). We distinguish two types of innovation, new-to-market and new-to-firm, respectively. To analyze the indirect effect of open innovation (measured via the depth and breadth of knowledge acquisition and innovation cooperation), we use structural equation models. The results confirm that open innovation mediates the effect of public support (mainly government and EU) on innovation activity. We also show that firms oriented on new-to-market innovation report significantly higher breadth of open innovation. We propose several important implications for innovation policymakers, stressing that the variety (breadth) of open innovation is the critical determinant to promote innovation activity.

Keywords Open innovation • Collaboration • Knowledge acquisition • Structural model • Czech manufacturing firms

---

P. Hajek (✉)
Faculty of Economics and Administration, Institute of System Engineering and Informatics, University of Pardubice, Studentska 95, Pardubice 53210, Czech Republic
e-mail: petr.hajek@upce.cz

J. Stejskal
Faculty of Economics and Administration, Institute of Economic Sciences, University of Pardubice, Studentska 95, Pardubice 53210, Czech Republic
e-mail: jan.stejskal@upce.cz

A. Dias et al. (eds.), *Modeling Innovation Sustainability and Technologies*, Springer Proceedings in Business and Economics,
https://doi.org/10.1007/978-3-319-67101-7_9

1 Introduction

Innovation is currently considered the most important long-term determinant of the economic performance of firms (see a review by Grossman and Helpman 1994). Innovative companies grow and develop more quickly by gaining benefit from the economies of scale (Almus and Nerlinger 1999) and other benefits. However, firms also have to invest in new technology, skilled employees and the acquisition of new knowledge. In other words, firms must ensure the so-called critical factors for successful innovations. Van der Panne et al. (2003) provide a list of firm related critical factors such as experience, R&D intensity, knowledge and innovation strategy, etc. These factors cannot be easily attained at the firm level. Therefore, it is necessary to use the existence of various external interfaces, which include information gathering, collaborations and alliances with actors ranging from universities to suppliers and customers (Ebersberger et al. 2012).

Firm management must ensure synthesizing the necessary internal and external production factors and ensure the involvement of the company in the new knowledge creation. In practice, the firms are involved in business alliances (George et al. 2002), knowledge networks (Tsai 2001), regional innovation systems (Asheim et al. 2011) and industrial clusters (Stejskal 2010; Stejskal and Hajek 2012), depending on the potential of the region (Uramova and Koziak 2008), type of industry, etc. When connecting to networks, firms find that it is not possible to only receive from other actors in the network, but also provide (usually knowledge, experience, know-how and contribute financially to the purchase or development of new technologies). Openness is a natural consequence of the firm's involvement in the network. According to Ebersberger et al. (2012), interfacing or "openness" cannot be reduced to formal ties or to one single dimension, such as alliances, purchases of patents or collaboration; and that these dimensions should not be considered in isolation from each other. Firm's openness should not only be the result, but a means to get new ideas and increase capacity to absorb innovation. The strategic intent of the firm should be to create a suitable environment for the creation of open innovation.

The concept of open innovation was introduced by Chesbrough (2003). This concept suggests that the advantages that firms gain from internal R&D expenditure have declined (Laursen and Salter 2006). The firms seeking for innovation in the open innovation system spend on R&D smaller funds, because new knowledge, information, ideas and experiences they obtain from external cooperating chains. The substitution (not total) of internal information sources for external sources needs the changes in management, typically in HR management, communication and firm strategies (Gavurova 2012). The firms, in practice, obtain new findings from external sources and cooperation networks. Based on them they create the innovation and offer it in the market. This does not include all firms, but only those who are able to recognize the opportunity. Therefore, it is needed to have an innovative management with know-how that will help to identify seemingly unrelated opportunity, to combine them and to expand their absorption and potential. This approach may also bring some loss; respectively require additional investment to the technologies. The change in firm orientation from internal to external knowledge is sometimes referred to as the 'connect and develop' model (Sakkab 2002). The firm itself as well as its environment is changed with focus on external stimuli. The increased interest of companies about cooperation is forced to change their strategies and market behaviour.

Previous research has shown that both the breadth and depth of open innovation is an important determinant of innovation activity (Ebersberger et al. 2012; Laursen and Salter 2006; Van de Vrande et al. 2009). However, more complex models are fundamental to enhance understanding of the role of open innovation in innovation activity. Here we aim to investigate the indirect role of open innovation, mediating the effect of (1) market and industry level, and (2) internal R&D expenditure and public support.

The remainder of this paper has been divided into four sections. The paper first gives a brief overview of the evidence for the positive effects of open innovation activities on firms' innovativeness. The next section lays out the description of the data and research methodology. Section 4 provides the results of the modelling. The final section concludes the paper and discusses its implications.

2 Effects of Open Innovation on Innovation Performance

Recently, there has been much effort in finding adequate strategies for innovative firms. The essence of the changes has been the shift away from investment to intramural R&D and supplement (or substitute) them by extensive use of external knowledge sourcing and external paths to commercialization (Cassiman and Veugelers 2006). The technological progress and new communication elements are the important prerequisites for creating innovation based on cooperative open systems. It brings with it the requirement of demand for new products, innovation systems, sales, marketing, etc. Similar changes in development, however, can be observed already during the twentieth century. There is a clear shift away from innovation creation by alone entrepreneur (Schumpeter 1942) to innovation generating by cooperating entities (numerous interactions and the knowledge spill-over effects are arisen spontaneously among them; Von Hippel 1988; Lundvall 1992). Recent studies conclude that the innovation process is dependent on the number, nature and openness of the interactions between different entities. The speed of innovation production depends inter alia on "swift trust," which determines the transfer of knowledge and is a prerequisite for the development of knowledge spill-over effects and the innovations (Brown and Duguid 2000).

Many studies have shown that the application of the open innovation model has three separate processes (Gassmann and Enkel 2004; Dittrich and Duysters 2007). Each firm can decide on which one to focus on. The first is outside-out process that targets the use and application of new ideas, thoughts and knowledge gained outside of companies and their commercialization into new products and services.

The second, inside-out process focuses on the use of ideas arising out of the company through patents, licensing or divestment knowledge in the market. The third, coupled process captures activities that have arisen as a result of cooperative linkages (other than ordinary business). Individual processes utilize various information sources such as customers, competitors, suppliers, customers, universities, research institutes etc. Search processes can therefore be seen as a dynamic capability that allows firms to sustain their competitive advantage over time (Eisenhardt and Martin 2000).

Laursen and Salter (2006) found that the innovation performance increases with the breadth and depth of external search, with a variety of external information sources and with the intensity of their use. These relationships are found take on inverse U-shapes, indicating the possibility of excessive dependence on external information sources (Ebersberger et al. 2012). External sourcing can be realized in two ways: on the market principle and based on collaboration, respectively. The former approach has several specifics: since firms buy the knowledge or innovation, the demand for these final products is increased (Lichtenthaler and Ernst 2007). However, the knowledge spill-over effects are limited, no matter whether it is bought from other firms or research organizations/universities. The prerequisite for this knowledge acquisition is the good protection of intellectual rights and the use of licenses and patents. Only then the commercialization of external technology is frequently utilized.

The latter approach is to acquire knowledge in collaboration processes towards the creation of new knowledge and innovation. First, it is necessary to establish a relationship with an organization that generates knowledge; this is with R&D organizations (Herstad et al. 2014), universities (Abramovsky and Simpson 2011), suppliers or customers (Greer and Lei 2012), etc. Firms may also enter into strategic alliances with other companies (Lew and Sinkovics 2013), associations or cooperatives (Lin et al. 2012), often in industrial clusters (Kesidou and Snijders 2012) and open innovation networks (Love et al. 2014).

Many researchers are concerned with the issue from different views. They often analyse the case studies related to changes in firm strategies, as well as the theoretical role of open innovation and the models of open innovation processes, respectively. However, few studies deal with the fact that open innovation may have different effect on innovative performance for enterprises of different industries (Sofka and Grimpe 2010; Schroll and Mild 2011). Several studies have investigated the role of open innovation in individual countries. For example, Spithoven et al. (2011) dealt with the formation of absorption capacity in the case of inside-out processes in traditional industries (Belgium), Chiarone et al. (2010) examined asset-intensive industries (Italy), and Bianchi et al. (2011) studied organisational modes for open innovation in the bio-pharmaceutical industry (Italy). Even less attention has been paid to the technology-intensive production. Martín-de Castro (2015) studied the role of openness and absorptive capacity in knowledge-based and high-tech industrial markets; Alberti et al. (2014) examined the Italian mid-high tech SMEs; and Park et al. (2014) were looking for the competitive dynamics and the knowledge seeking behaviour of high-tech firms (South Korea).

3 Data and Research Methodology

To examine the role of open innovation in the Czech manufacturing industry, we collected data from the Community Innovation Survey (CIS), which is based on a harmonized questionnaire of EU Member States. The survey was carried out in the Czech Republic for the period 2008–2010 by combining sample (stratified random sampling) and exhaustive surveys taking into account the regional dimension of NUTS3. In total, data on 5151 Czech firms with at least ten employees was obtained (response rate greater than 60%). The CIS is regarded as a reliable source of innovation statistics in the EU owing to comprehensive data validation and measurement error reduction procedures (see Eurostat for details). After discarding the firms without variables necessary for the calculation of open innovation depth and breadth, we obtained $N = 1318$ manufacturing firms.

The basic characteristics of the dataset are given in Tables 1 and 2. The innovation activity of the firms was estimated by calculating the number of companies that introduced a new product or process to the firm/market. Table 1 shows that low-tech firms dominated in the dataset with 40.12%, most of the firms were oriented on the national market (48.36%), and only 19.12% received public financial support for innovation activity from the EU.

In calculating the variables for open innovation, we adopted the approach used by Laursen and Salter (2006), Van de Vrande et al. (2009) and Ebersberger et al. (2012), where open innovation is categorized in two ways, namely into knowledge sources for innovation/collaboration on innovation and depth/breadth. Thus, we obtained four indicators of open innovation: (1) depth of knowledge sources, (2) breadth of knowledge sources, (3) depth of collaboration, and (4) breadth of collaboration. Regarding the knowledge sources, each firm was asked to indicate on a 0–3 scale the degree of use for each of the nine sources (market source: suppliers, clients, competitors, and consultants; research institutes: universities, and government research institutes; other: conferences, scientific journals, and professional associations). To obtain the breadth of knowledge sources, the nine sources were simply added up, this is the overall score was on the scale 0–9. For the depth of knowledge sources, a similar approach was used, but only those sources were added up that were used to a high degree (this is 3 on the 0–3 scale).

Similarly, the breadth of collaboration was calculated as the sum of collaborating partners (other firms, suppliers, clients, competitors, consultants, universities, and government institutes), leading to the 0–7 scale. To calculate the depth of collaboration, only those partners were added up, when the firm used both domestic and international partners for innovation collaboration. Table 3 shows the descriptive statistics for these variables. The sophistication of market represents a strong

Table 1 Relative frequencies of categorical variables

Industry	Low-tech	Medium low-tech	Medium high-tech	High-tech	
Rel. freq.	40.12	26.25	27.56	6.07	
Market	Local	Regional	National	EU	Other
Rel. freq.	13.62	11.09	48.36	22.87	4.07
Funding	Local	Govern.	EU		
Rel. freq.	4.17	17.30	19.12		
Innovation	New-to-firm	New-to-market			
Rel. freq.	79.71	71.19			

Table 2 Average values and standard deviations of numerical variables

	Mean	St. dev.
TURN10	1,132,335	7,584,428
EMP10	223	1057
RTOT10	33,614	275,921

TURN10 total turnover in 2010, *EMP10* average number of employees in 2010, *RTOT10* total innovation expenditure in 2010

Table 3 Descriptive statistics of open innovation variables

	Breadth of knowledge sources	Depth of knowledge sources	Breadth of collaboration	Depth of collaboration	*N*
Industry					
High-tech	7.90 ± 2.65	1.78 ± 1.71	3.15 ± 1.97	**0.97 ± 1.33**	73
Medium high-tech	**8.03 ± 2.61**	**2.07 ± 1.61**	**3.40 ± 1.87**	0.71 ± 1.06	370
Medium low-tech	7.51 ± 2.78	1.62 ± 1.41	3.04 ± 1.97	0.66 ± 1.24	279
Low-tech	7.89 ± 2.58	1.98 ± 1.60	3.16 ± 1.84	0.72 ± 1.13	576
Market					
Local	7.45 ± 2.82	1.79 ± 1.64	3.31 ± 1.83	0.42 ± 1.03	104
Regional	7.27 ± 2.97	1.68 ± 1.60	2.95 ± 2.10	0.63 ± 1.20	99
National	7.86 ± 2.56	1.91 ± 1.59	3.24 ± 1.89	0.65 ± 1.13	679
EU	7.91 ± 2.67	1.92 ± 1.44	3.05 ± 1.80	0.78 ± 1.09	358
Other countries	**8.83 ± 2.19**	**2.56 ± 1.86**	**3.96 ± 1.83**	**1.06 ± 1.31**	78
Innovation					
New-to-firm yes	8.20 ± 2.48	2.04 ± 1.57	3.33 ± 1.84	0.81 ± 1.18	735
New-to-firm no	7.97 ± 2.49	2.15 ± 1.74	3.25 ± 1.92	0.68 ± 1.18	251
New-to-market yes	**8.39 ± 2.45**	**2.22 ± 1.67**	**3.58 ± 1.83**	**0.89 ± 1.27**	587
New-to-market no	7.77 ± 2.49	1.84 ± 1.50	2.78 ± 1.81	0.56 ± 0.95	399

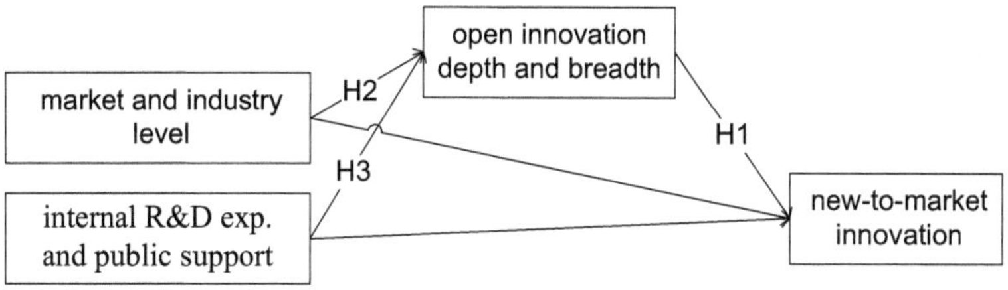

Fig. 1 Conceptual model

determinant of open innovation. Further, firms introducing new-to-market innovations strongly depend on both depth and breadth of open innovation.

To study the effect of open innovation variables on innovation activity, we constructed a structural equation model. In this model, the depth and breadth of knowledge sources/collaboration represent mediator variables, causally located between the input variables (technology level of industry, market sophistication, R&D expenditure, and public support) and output (new-to-market innovation activity). In other words, we tested both the direct and indirect (via open innovation variables) effect of the input variables on innovation activity. Previous research has shown significant positive effects of open innovation variables on innovation activity in manufacturing industries (Ebersberger et al. 2012; Laursen and Salter 2006).

However, the strength of the effect seems to be different across EU countries (Ebersberger et al. 2012). The indirect effect, on the other hand, has not been investigated so far, although Abramovsky et al. (2009) reports that both internal R&D expenditure and the receipt of public support may be positively related to open innovation variables (Stejskal and Hajek 2015). In addition, here we also examine the effect of industry technology level and market sophistication on open innovation variables. As indicated in Table 3, both the (medium) high technology firms and the firms oriented on non-EU markets show higher values of open innovation variables. Therefore, we hypothesize that (see Fig. 1):

H1: Open innovation depth and breadth positively affect new-to-market innovation activity.
H2: Market and industry level positively affect open innovation depth and breadth.
H3: Internal R&D expenditure and public support positively affect open innovation depth and breadth.

4 Empirical Results

To test the hypotheses posed in the previous section, the modeling was performed using the structural equation models in the Process tool developed by Hayes (2013) for the SPSS statistical software package. Figure 2 shows the results of the modelling.

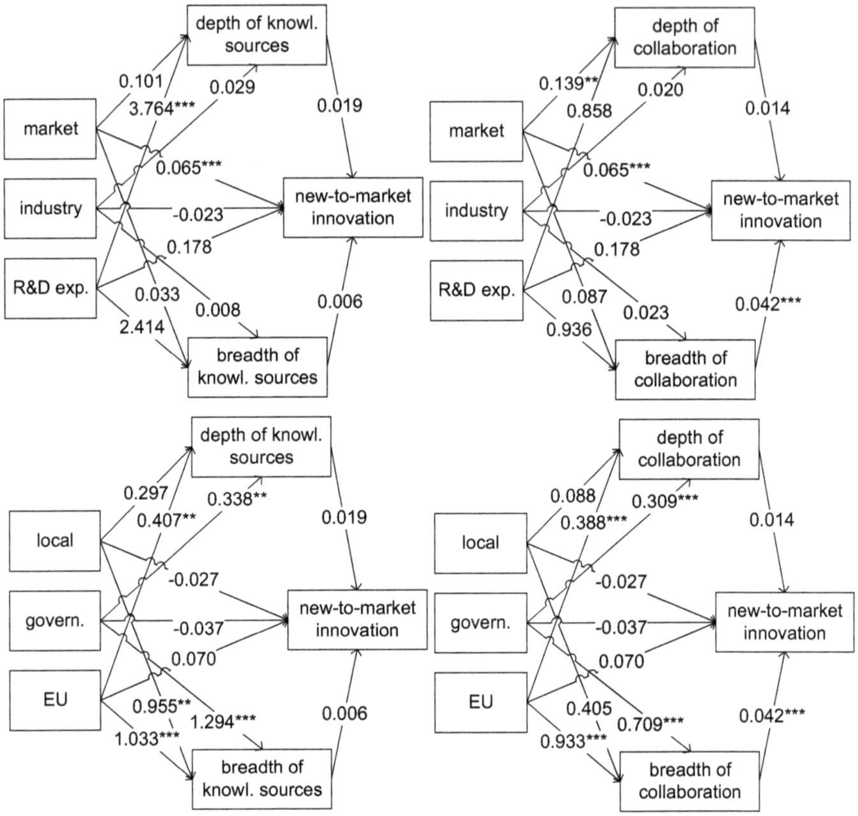

Fig. 2 Structural equation models. *Significant effect at $p = 0.10$, **Significant effect at $p = 0.05$, and ***Significant effect at $p = 0.01$

The results of the modelling confirmed that private R&D expenditures affect the depth of knowledge sources (public financing is viable due to significant differences in knowledge resources needs in individual firms). In contrast, private firms do not fund the establishment of collaboration activities. The sophistication of the target market forces firms to innovate and collaborate on a much broader level than if their sales market is "only" a local one. On the other hand, the results did not confirm that the type of industry (according to the necessity of knowledge necessary for production) significantly affects the depth of knowledge sources and depth of collaboration.

Regarding the public funding of innovative activities: EU funds (0.407**) and government support (0.338**) have the strongest impact on the depth of knowledge sources. The results are similar for the breadth of knowledge sources, with the significant effects of EU (1.033***), government (1.294***) and local budgets (0.955**). The direct effects of various sources of public support on innovation activity were not confirmed. Similar results were observed for the depth and breadth of collaboration, with the strongest impact of the EU budget (0.388*** for the depth of collaboration and 0.933*** for the breadth of collaboration, respectively).

Taken together, market sophistication directly affects new-to-market innovation activity and the depth of collaboration, respectively. The depth of collaboration was however not a significant determinant of the innovation activity. The effects of industry level were not significant at all. Internal R&D expenditure showed strong effect on the depth of knowledge sources. Again, this effect did not lead to innovation activity owing to the weak effect of the depth of knowledge sources. In fact, the breadth of collaboration was the only significant open innovation determinant of innovation activity. It was the government and EU public support that mainly affected the breadth of collaboration.

5 Discussion and Conclusion

New-to-market innovation substantially contributes to the firm performance and may also ensure the firm's long-term competitiveness. This type of innovation is the result of available knowledge, experience and know-how application. In order to achieve the new-to-market innovation it is also necessary to utilize spill-over effects present both in the firm and in its innovative ties. New-to-market innovation is also the common aim of public policies. In the production of these innovations, a number of positive effects (often spill-overs) is formed, which ultimately enhance societal welfare.

Given the importance of this type of innovation, it is becoming essential that researches must help identify the determinants of the innovation environment in order to ensure maximum efficiency. These efforts also facilitate the concept of open innovation, namely the analysis of the depth and breadth of open innovation.

In our study, we present results, which confirm that the breadth of collaboration is a key factor for new-to-market innovation activity. Therefore, our results suggest that the new-to-market innovation depends on a variety of subjects that are involved in innovation (knowledge) chains. This fact may imply that the innovation is conditioned by the spill-over effects that occur naturally in these collaborative knowledge networks. The results also show that the breadth of collaboration is affected the most with the EU and government financial support. The results confirm that the resources of regional and local budgets have less effect, which may be primarily due to the budget limits. Next we examined the role of industry and market sophistication. Although these determinants have a positive effect on open innovation activities, it was not significant (however, market sophistication proved to be the strongest direct determinant of the innovation activity).

Our results confirm the results by Ebersberger et al. (2012). Their study examined the effect of open innovation activities on new-to-market innovations in four countries (Austria, Belgium, Denmark, and Norway). They demonstrated that new-to-market innovation is mostly influenced with the breadth of open innovation (from 0.124*** to 0.229***). The depth of open innovation did not show the significant effect for all countries, respectively, their analysis showed different results, including negative effects. Namely, the breadth of collaboration showed significant influence in Belgium and Norway, whereas the depth of collaboration has seen only negative effects.

Returning to the hypotheses posed in this paper, it is now possible to state that: (1) open innovation activities have positive effect on new-to-market innovation (but significant only in the case of collaboration breadth); (2) market level is more important than industry level for both open innovation and new-to-market innovation activity; and (3) public support is critical for open innovation depth and breadth.

Based on our results, we can imply that private investment should be better used for other purposes (for example, investments to technology or staff training), rather than to promote collaboration in knowledge networks. From our results, it is also possible to deduce the implications for public decisions and policies. We unequivocally support the increase in collaboration breadth. Governments should promote collaboration between firms, market actors, as well as other knowledge-based organizations. Financial support from the EU and national governments should be directed primarily on collaborative activities. Governmental organizations should be facilitators or moderators in industry-university cooperation. Properly formed government spending programs may also help increase the interest in collaboration and help remove the obstacles in initiation phase of relationships between different organizations. Consequently, this will help produce spill-over effects and enhance societal welfare.

Acknowledgments This article was created as part of the resolution of the research task No. 17-11795S financially supported by the Grant Agency of the Czech Republic.

References

Abramovsky L, Simpson H (2011) Geographic proximity and firm–university innovation linkages: evidence from Great Britain. J Econ Geogr 11(6):949–977

Abramovsky L, Kremp E, López A, Schmidt T, Simpson H (2009) Understanding co-operative innovative activity: evidence from four European countries. Econ Innov New Technol 18 (3):243–265

Alberti FG, Ferrario S, Papa F, Pizzurno E (2014) Search breadth, open innovation and family firms: evidences in Italian mid–high tech SMEs. Int J Technol Intell Plan 10(1):29–48

Almus M, Nerlinger EA (1999) Growth of new technology-based firms: which factors matter? Small Bus Econ 13(2):141–154

Asheim BT, Smith HL, Oughton C (2011) Regional innovation systems: theory, empirics and policy. Reg Stud 45(7):875–891

Bianchi M, Cavaliere A, Chiaroni D, Frattini F, Chiesa V (2011) Organisational modes for open innovation in the bio-pharmaceutical industry: an exploratory analysis. Technovation 31 (1):22–33

Brown JS, Duguid P (2000) Mysteries of the region: knowledge dynamics in Silicon Valley. In: Lee CM, Miller WF, Hancock MG, Rowen HS (eds) The Silicon Valley edge. Stanford University Press, Stanford, pp 16–39

Cassiman B, Veugelers R (2006) In search of complementarity in innovation strategy: internal R&D and external knowledge acquisition. Manag Sci 52(1):68–82
Chesbrough H (2003) Open innovation. Harvard University Press, Cambridge, MA
Chiaroni D, Chiesa V, Frattini F (2010) Unravelling the process from closed to open innovation: evidence from mature, asset-intensive industries. R&D Manag 40(3):222–245
Dittrich K, Duysters G (2007) Networking as a means to strategy change: the case of open innovation in mobile telephony. J Prod Innov Manag 24:510–521
Ebersberger B, Bloch C, Herstad SJ, Van De Velde E (2012) Open innovation practices and their effect on innovation performance. Int J Innov Technol Manag 9(06):1–25
Eisenhardt KM, Martin JA (2000) Dynamic capabilities: what are they? Strateg Manag J 21 (10–11):1105–1121
Gassmann O, Enkel E (2004) Towards a theory of open innovation: three core process archetypes. Mimeo, Institute of Technology Management, University of St. Gallen, Switzerland
Gavurova B (2012) Source identification of potential malfunction of balanced scorecard system and its influence on system function. E+M Ekon Manag 15(3):76–90
George G, Zahra SA, Wood DR (2002) The effects of business–university alliances on innovative output and financial performance: a study of publicly traded biotechnology companies. J Bus Ventur 17(6):577–609
Greer CR, Lei D (2012) Collaborative innovation with customers: a review of the literature and suggestions for future research. Int J Manag Rev 14(1):63–84
Grossman GM, Helpman E (1994) Endogenous innovation in the theory of growth. J Econ Perspect 8(1):23–44
Hayes AF (2013) Introduction to mediation, moderation, and conditional process analysis: a regression-based approach. Guilford, New York, NY
Herstad SJ, Aslesen HW, Ebersberger B (2014) On industrial knowledge bases, commercial opportunities and global innovation network linkages. Res Policy 43(3):495–504
Kesidou E, Snijders C (2012) External knowledge and innovation performance in clusters: empirical evidence from the Uruguay software cluster. Ind Innov 19(5):437–457
Laursen K, Salter A (2006) Open for innovation: the role of openness in explaining innovation performance among UK manufacturing firms. Strateg Manag J 27(2):131–150
Lew YK, Sinkovics RR (2013) Crossing borders and industry sectors: behavioral governance in strategic alliances and product innovation for competitive advantage. Long Range Plann 46 (1):13–38
Lichtenthaler U, Ernst H (2007) External technology commercialisation in large firms. R&D Manag 37(5):383–397
Lin C, Wu YJ, Chang C, Wang W, Lee CY (2012) The alliance innovation performance of R&D alliances—the absorptive capacity perspective. Technovation 32(5):282–292
Love JH, Roper S, Vahter P (2014) Learning from openness: the dynamics of breadth in external innovation linkages. Strateg Manag J 35(11):1703–1716
Lundvall BA (ed) (1992) National systems of innovation: towards a theory of innovation and interactive learning. Pinter, London
Martín-de Castro G (2015) Knowledge management and innovation in knowledge-based and high-tech industrial markets: the role of openness and absorptive capacity. Ind Mark Manag 47:143–146
Park NK, Mezias JM, Lee J, Han JH (2014) Reverse knowledge diffusion: competitive dynamics and the knowledge seeking behavior of Korean high-tech firms. Asia Pac J Manag 31 (2):355–375
Sakkab NY (2002) Connect and develop complements research and develop at P&G. Res Technol Manag 45(2):38–45
Schroll A, Mild A (2011) Open innovation modes and the role of internal R&D: an empirical study on open innovation adoption in Europe. Eur J Innov Manag 14(4):475–495
Schumpeter JA (1942/1987) Capitalism, socialism and democracy. Unwin, London

Sofka W, Grimpe C (2010) Specialized search and innovation performance–evidence across Europe. R&D Manag 40(3):310–323
Spithoven A, Clarysse B, Knockaert M (2011) Building absorptive capacity to organise inbound open innovation in traditional industries. Technovation 31(1):10–21
Stejskal J (2010) Comparison of often applied methods for industrial cluster identification. In: International Conference on Development, Energy, Environment, Economics—Proceedings, pp 282–286
Stejskal J, Hajek P (2012) Competitive advantage analysis: a novel method for industrial clusters identification. J Bus Econ Manag 13(2):344–365
Stejskal J, Hajek P (2015) The influence of public expenditure on innovation activity in Czech manufacturing industry. In: Proceedings of the 25th international business information management association conference—innovation vision 2020: from regional development sustainability to global economic growth, IBIMA 2015, pp 1820–1827
Tsai W (2001) Knowledge transfer in intraorganizational networks: effects of network position and absorptive capacity on business unit innovation and performance. Acad Manage J 44 (5):996–1004
Uramova M, Koziak R (2008) Regional disparities in Slovakia from the aspect of average nominal wage. E+M Ekon Manag 11(2):6–17
Van de Vrande V, De Jong JP, Vanhaverbeke W, De Rochemont M (2009) Open innovation in SMEs: trends, motives and management challenges. Technovation 29(6):423–437
Van der Panne G, Van Beers C, Kleinknecht A (2003) Success and failure of innovation: a literature review. Int J Innov Manag 7(03):309–338
Von Hippel E (1988) The sources of innovation. Oxford University Press, New York, NY

Petr Hajek is an associate professor with the Institute of System Engineering and Informatics, Faculty of Economics and Administration, University of Pardubice, Czech Republic. He deals with the modelling of economic processes (especially in the field of public economics and public finance).

Jan Stejskal is an associate professor with the Institute of Economics, Faculty of Economics and Administration, University of Pardubice, Czech Republic. His domain is connection of the public economy in the regional scope and view. Especially, he analyzes regional policy, tools of the local and regional economical development, and public services.

German Law Covering the Public Participation in Planning and Building Infrastructure Projects

Anja Bothe

Abstract The analysis presented here focuses on the latest German legislation changes introduced in order to make the planning and building of infrastructure projects more harmonious and more effective.

From a due-process perspective, International and European Environmental law played a pioneer role by leading the German legislators to introduce an earlier public participation in its extremely complex system of spatial and sectorial planning. At first it mainly addressed sustainability. Then, after a few major infrastructure projects were confronted with massive resistance by the population affected, a rethinking about the citizens' involvement at an early stage became necessary, aiming at streamlining and speeding up legally binding decisions on the more controversial infrastructure projects. As a result, the Administrative Procedure Code was expanded to incorporate the "Law for broadening the public participation and for the standardization of the procedures for determining sectorial plans."

In order to debate, in an interdisciplinary way, the role that the law is able to play in communication procedures, this article discusses the essential factors impacting the different types of plans and planning. It then questions the strategies leading to acceptance of certain projects and the trade-offs between speed of implementation and the respect for public participation rights.

Keywords Public participation • Spatial planning • Licensing procedure of infrastructure projects • Sectorial planning • Communication procedures

1 Introduction

Large size infrastructure projects, such as airport, rail and road extensions, are particularly likely to encounter frequent citizens' protests. The experience with the enhancements of the Stuttgart train station is an especially relevant example, since

A. Bothe (✉)
Department of Law, Universidade Autónoma de Lisboa, Lisbon, Portugal

Department of Management Sciences, Universidade Atlântica da Barcarena, Oeiras, Portugal
e-mail: abothe@ual.pt

A. Dias et al. (eds.), *Modeling Innovation Sustainability and Technologies*,
Springer Proceedings in Business and Economics,
https://doi.org/10.1007/978-3-319-67101-7_10

it has motivated the economic stakeholders, as well as the German federal and regional legislatures and administrations, to rethink the project. In this case, which has become known as "*Stuttgart 21*," massive resistance only materialized **after** the planning approval procedure. As a consequence, the "*Law for broadening the public participation and for the standardization of the procedures for determining sectorial plans*" was introduced in the German legislature in 2012. Among other elements in this law, public participation is now required prior to the formal opening of the procedure for planning and approval of sectorial plans, as described in Article 25, paragraph 3, of the Administrative Procedure Code. In 2013, the Association of German Engineers developed standards governing the communication and public participation in planning and building infrastructure projects. These standards are directed at project sponsors, general and specialist design contractors, project managers, executing companies and their officials. In the same year, regional governments produced binding administrative provisions to intensify public participation in the planning and licensing procedures of large infrastructure projects. These standards and administrative provisions establish several mandatory steps and procedures within the regional planning and approval processes: these mandatory steps include scoping the previous and present participation of the public and their interaction, both at formal and informal levels, conducting an official investigation to show just cause for the project, establishing internet access to the project allowing for on-line search of the detailed plans by the public, a requirement to request and enable public participation periodically as the planning process develops, and a need to re-evaluate these new rules as appropriate.

The main objective of these mandatory steps is a more efficient harmonization of the different interests involved in spatial planning. The International, European and German legislation on Environmental Impact Assessment, as well as public participation concerning sustainability, have taught us about the positive impacts of these procedures. This in turn has raised the possibility that we are in fact headed towards a progressively earlier inclusion of the public in the planning process.

In order to establish a more solid foundation for a critical analysis of several of the elements in these rules, this article discusses the essential factors impacting these types of plans, the skills required in their preparation, the strategies leading to acceptance of those plans, and the possible dynamics affecting their implementation.

2 Types of Plans

Only since 2009 has a Federal Plan of Spatial Development been formally in place. However, that new legal instrument does not cover all aspects of Spatial Development, but rather it only addresses eight areas of policy (Article 2, paragraph 2, of the Spatial Planning Federal Law, the followings are the eight areas of possible Federal Spatial Plans: 1. Sustainability, 2. Urbanization structures, 3. Infra-structure and traffic, 4. Economy, 5. Cultural landscape, 6. Environment and climate protection, 7. Defence and civil protection, 8. European Cooperation). Several major studies of this new instrument have concluded that it is most appropriate for the organization of urbanization structures, infra-structure and traffic communications. All-encompassing federal spatial plans, which would cover the coordination of the different affected powers and which would be focussed on implementation, are not foreseen by law. They are only addressed by a set of Federal guidelines that have been produced since 1975. The essential purpose of enacting a Spatial Planning Federal Law is the harmonization of the different interests involved (Article 1, paragraphs 1 and 3, of the Spatial Planning Federal Law).

Before defining the federal and regional skills participating in spatial planning, we need to distinguish between legislation and administration. The **administration** of spatial planning is a regional skill that resides at the state level, while its **legislation** perspective is a federal one (Articles 83 and 87–90 of the German Constitution). As mentioned, the federal administrators decided not to make use of their powers, but rather to leave that responsibility to the States. The States are thus obliged to produce two levels of spatial plans: one level covering the entire territory of the state and the other level covering only specified portions of the state. Those types of plans need to be distinguished from the planning of certain policy areas. In German law policy planning is separate from spatial planning, because the definition of spatial planning usually relates to several precisely defined policy areas and not just to one. (Bothe 2014b, p. 306) In order to understand the challenge of coordinating between spatial planning and single-section sectorial planning we need to define the different forms of legal binding.

3 Legal Types of Binding Affecting State Plans of Spatial Development

Three different types of legal binding (Article 3, paragraph 1, and article 7, of the Spatial Planning Federal Law) are foreseen when formulating State plans for Spatial Development:

- "Goals" must be determined or determinable in spatial terms and content. "Goals" include a complete weighting done by the holder of spatial territory.
- "Principles" of spatial planning are affirmations controlling the development, order and security of space, in the form of demands addressed to the weighting of decisions and the discretionary process that follows. "Principles" can arise as a result of laws or of decisions taken as part of a state's plan of spatial development.
- "Other requirements" of spatial development can be either goals that are in the process of being integrated into a plan, or the result of a formal state's planning process, or the statement of planning intentions made by a State's authority.

"Goals" are the strongest form of legal binding: they must be implemented. The other forms of binding, "principles" and "other requirements," are to be respected when possible, but they may not be implemented if other opposing interests are given a higher weight.

The types of legal binding described above are the most important elements affecting the coordination between spatial planning and planning of just one sectorial concern.

4 Possible Dynamics Affecting the Implementation of Plans

In certain limited circumstances public entities which are part of the Federal State can initiate changes to the goals integrated in the plans (Article 5 of the Spatial Planning Federal Law). Other holders that intend to implement the plans have different mechanisms for exerting that dynamic: during the process of elaboration of the plan, they may propose that exceptions be allowed. If that is not successful, "goals" may also be removed after concluding the planning process, if such deviation is justified by regional planning aspects and if the essential of the planning is not affected (Article 6 of the Spatial Planning Federal Law). These exceptions or deviations do not include an outright **change** of the "goals", but rather a selective modification in order to implement a certain construction or other concretization. This system of exceptions and deviations was introduced in the Spatial Planning Federal Law in 2008, but it had already been widely used by administrations before that date because it was considered legal by jurisprudence (Federal Administrative Court decision of 18th of September of 2003, Proc. 4 CN 20.02—*BVerwGE* 119, 54; in the reasoning of the chances in the Federal Spatial Law the legislator affirms the only declarative effect of the new article 6, paragraph 2: Journal of the Federal Parliament—*BT-Drs.*—16/10292, p. 23). Several authors consider that the system of exceptions and deviations lowers the value of regional planning (Kment and Grüner 2009, pp. 93–98). Moreover, these authors criticize this practice on the basis that the procedural and content demands for incorporating exceptions and deviations are not defined with sufficient precision (Müller 2008, pp. 360–363).

The process of definitely removing a "goal" from a plan is the same as the process for its initial introduction (Article 7, paragraph 7 of the Spatial Planning Federal Law).

5 Regional Planning Procedure

One of the most important instruments of planning above local planning is the regional planning procedure (*Raumordnungsverfahren*—for its definition see article 15 and 16, of the Spatial Planning Federal Law). As mentioned previously, legislation of spatial development issues is a federal skill, but frequently, and also in the case of the regional planning procedure, the Spatial Planning Federal Law just determines the bases of that instrument, and leaves its concretization to the State's legislator.

Concerning the implementation of major infra-structure projects, it is frequently debatable whether certain decisions are part of the spatial development planning or part of the sectorial planning.

The Spatial Planning Federal Law leaves it to the State's legislator to decide whether public participation is or is not demanded in the regional planning procedure. Six States have established public participation as an obligation on the local administration. The other States leave that decision to the project administration, on a case-by-case basis.

6 Distinction Between Spatial and Sectorial Planning

The previously mentioned *"Law for broadening the public participation and for the standardization of the procedures for determining sectorial plans,"* which was introduced in the German legislature in 2012, regulates sectorial planning procedures, which are to be distinguished from regional planning procedures. The result of the regional planning procedure is the determination of "other requirements," which are defined in article 3, number 4 of the Spatial Planning Federal Law: they are to be respected when possible, but they may not be implemented if other opposing interests are given a higher weight (see above Chap. 3).

It is up to the State Authority for Spatial Planning to decide on the necessity of a regional planning procedure. After receiving all the documentation pertaining to that procedure, it has 4 weeks to decide about the necessity of the regional planning procedure, and 6 months to conclude on it. The demand for a regional planning procedure does not exist in the three German States that are also cities: Berlin, Bremen and Hamburg.

The objective of the regional planning procedure is a concrete project, and that is why it is difficult to distinguish the regional planning procedure from the determination of a sectorial plan (Bothe 2014a, p. 275; 2014b, p. 297). The latter is a different instrument, which is not integrated in the spatial planning, but rather in the subsequent sectorial planning. We list below four judged cases in which courts decided that the debatable determinations were part of the regional planning procedure and not part of sectorial planning:

- The location of the new Berlin airport, including closing the old airport and an assessment of the noise levels involved. Only questions about technical feasibility are part of the sectorial planning (Federal Administration Court 2006, Judgement 125, pp. 116–167).
- Limitations on night-time airplane traffic at the Frankfurt airport: they were considered a "goal" of spatial planning, meaning that all weighting of interests had been completed, see article 3, number 4 of the Spatial Planning Federal Law (Administration Court of the State of Hesse Aug 21 2009).

- Also the location of part of a highway was considered a legal "goal" of spatial planning (Constitutional Court of Bavaria July 15 2002).
- In the State of North Rhine-Westphalia, requirements for greenhouse gas reduction in spatial planning were considered mandatory before closing central power plants based on their carbon footprint (Administration Court of Münster Sept 3 2009).

7 Procedure for Determining Sectorial Plans

Regarding major infra-structure projects, the determination by spatial planning already allows for addressing, at least partially, any concerns regarding soil use. Thus the question arises as to what improvements are still needed to the procedure for determining sectorial plans, in addition to the recent legislation changes:

"*Included in the procedure for determining sectorial plans is a decision on the admissibility of the project, whereby the necessary follow-up actions addressing other systems are taken, and all affected public concerns are identified. Once the procedure for determining sectorial plans has been followed, other regulatory decisions such as public approvals, awards, licenses, authorizations, consents and plan approvals are not required. All public relations between the holder of the project and those parties affected by it are defined by the procedure for determining sectorial plans, thus producing the associated legal binding rights.*" (Article 75, paragraph 1, of the Federal Administrative Procedure Code. "*Durch die Planfeststellung wird die Zulässigkeit des Vorhabens einschließlich der notwendigen Folgemaßnahmen an anderen Anlagen im Hinblick auf alle von ihm berührten öffentlichen Belange festgestellt; neben der Planfeststellung sind andere behördliche Entscheidungen, insbesondere öffentlich-rechtliche Genehmigungen, Verleihungen, Erlaubnisse, Bewilligungen, Zustimmungen und Planfeststellungen nicht erforderlich. Durch die Planfeststellung werden alle öffentlich-rechtlichen Beziehungen zwischen dem Träger des Vorhabens und den durch den Plan Betroffenen rechtsgestaltend geregelt.*")

Since the dawn of spatial planning one of the major challenges has been how to reach the desired legitimate results in the most effective yet fairest way. Included in those results must be one of the essential characteristics of the procedure for determining sectorial plans: the production of binding rights for everyone involved. The principal motivation for legislators has so far been the speedy achievement of the required results, often lowering the standards of public participation (Journal of the Federal Parliament 17/9666, May 16 2012: proposal of Law for broadening the public participation and for the standardization of the procedures for determining sectorial plans.). In the beginning of the 1990s, legislation for a speedier production of plans for infra-structure implementation, including all binding rights, was justified by special needs and urgency associated with East Germany. When this special legislation, addressed to the needs of the five States of the former German Democratic Republic, was extended to the rest of the country, the criticisms of the resulting reduction in public participation and legal protection became louder and more frequent. Rather extreme examples of citizen protests include those organized to oppose the extensions of the Frankfurt and Berlin airports, and the construction of interim storage facilities for radioactive waste and waste incinerators. These incidents serve as lessons pertaining to several aspects of spatial planning. The federal law, enacted in 2006, for accelerating the planning of infra-structure projects, brought among others the following innovations:

- The Federal Administration Court became the first and last resort for contesting any decisions produced by a procedure for determining sectorial plans
- Public participation became optional,
- Associations formed to deal with spatial impact concerns lost the right to automatically receive information about projects, so that they need to actively maintain themselves up to date like any other citizen,
- Anyone who does not present objections during the public participation period cannot do so later in court, except when that entity has an overriding special subjective civil right.

These innovations were introduced into six special laws in 2006 and were transferred into the general Federal Administration Process Law in 2013.

8 Prior Public Participation

The above-mentioned conflicts over the implementation of major infra-structure projects like airports, highways, etc., have led many to conclude that it is imperative to introduce changes adding citizens' rights instead of lowering their standard. The previously referred *Law for broadening the public participation and for the standardization of the procedures for determining sectorial plans*, for instance, has introduced public participation prior to the formal opening of the procedure for planning and approval of sectorial plans. Towards that objective, this law states that "*The authority must do the utmost to ensure that the holder, who is planning a project that might have impacts essential to the interests of a large number of citizens, must inform the affected public about the goals of the project, the way the project is to be realized, and the probable effects of that project (prior public participation).*" (Article 25, paragraph 3, of the Administrative Procedure Act: "*Die Behörde wirkt darauf hin, dass der Träger bei der Planung von Vorhaben, die nicht nur unwesentliche Auswirkungen auf die Belange einer größeren Zahl von Dritten haben können, die betroffene Öffentlichkeit frühzeitig über die Ziele des Vorhabens, die Mittel, es zu verwirklichen, und die voraussichtlichen Auswirkungen des Vorhabens unterrichtet.*"). This new legal instrument became generally effective in June 2013, although for some federal sectorial acts it only became effective in June 2014. In March 2015 the Federal Government was the subject of an inquiry in the Federal Parliament about its accumulated experience with this mode of prior public participation. The Federal Government informed the Parliament of 36 cases of road construction projects including prior public participation, as well as 25 railway projects and 19 water infra-structure projects. The inclusion of prior public participation in several of these projects actually happened independently of the new legislation, because when the initiatives started the States had not yet incorporated the Federal Administrative Procedure Code into their State Administrative Procedure Codes. While the Government advocated that "*prior and continuous public participation is a central part of the successful realisation of traffic infra-structure projects,*" (Journal of the Federal Parliament 18/4159, March 2 2015) other parties supported an even earlier public participation. According to these parties, prior public participation should not only be part of the sectorial planning procedure, but should also be included in the Regional planning procedure, because spatial planning usually takes place before a concrete plan has been produced. Continuing along the same logic—that the earlier the public is involved the more effective is the harmonization of the process—and analysing the evolution of the European and national legislations regarding the evaluation of environmental impacts of territorial projects, (Directive 2003/35/EC of the European Parliament and of the Council of 26 May 2003 providing for public participation in respect of the drawing up of certain plans and programs relating to the environment and amending with regard to public participation and access to justice Council Directives 85/337/EEC and 96/61/EC—Statement by the Commission.) it may be concluded that public participation has happened progressively earlier in the project timeline after each legislation change. However, it must be recognized that many essential decisions are made long before both of the referred procedures, the regional planning procedure **and** the procedure for determining sectorial plans, take place. The decision regarding the **if** of the project, in particular, is usually made during the federal planning of traffic routes (*Bundesverkehrswegeplanung*). This basic decision about the need for the project establishes early legal binding affecting all the subsequent planning procedures. This binding includes a rough definition of the communication process and the classification of the type of route. All these essential steps are thus legally considered to be an internal instrument of the Federal Government. They are usually integrated in the Federal laws ruling the extension of highways or railways. Although these decisions are binding and address many essential elements, they only impose an analysis of alternatives and public participation, the latter only since 2006 when the Public Participation law became effective. However, the objective of the public participation is limited to environmental concerns.

9 International Sustainability Concerns as a Valuable Forerunner in Public Participation Tools

The integration of ecological concerns in the planning process has been a reality in the European Union since 1985, as a consequence of the Directive on Environmental Impact Assessment (Directive 85/337/EEC on the assessment of the effects of certain public and private projects on the environment, EIA). This is the first European Union instrument concerning all environmental sectors simultaneously, and it reflects the regulatory philosophy of Niklas Luhmann, namely the "*legitimation by due process.*" It is all about process, as expressed by the text of the law: "*The environmental impact assessment shall identify, describe and assess in an appropriate manner, in the light of each individual case ... the direct and indirect effects of a project on the following factors:—human beings, fauna and flora;—soil, water, air, climate and the landscape;—material assets and the cultural heritage;...*" (Article 3 Council Directive 97/11/EC, March 3 1997 amending Directive 85/337/EEC on the assessment of the effects of certain public and private projects on the environment).

Until 2001, the public participation was supposed to be introduced before the realization of certain projects, but not before the beginning of the planning stages, which we nowadays consider to be too late: "*1. When a decision to grant or refuse development consent has been taken, the competent authority or authorities shall inform the public thereof in accordance with the appropriate procedures and shall make available to the public the following information:*

- *the content of the decision and any conditions attached thereto,*
- *the main reasons and considerations on which the decision is based,*
- *a description, where necessary, of the main measures required to avoid, reduce and, if possible, offset the major adverse effects.*" (Article 9 Council Directive 97/11/EC, March 3 1997)

In 2001 was approved its "sister directive" (Directive 2001/42/EC of the European Parliament and of the Council of 27 June 2001 on the assessment of the effects of certain **plans and programmes** on the environment.) "*This Directive requires certain plans and programmes, which are likely to have significant effects on the environment, to be subject to an environmental assessment. This assessment specifically enables environmental considerations to be integrated in the preparation and adoption of these plans and programmes. It (...) includes the introduction of an environmental report (...), as well as carrying out consultations (with the public, the authorities with environmental responsibilities...).*" (The official summary of the directive can be found in http://eur-lex.europa.eu/legal-content/EN/TXT/HTML/?uri=URISERV:l28036&from=DE). The principal lesson learned from the experience with the first directive is that all evaluation including public participation **needs to occur much earlier**: not only when the **project** is to be implemented, but in the early stages of the elaboration of the **plan**.

In the meanwhile, the European Union was already obliged to realise the mentioned revisions because it signed the Aarhus Convention, the United Nations Economic Commission for Europe (UNECE) Convention on Access to Information, Public Participation in Decision-Making and Access to Justice in Environmental Matters (adopted on 25 June 1998 in the Danish city of Aarhus at the Fourth Ministerial Conference as part of the "Environment for Europe" process. It entered into force on 30 October 2001). The Convention provides for the following three rights: 1. "access to environmental information," which means that it is to be made possible for everyone to receive environmental information that is held by public authorities in order to participate in environmental decision-making. 2. "public participation in environmental decision-making": "*Arrangements are to be made by the public authorities to enable the public affected and environmental non-governmental organisations to comment on, for example, proposals for projects affecting the environment, or plans and programmes relating to the environment. These comments are to be taken into due account in decision-making, and information is to be provided on the final decisions and the reasons for it.*" (European Commission, http://ec.europa.eu/environment/aarhus/). 3. "Access to justice," the right to review procedures to challenge public decisions that have been made without respecting the two aforementioned rights or the environmental law in general (http://ec.europa.eu/environment/aarhus/).

Therefore we can conclude that the European Union Environmental Law has been a forerunner or pioneer in public participation: It first obliged the national legislators to introduce public participation at progressively earlier stages of each project. One of the latest developments came from the Grid Expansion Acceleration Act transmission network (NABEG, Article 20 Application conference) establishing the need for an early public meeting for defining the scope of the project. The mandated application conference is unlike a conventional scoping appointment, since it is not limited to the scope and methods of environmental impact assessment, but extends instead to all of the planning approval process, considering a large range of issues related to the intended project (paragraph 1 sentence 2). And while the scoping concerning the environmental impact assessment is to be realized with participation of the public authorities, the referred Grid Expansion Acceleration Act transmission network (NABEG) includes the public in the full scoping process. "*Thus, the application conference is a crucial building block to further develop the existing participation rights in major projects through confidence and acceptance-enhancing measures. Conflicting public and private interests come to the fore at an early stage of the procedure and thus ensure full consideration of the needs of the general public. The application conference thus contributes to process transparency, acceptance and peace. It also serves to accelerate the planning approval procedure.*" (Federal Parliament Publication 2011, p. 28) The reflections described above are currently limited to the planning process, according to the Grid Expansion Acceleration Act transmission network. Will they in the future expand to general sustainability concerns, or even to all types of planning and all types of interests presented by the public?

10 Assessing the New Legislation

Summarizing the elements introduced by the new legislation, incorporated in the Federal Administrative Procedure Code, that are aimed at speeding up the process of producing binding rights, we must consider the following aspects:

1. The authorities in charge of the procedure for determining sectorial plans have been given more power of discretion to decide what communities must be involved in the public participation process. (Article 73, paragraph 2, Federal Administrative Procedure Code.)
2. The participation of special-interest associations occurs right at the beginning of the decision process, together with the public entities involved. (Article 73, paragraph 4, Federal Administrative Procedure Code.) If these associations do not present a particular opinion, they lose their right to do so afterwards. This is advantageous for the authority in charge of the process since they do not need to worry about new opinions being introduced at a later stage. For the associations, however, it means that they lose their right of participation throughout the process unless they manage to be well prepared and organized during the very early stages of the public participation process.
3. The legislation introduces deadlines aimed at forcing the authority in charge of the procedure to establish a clear schedule and puts some pressure on them to respect it. (Article 73, paragraph 6, *in fine,* Federal Administrative Procedure Code) Nevertheless, these deadlines are weakened by the fact that there are no consequences if they are not met. Moreover, subsequent legislation changes lowered the bar further by limiting even more the types of procedure failures that carry legal consequences. (Article 75, paragraph 1a, and articles 45 and 46, Federal Administrative Procedure Code)
4. To recap, how would we rate the new process of public participation prior to the formal opening of the procedure for planning and approval of sectorial plans? As mentioned above, (see Chap. 8) one of the most fundamental criticisms of this new instrument is the exclusion of public participation in the initial decision about whether or not to pursue the project, except for consultation of the public regarding ecological concerns. The German legislators were forced to introduce early public participation by the European Directive 35/EC of 2003. However, the new public participation prior to the formal opening of the procedure for planning and approval of sectorial plans is optional in all of its aspects because the legal expression is not sufficiently precise and thus admits several interpretations. It stipulates that the authority must do the utmost to ensure that the holder, who is planning a project that might have impacts essential to the interests of a large number of citizens, must inform the affected public about the goals of the project, the means to achieve those goals, and the expected impacts of the project (prior public participation, Article 25, paragraph 3, Federal Administrative Procedure Code). But it does not define precisely what doing the utmost means. Therefore, the new legal instrument is for all purposes optional for the authority, as well as for the investor, who can choose the content of the information released to the public and the way to present it.

11 Conclusion

In conclusion, does the new legislation achieve its aim of reaching a result with binding rights in a faster and more harmonious way?

Empirical studies show that non-governmental organizations, NGOs, face very few legal processes, and when they do so the processes are concluded quite fast and in 40% of them the NGO wins the process. (The same percentage of individual legal processes is about 10–12%, see Independent Institute for environmental questions, Lawsuits of associations in the law of nature and environmental protection 2013, p. 3) These organizations consider that their rights are very dependent on the information they receive and also on the possibility or not of participating at a later stage, even if they failed to present opinions at an earlier stage of the process. (Schmidt et al. 2011, p. 6; 69th Meeting of German Lawyers 2012, p. 51) Federal departments have analysed the time management of the implementation of major infra-structure projects, and they concluded that many months usually elapse between the conclusion of the administrative process and the beginning of the work. (Federal Government; German Parliament, Circular 18/4159, 2015). These conclusions lead us to believe that the main reasons behind delays in starting the work are not legal ones.

A clear and efficient harmonization of the interests involved promotes a faster start of the work. That is why several entities published recommendations for the procedure of public participation (Association of German Engineers: Standard AGE 7000 about prior public participation for industry and infra-structure projects 2015; European Commission: Guidance on Integrating Climate Change and Biodiversity into Strategic Environmental Assessment 2013; University of applied sciences: Final Report on the analyses: Evaluation of the Federal environmental impact assessment act 2008; Federal Authority for Environment: Guides for authorities and legal recognition 2015; Federal Ministry for traffic and digital infrastructure, Manual for a good public participation 2012). The following are some of their reflections, conclusions and experiences:

11.1 More or Less Legal Demands?

There are pros and cons concerning the extent of mandatory legal demands: on the one hand they aim at facilitating communication, which has raised concerns with some authors about this not being a suitable object of legal rules (69th Meeting of German Lawyers 2012, p. 44). Each case is very specific, (69th Meeting of German Lawyers 2012, p. 49), and often the parties being legally mandated to this enhanced communication may tend to see the other parties as an opposition. On the other hand, legal rules that are only voluntary end up bypassing discussion of uncomfortable matters, which will lead to increased conflicts during the later stages of the project (69th Meeting of German Lawyers 2012, p. 41).

11.2 Common Plan About the Exact Procedure of Planning and the Coordinating Party

From the very beginning of the process, technical aspects, communication, legal aspects and financing considerations must all be taken into account (Association of German Engineers 2015a, b). A neutral authority, working together with all parties involved, should establish a plan about the proper procedure, especially about the liability of the decisions taken in each stage of planning (69th Meeting of German Lawyers 2012, p. 42). However, while some consider essential that this coordination be carried out by a neutral authority (69th Meeting of German Lawyers 2012, p. 42), others point out the disadvantages that arise from the fact that the procedure coordinating authority and the decision making authority are not the same:

- the waste of material resources,
- the increased requirements of communication
- the blurring of responsibilities (69th Meeting of German Lawyers 2012, p. 51)
- the less concentrated information level (69th Meeting of German Lawyers 2012, pp. 12–45).

For the holder of the project an integrated project management is usually recommended, as opposed to a complete outsourcing of the communication process.

In order to form engineers possessing the necessary communication sensibility, universities should integrate the necessary subjects into their education program. Notwithstanding, for maximum effectiveness communication teams should be interdisciplinary.

11.3 Positive Experiences with Accompaniment in Difficult Periods

There have been a few positive experiences with sensitive practical problems being introduced and discussed progressively in stages, overseen by a neutral authority, like in the following examples related by the Association of German Engineers:

- During the reconstruction of the Wiener Central Station an ombudsman was always present to oversee the needs and concerns of frustrated neighbours.
- The 2 year mediation process for the extension of the Frankfurt airport interlinked the extension work with an anti-noise treaty, the prohibition of night flights and the establishment of a regional dialog forum.

11.4 Assessing the Potential for Influencing Decisions and Its Limitations

It is difficult to select the right moment to introduce public participation in the planning process: when it occurs too early the issues may be too abstract for the public, which does not yet feel directly affected. Nevertheless, the initial decision about the need for the project is obviously central in the planning procedure, and public participation in that decision is highly desirable (Federal Ministry for traffic and digital infrastructure 2012, p. 27). When public participation is introduced too late, there is usually no openness on the part of the authorities and the project holder to accept any changes to the decisions that have already been made.

When there are no viable alternatives to the project, it is not correct nor effective to pretend that there might be one. In these situations, the environmental associations in particular are seen by the authorities as being the enemy standing in the way of an effective implementation of a necessary project (Fehling 2012). In these cases it is often better to assume that public participation is to be meant as participation in the procedure, but not as participation in the decision which is seen as inevitable (69th Meeting of German Lawyers 2012, p. 44).

Naturally, questions about the kind of democracy we want and what are to be its basic values have been the motivation behind several deep conflicts. They need to be confronted, but the planning procedure itself has different objectives (69th Meeting of German Lawyers 2012, p. 46).

11.5 Deterioration of Material Conditions and Motivation

Environmental associations tend to have increasingly less material conditions to realise their objectives: they have progressively fewer personnel and weaker financial capacities. Consequently, they often do not have much faith in the effectiveness of their work. In a recent opinion survey, 79% of the people participating in these associations stated that they consider that their work has no influence, or an extremely small influence, on the decisions (University of applied sciences 2008). These facts lead to a lack of balance between the investor and the public. In order to even out the playing field, it would be necessary to financially support at least a portion of the costs of the citizens' associations (Independent Institute for Environmental concerns 2013; 69th Meeting of German Lawyers 2012, p. 52).

All in all we conclude that the controlling force of the law is weaker than the combined power of all other influences.

References

69th Meeting of German Lawyers—69. Deutscher Juristentag (2012) Theses of the experts and speakers, Thesen der Gutachter und Referenten, Ziekow, Jan, pp 41–43; Dolde, Klaus-Peter, pp 43–45; Gabriel, Oscar W., pp 46–49; Wegener, Bernhard W., pp 49–52

Administration Court of Münster (2009) OVG NRW, Sep 3, 2009. In: ZNER, Review for the new energy law, vol 3, pp 284–289

Administration Court of the State of Hesse (2009) HessVGH, Aug 21, 2009. In: LKRZ, Review for the law of the states and the municipalities of Hesse, Rhineland Pfalz, e Saarland, vol 2, 2010, pp 66, 69 and following

Association of German Engineers (2015a) Verein Deutscher Ingenieure, Standard AGE 7000 about prior public participation for industry and infra-structure projects. http://www.vdi.de/7000/. Accessed 25 Sep 2015

Association of German Engineers (2015b) AGE news, Vdi nachrichten. http://www.vdi-nachrichten.com/Technik-Gesellschaft/Wir-voellig-neue-Dialogkultur/. Accessed 25 Sep 2015

Bothe A (2014a) Classification and qualification of the ground on the right of the German urbanism, A classificação e a qualificação do solo no direito do urbanismo alemão. In: JURISMAT, Legal Magazine of the Institute Manuel Teixeira Gomes, n. 5, pp 271–285

Bothe A (2014b) Law of urban planning, spatial planning and the land in Portugal and Germany, Direito do Urbanismo, do Ordenamento do Território e dos Solos em Portugal e na Alemanha. In: JURISMAT, Legal magazine of the institute manuel teixeira gomes, n. 4, pp 289–319

Constitutional Court of Bavaria (2002) *BayVerfGH*, July 15, 2002. In: DÖV, Review of public administration and administration science 2003, pp 78–80

European Commission (2013) Guidance on integrating climate change and biodiversity into strategic environmental assessment

Federal Administration Court, Judgement (2006) 125, pp 116, 167, note 155. http://www.bverwg.de/entscheidungen/entscheidung.php?ent=160306U4A1075.04.0. Accessed 27 March 2014

Federal Authority for Environment (2015) *Bundesumweltamt*, Innovative public dialogue process with environmental assessment using the example of certain types of projects/project properties—Guides for authorities and legal recognition, Neuartiger Öffentlichkeitsdialog in Verfahren mit Umweltprüfung am Beispiel bestimmter Vorhabentypen/Vorhabeneigenschaften—Leitfäden für Behörden und rechtliche Verankerung

Federal Government (2015) Experience report, *Erfahrungsbericht*,—German parliament, Circular 18/4159, March 2, 2015. http://dip21.bundestag.de/dip21/btd/18/041/1804159.pdf /. Accessed 10 Oct 2015

Federal Ministry for Traffic and digital Infrastructure (2012) Manual for a good public participation, Handbuch für eine gute Bürgerbeteiligung

Federal Parliament (2011) Fundamentation of the law, Federal parliament publication, BT-Drucks. 17/6073, 6.6.2011. http://dip21.bundestag.de/dip21/btd/17/060/1706073.pdf. Accessed 20 Sep 2015

Fehling M (2012) Reform of public participation for plan approval of infrastructure projects, Reform der Bürgerbeteiligung für die Planfeststellung von Infrastrukturvorhaben. In: Transport scientific seminar, Verkehrswissenschaftliches Seminar, Freiburg, 20 September 2012

Independent Institute for Environmental concerns (2013) Unabhängigies Institut für Umweltfragen, Public participation in environmental protection in Germany—Status Quo and new ways, Öffentlichkeitsbeteiligung im Umweltschutz in Deutschland—Status Quo und neue Wege

Kment M, Grüner J (2009) Exceptions of the goals of spatial planning—about the new concept of the spatial planning federal law, Ausnahmen von Zielen der Raumordnung–zur Neufassung des Raumordnungsgesetzes. In: UPR, Magazine for the environmental and planning law, pp 93–111

Müller C (2008) The proposal of the spatial planning federal law, Der Referentenentwurf des Raumordnungsgesetzes 2008. In: RUR, Magazine for Spatial Science and Spatial Order, pp 360–374

Schmidt A et al (2011) The development of collective action in nature and environmental protection law 2007 to 2010—an empirical study commissioned by the Federal Agency for Nature Conservation, Die Entwicklung der Verbandsklage im Natur- und Umweltschutzrecht von 2007 bis 2010—eine empirische Untersuchung im Auftrag des Bundesamtes für Naturschutz. https://www.bfn.de/fileadmin/MDB/documents/themen/recht/BfN-Verbandsklagen-Studie-2011.pdf. Accessed 25 Sep 2015

Spacial Planning Federal Law (2008) BGBl. I, p 2986, last changed at 20th of July 2017, BGBl. I, p 2808. www.gesetze-im-internet.de/rog_2008/BJNR298610008.html. Accessed 25 Sep 2017

University of applied sciences, Hochschule Darmstadt et al. (2008) Final report on the analyses: evaluation of the federal environmental impact assessment act, Abschlussbericht zum Vorhaben, Evaluation des UVPG des Bundes

Anja Bothe is an assistant professor in the Department of Law at the Autonomous University of Lisbon, where she is the Head and Scientific Coordinator of the first cycle of law program, which leads to the award of the first degree in law. Dr. Bothe is also an assistant professor in the Department of Business Management at the Atlantic University of Barcarena, in Oeiras, Portugal. She specializes in City Planning Law, and she teaches classes in Constitutional Law, European Law, Health Law, Law of the Information Society and Commercial Law. Within the Portuguese Institute for Quality and its National Standardization Department, Dr. Bothe holds a voting position in the Technical Commission that develops legal standards ruling software systems in the area of Health, being part of working group 3 focused on safety and quality. She has published several papers in highly-regarded peer-reviewed international, German and Portuguese journals, she has authored and co-authored several books and has participated in a number of conferences.

The Critical View on Innovation Activity in SME's Sector in Slovakia

Ľubica Lesáková

Abstract Innovations are the drive of economic development advancing the possibilities of future competitiveness in the form of new knowledge and increasing economic efficiency and performance, particularly within small and medium enterprises (SME). To strengthen innovation activities is one of the main tasks of SME nowadays. The objective of this chapter is to give a critical view on innovation activity in the SME sector in Slovakia. Because of the need to use a systemic approach the stress will be given to identifying main factors influencing innovation activity of Slovak SME. The results of our own empirical research are devoted to the identification of the main barriers to develop innovation activity identified among SMEs. Recommendations for relevant public institutions as well as for SME to overcome the barriers will be discussed.

Keywords Innovations • Small and Medium Enterprises • Slovak Republic

1 Introduction

Current economy tends to be characterized as a new, global a knowledge-based economy. The new, global economy is the economy of knowledge and ideas, where innovative ideas and technologies fully integrated in services and products became a key to generation of new working positions and higher life standard. Only those businesses that are dynamic are able to respond to the market demand swiftly and are capable of research and development of new products, innovations and technological changes.

In the Europe 2020 Strategy is the area of innovations one of the initiating policies to start the EU potential for growth (Europe 2020. A strategy for smart, sustainable and inclusive growth). Innovations are the drive of economic development advancing the possibilities of future competitiveness in the form of new knowledge and increasing economy's efficiency and performance (European

Ľ. Lesáková (✉)
Faculty of Economics, Matej Bel University, Banská Bystrica, Slovakia
e-mail: lubica.lesakova@umb.sk

A. Dias et al. (eds.), *Modeling Innovation Sustainability and Technologies*,
Springer Proceedings in Business and Economics,
https://doi.org/10.1007/978-3-319-67101-7_11

Commission, 2010, 2013). To strengthen innovation activities is one of the main task of small and medium enterprises nowadays.

Small and medium enterprises (SME) are forced to make innovations, because they are under permanent pressure of competitors at the market. From this point of view the ability to compete in innovations plays very important role as a factor of their competitiveness (Bessant and Tidd 2009).

The aim of the article is to give the critical view on innovation activity in SME's sector in Slovakia. Because of the need to use a systemic approach the stress will be given to identifying main factors influencing innovation activity of Slovak SME. Presented will be results of own empirical research devoted to identification of main barriers to develop innovation activity identified among SME. Formulated will be presumptions and recommendations for relevant public institutions as well as for SME to overcome the barriers. For the research purposes we will utilize evaluation of relevant secondary data as well as results of own empirical research.

The paper was elaborated as a part of the VEGA project 1/0494/15 "The research of factors influencing the successfulness of innovative small and medium enterprises in the Slovak Republic."

2 Innovation Performance of Slovakia

Innovation performance of the country is important indicator, which describes situation and effectiveness of the state policy and other tools supporting the innovation performance of domestic and foreign enterprises. In order to secure international comparison of success of innovations support within European Union, the Summary Innovation Index (SII) was created. Summary Innovation Index takes into account a score of 25 indicators, which are divided into five groups.

The first three groups of indicators include innovative inputs and the last two groups include innovative outputs (Innovation Union Scoreboard):

1. Innovation enablers (five indicators), which measure the structural conditions required for innovation potential.
2. Knowledge creators (four indicators) which present the value of investment into research and development activities, which are considered to be a key determinant of the knowledge-based economy development.
3. Innovation and entrepreneurship (six indicators) which measure the effort of enterprises on innovations, the effort of SME, private and cooperative, spending on research and development, venture capital and GDP.
4. Application of innovations in practice (five indicators) which measure innovation performance expressed through business activities and participation in employment and added value in innovative sectors.
5. Intellectual property (five indicators) which measures the achieved results in terms of successful know-how such as patents, trademarks and new design.

Classification of countries is consequently made upon the results of measurement for the 25 indicators. The countries in EU are classified into four categories: innovation leaders, innovation followers, moderate innovators, modest innovators. The position of Slovakia between the EU member countries according to Summary Innovation Index (in the year 2014) presents Fig. 1.

From a long-term point of view, Slovakia belongs according to the Innovation Union Scoreboard international comparison to the EU countries which lag behind the EU average considerably in the innovation performance. Slovakia ranks with regard to the SII into the group of moderate innovators (together with the Czech Republic, Hungary, Poland and other nine countries). Innovation performance of the Slovak Republic (Innovation index relative to EU average) in the years 2007–2014 is presented in Fig. 2.

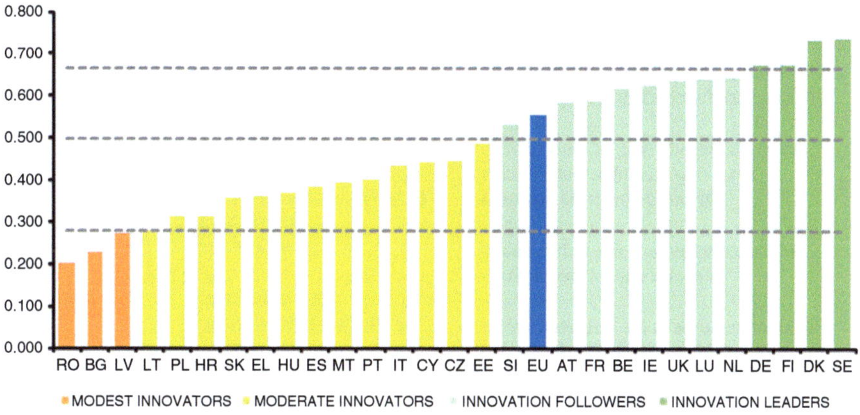

Fig. 1 Summary innovation index of EU countries in the year 2014 (Source: Innovation Union Scoreboard 2015)

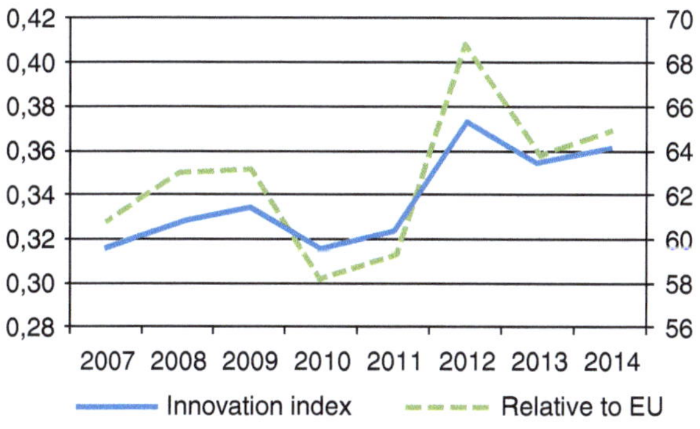

Fig. 2 Innovation performance of the Slovak Republic—Innovation index relative to EU average (Source: Innovation Union Scoreboard 2015)

From Fig. 2 is evident, that innovation performance in Slovakia has increased between 2007 and 2014, but declined in 2010 and in 2013. The performance relative to the EU has had more fluctuations but over time has increased significantly. Performance relative to the EU reached a peak in 2012 at 69% of the EU, but fell to 64% in 2014. Based on the obtained values of the Summary Innovation Index 2014 for Slovakia it was 0.35, while the EU average was 0.55 (Innovation Union Scoreboard 2015).

Detailed view on SII indicators value of the Slovak Republic relative to the EU 27 average in the year 2014 is presented in Fig. 3.

Figure 3 shows that Slovakia performs below the EU average for many dimensions. As to the year 2014 large relative strengths in terms of indicators are in Sales share of new innovations (158% to average of EU member states—large enterprises

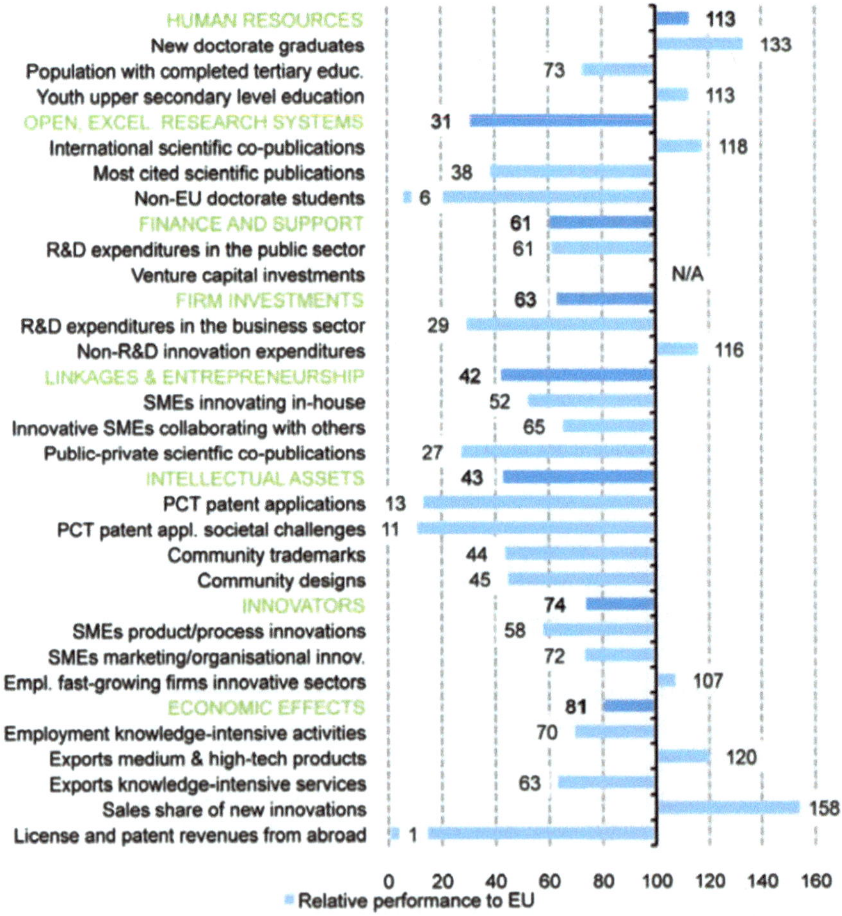

Fig. 3 SII indicators value of the Slovak Republic relative to the EU 27 average in the year 2014 (Source: Innovation Union Scoreboard 2015)

in automobile industry and machinery industry) and New doctorate graduates (133% of the EU average). Large relative weaknesses are in License and patent revenues from abroad (indicator hardly reaches 1.5% of the average of EU member states!), in Non-EU doctorate students (5.8% of the average of the EU member states) and in PCT patent applications in societal challenges (11% of the EU average) and PCT patent applications (13.2% of the EU average).

According to indicators given in European Innovation Scoreboard the financial investment in innovation is in Slovakia far below the average of the member states of the European Union. Indicator R&D expenditure in the business sector was in Slovakia in the year 2014 only 0.38–1.29 in the EU (that means 29% of the average in EU). (Better situation was in Non R&D innovation expenditure—acquisition of machinery, equipment and software, investments into training, acquisition of external knowledge, investment for introducing innovation on the market—that was 11.5% of the EU average).

As to the R&D expenditure in the public sector, it is for a long time very low (in the year 2014 it was 61% of the average of EU member states), the situation in venture capital investments is much worse (indicator is not given).

The two main indicators concerning the innovations in SME (SME introducing product or process innovations and SME introducing marketing/organizational innovations) are both below the EU member states. A detailed analysis within the companies is necessary to find out what are the main barriers to innovation activity in SME's sector.

3 Innovations in Small and Medium Enterprises in the Slovak Republic

From a long-term point of view the innovation activity of Slovak SME is according to the Innovation Union Scoreboard behind the EU average. Both main indicators concerning the SME are for a long time below the average of EU member states. The indicator on "SME introducing product or process innovations" reached in the year 2014 the value 58% of the EU average, the indicator "SME introducing marketing/organizational innovations" reached 72% of the EU average (Tables 1 and 2).

It is evident that all indicators (SME introducing product or process innovations and SME introducing marketing/organisational innovations) decreased during the last 3 years significantly. The situation is critical especially in indicator SME introducing product or process innovations—it has decreased in the year 2014 according to the year 2013 by 32% (!).

The decline is evident also in the indicator SME innovating in-house (in the year 2014 it was 31% less to the year 2013) as well as by the indicator Innovative SME collaborating with others (decrease by 20%) (Statistical Yearbook of the Slovak Republic, 2014).

Table 1 Innovating small and medium enterprises in Slovakia (Source: Innovation Union Scoreboard 2015)

Indicator	2011		2012		2013		2014	
	EU27	Sk	EU27	Sk	EU27	Sk	EU28	Sk
SMEs introducing product or process innovations	34.2	19.0	38.4	26.0	38.4	26	30.6	17.7
SMEs introducing marketing/ organisational innovations	39.1	28.3	40.3	27.2	40.3	27.3	36.2	26.2
SMEs innovating in-house	30.3	14.9	31.8	21.8	31.8	21.8	28.7	15.0
Innovative SMEs collaborating with others	11.2	5.8	11.7	8.3	11.7	8.3	10.3	6.7

Table 2 Division of enterprises according to size structure and realization of innovations

Enterprise/Realization of innovation	Yes	No
Small enterprises	112	272
Medium-sized enterprises	62	81

To identify the main reasons of low innovation activity we have conducted in the year 2014 the large empirical research aimed at evaluating innovation activities in SME in Slovakia. One part of the research was oriented on identifying the main barriers of innovation activities in SME in Slovakia.

4 Results and Discussion

The research conducted at our school in the year 2014 was aimed at three areas: evaluation of innovation activities of SME in Slovakia, identifying the barriers of their development and formulating the presumptions and recommendations to elimination of identified barriers. The research was conducted by the questionnaires distributed via electronic mail. Selective sample was created by 527 enterprises, 384 of them were small and 143 medium-sized enterprises. Representativeness of the sample was verified statistically by means of non-parametric test—chi-square test. The test confirmed a representative sample of selected set. From the overall number of enterprises more than 40% of enterprises worked in industrial branches (engineering, woodworking, electro technology, chemistry, and rubber industries). A third of enterprises were active in the sector of market services, 20% were from building industry and 10% acted in information-communication technologies.

For the question, if SME practice innovation activities, 174 enterprises responded positively from the overall number of 527 enterprises (33.02%), 112 of which were small and 62 medium-sized enterprises. Together 353 enterprises (272 small and 81 medium-sized) stated that they do not practice any type of innovation activities, while they try to act in the market without changes in entrepreneurial activity and they trust to their entrepreneurial strategy.

The question on identifying main barriers of innovation activity in small and medium-sized enterprises in the Slovak Republic was answered and discussed not only by SME that realize innovations, but as well by those enterprises without innovation activity. Innovative enterprises can provide information on barriers; they really meet when realizing innovation activities. On the other hand innovatively inactive enterprises will provide the reasons for innovation activities they do not perform.

We divided the barriers of innovation activity in SME into three groups from the viewpoint of their significance and influence upon innovation activities of SME. The division of barriers according to significance was carried out by means of statistical program of SPSS.

The five main significant barriers to innovations in SME in Slovakia were identified: lack of financial means for innovation, high costs for innovation, quality of innovative environment, lack of qualified employees and absence of cooperation with other subjects in the field of innovation activities.

The main highly important factors that limit the possibilities of further innovation of Slovak SME are cost based factors: an insufficiency of resources within the enterprise and high costs for innovation (more than 80% of respondents consider these two barriers as most significant). The main barrier within the group of barriers with a significant influence on innovation activities is the lack of financial sources for innovation in an enterprise. Financing of R&D activities in Slovakia is strongly below the average of the EU 27 countries. The indicators on R&D expenditure in the business sector are only 29% of the average of the EU, the indicator on venture capital investments is very low (is not given). The majority of SME in Slovakia considers the problems with the accessibility of financial sources for the most expressive factors limiting their innovation activities. The main external source of funding the innovation activities remain the structural funds through the priority axes of the Operational Program Competitiveness and Economic Growth (Ministry of Economy) and the Operational Program Research and Development (Ministry of Education). The two Ministries and their agencies (due to strict implementation of the Competence Act) cooperate insufficiently, which leads to fragmentation and implementation deficiencies. Enterprises introduced negative experience when they were acquiring means from the funds of the European Union, structural funds, or other public financial sources (bureaucratic demand, administration, corruption, ineffective redistribution of means, as well as ignorance of their drawing). The problems with acquiring the financial means force SME to innovate predominantly from their own financial sources.

High costs for innovation activities are the second main barrier belonging to this group. Realization of innovation activities is connected with high costs. Eighty one percent of respondents evaluate them as a significant barrier. Nevertheless, managers of enterprises should take into consideration that innovation is a prerequisite for obtaining a favorable position in the future (Cameron and Green 2006).

The third important barrier is the quality of innovative environment and infrastructure for innovations. More than 70% of respondents is critical to the quality of innovative environment. Respondents expressed critical attitude to the existence and activities of institutions supporting innovation activities as well as to the support of the rise and development of innovative SME from the side of the state. Critical is viewed that in Slovak regions the higher territorial units (VÚC) do not have innovation structures; there is no scheme for effective management of the state innovation policy and regional innovation strategies. An institutional framework for a more efficient connection between industry, results of R&D and practice is missing. The intention to create the regional innovation centers was to ensure implementation of the regional and state innovation policy in regions in order to assure the growth of competitiveness, reduction of regional disparities and growth of regional employment. Slow implementation, lack of coordination and consensus among the relevant ministries appears to be critical. Respondents were critical to the long-term absence of creating regional innovation centers, which should help to start the cooperation between SME on the one side and universities, research centers, technological parks on the other side, as well as to be helpful in the process of establishing the clusters.

Qualified human resources belong to one of the most important factor determining innovation activity. Managers of SME identified the lack of qualified employees as one of the significant barrier. In the Innovation Strategy of the Slovak Republic for the years 2014–2020 was indicated as the second priority “High-quality human resources.” A special measure in the above mentioned strategy is oriented on innovation education for SME. The objective is to provide education and training to firms and entrepreneurs in the area of innovation activities. The reason for the adoption of this measure is a low level of innovation activities and creativity of businesses with SME falling in the category of low innovative enterprises (this is also about the motivation of employees to develop innovation activities). A series of special training courses on innovative activities and special practices and procedures has to be organised under this measure. Educational activities have to be carried out in cooperation with cluster organisations, industrial chambers and associations operating in Slovakia, as well as with higher territorial units and municipalities.

According to results of our research the absence of cooperation with other subjects in the field of innovation activities could be included into the group of averagely significant barriers. Slovak enterprises are dependent on innovation cooperation. When we look into the indicator of innovative SME collaborating with others the value of this indicator was in the year 2014 only 65% of the average of EU member states. As barriers to cooperation within the SME subjects is in many cases lack of interest, low motivation, insufficient financial sources and communication problems. The necessity to develop innovation activity calls for the cooperation in this area. The cooperation of SME with other subjects in the field of innovation activities brings several synergic effects to the enterprise (Kressel and Lento 2012). The most important of them is common sharing of knowledge and simpler approach to the latest know-how, common sharing of capacities, lower demands for financial sources, etc.

5 Conclusion

Small and medium enterprises sector is of paramount importance to improve innovativeness of the entire economy. Given the range of main barriers that have considerable restricting effect on innovation activities of SME, actions need to be urgently identified to overcome the barriers.

In the following part we conclude briefly results of the research aimed at identification of basic presumptions for the development of innovation activities in the SME in Slovakia. We aimed at inside and outside pre-conditions and by means of questions in the separate part of the questionnaire we revealed their importance for small and medium-sized enterprises. Through the analysis we summed them up as follows.

To the question, which basic presumptions have to be fulfilled so that the enterprises could realize the innovation activities, SME stated these six basic presumptions.

1. The main presumption that was stated by the respondents is the sufficient financial sources. The enterprises declared the need to simplify the approach to financial sources, liquidation of huge administrative demand and bureaucracy connected with acquiring financial means from the European funds (structural funds) or from other public sources. For the future it will be necessary to mobilize all financial sources in the area of innovation support in order to ensure that innovation activities performed by business entities receive the same level of funding as those in advanced EU countries (Inovačná stratégia Slovenskej Republiky na roky 2014 – 2020, 2013). In connection with efforts towards the most effective use of allocated financial resources, an indirect state aid has to be provided to profit-generating projects implemented by SME, i.e. financial engineering instruments such as guarantee funds, credit funds, venture capital funds and municipal development funds. There is an enormous interest of responsible institutions in coordination with the Ministry of Finance of the Slovak Republic to apply the upgraded model of usage of innovative financial tools in order to support innovation activities in SME. To support the financing of innovations the situation could be changed not only by one way financial support from state budget, but we see the solution also in overall improvement of the business environment (for example through a reduction of indirect taxes—especially VAT rate, in reduction of contribution to social and health insurance companies and in all other areas mentioned above).
2. As the second presumption for the development of innovation activity was stated the cooperation and participation of SME in networks and clusters. Positive examples from EU countries confirm, that the participation of small and medium enterprises in networks and clusters, support of partnership's building is the way, how to involve small and medium enterprises into innovation activities. Innovation process of a higher level calls for improvement of interaction between small and medium enterprises, research institutions and universities and for creation of various effective networks and partnerships (Lesáková 2009).

Building partnership is a way how to be involved into innovation activities. In the Innovation strategy of the SR for the years 2014–2020 was indicated as one of the main measures the support to innovative industrial cluster organisations. The purpose is to improve competitiveness through support to selected activities of industrial cluster organisations with a view to promote joint industrial activities in selected areas (Innovation Strategy of the Slovak Republic for the years 2014–2020). It could help to support cluster activities which contribute to increasing the competitiveness of the innovative cluster member companies.

3. As important presumption to develop innovation activity was indicated the high-quality human resources. Quality management, as well as employees able to think creatively and to implement innovations in their activities, represent one of the most important presumptions of the development of innovation activity of an enterprise. Management must be able to lead and direct the thoughts and ideas in the enterprise, to search and use talents, be aware of the fact that the enterprise will be successful due to being distinguished by the human resources (Frappaolo 2006). From the viewpoint of employees pro-active approach is expected, as well as the ability to learn and implement knowledge in the innovation activity. Remuneration of employees for their innovation ideas is a significant motivator and presumption for the increased effort of employees when searching for new, innovative solutions (Lesáková 2009).
4. According to our survey results, the fourth presumption is the suitable proinnovative environment. It is necessary to create an innovative environment in the SR that eliminates weaknesses in the area of research and innovation (R&I) and develops opportunities which create conditions for fundamental enhancement of innovative environment. Of special importance is the development of institutions supporting innovation activities on national and regional level. A critical element is above all the autonomous functioning of sectors of education, R&I and business practice, which results into different understanding of R&I. Of special importance is the creation of linkages between multinational corporation's R&I and domestic businesses R&I framework (including the SME), and increasing interest of businesses and industrial clusters in rebuilding of industrial R&I structures (entities). Successful implementation of innovation strategy requires a structural change of current competencies in the management of research and innovation in the SR and a principle manoeuvre in cultural change of innovative environment. According the research results government should pay much more attention to elimination of administrative barriers and create a systematic institutional support to SME on national and regional level.
5. SME's managers agreed on the fact that without well created vision and clearly formulated aims the innovation activity in SME is limited. The pre-condition for clearly formulated aims is the vision corresponding to the possibilities of an enterprise and responding to the situation on the market. Clear vision is a strong predictor of the success (Wagner and Hollenbeck 2012).
6. The important pre-condition identified on the basis of responses is the willingness of enterprises to innovate. This is inevitable, even if it is connected with certain risk. At present many innovative SME are successful and perspective,

and vice-versa many enterprises without innovative activity are getting into financial problems. The willingness to innovate should be accompanied by such an environment that will support the rise of innovation activities (Lesáková 2013). Due to this fact innovation activities may be introduced faster and at the same time several barriers that could retard the rise of innovation activities could be limited.

In today's entrepreneurial practice innovations must be natural part of any entrepreneurship. Permanent and regular innovation is becoming a competitive necessity; to be successful in the future requires interrupting conventions. There is a time of changes and the only way how enterprise can be successful is to accept these changes, adapt to them and utilize them.

References

Bessant J, Tidd J (2009) Innovation and entrepreneurship. Wiley, Chichester
Cameron E, Green M (2006) Making sense of change management. Kogan Page, London
European Commission (2010) Europe 2020. A strategy for smart, sustainable and inclusive growth. European Commission, Brussels
European Commission (2013) Innovation union scoreboard 2013. European Commission, Brussels
Frappaolo C (2006) Knowledge management. Capstone, Chichester
Inovačná stratégia Slovenskej Republiky na roky 2014–2020 (2013) Ministry of Economy of the Slovak republic, Bratislava
Kressel H, Lento T (2012) Entrepreneurship in the global economy. Engine for economic growth. Cambridge University Press, Cambridge
Lesáková Ľ (2009) Innovations in small and medium enterprises in Slovakia. Acta Polytech Hung 6(3):23–34
Lesáková Ľ (2013) Behavioural norms as an element supporting creativity and innovation in small and medium enterprises. In: Contemporary challenges towards management III. Wydawnictwo Uniwersytetu Slaskiego, Katowice, pp 99–107
Statistical Yearbook of the Slovak Republic (2014) Statistical Office of the SR, Bratislava
Wagner J, Hollenbeck J (2012) Organization behaviour. Securing competitive advantage. Taylor & Francis Group, Routledge
Innovation Union Scoreboard (2015) EU publications

Ľubica Lesáková—is Professor at the Department of Business Economics and Management at the Faculty of Economics of Matej Bel University in Banská Bystrica (Slovak Republic). She is the director of the magister and doctoral study programme "Business Economics and Management" at Matej Bel University in Banska Bystrica. In her research she was responsible for leading national research projects dedicated to SME's management as well as to the SME's innovation.

How to Measure Intermunicipal Cooperation in Conditions of the Czech Republic

Jiří Dušek

Abstract In the current concept of regional politics as an activity, the main purpose of which is to reduce disparities in development of individual regions and to ensure their harmonious development, emphasis is also laid on cooperation of towns and municipalities in development of a given area. Local initiative thus gains more and more on significance and under certain circumstances becomes the most important factor in regional development. This contribution analyses the ways to measure the intensity of intermunicipal cooperation in the individual regions of South Bohemia Region and it also attempts to describe all relevant relations between these variables in order to better understand the significance of processes and elements of cooperation using an originally created data-base of towns and municipalities in the South Bohemian Region after the year 2001.

Keywords Czech Republic • Intermunicipal cooperation • Municipality • Regional development

1 Introduction

A historic example of intermunicipal cooperation is the Hanseatic League. It was created by municipalities in Northern Europe and lasted from the thirteenth to the seventeenth century. Change came with the Industrial Revolution as cities grew rapidly and the requirements concerning public services increased. While cooperation was initially without obligation, with the beginning of the twentieth century IMC was increasingly legally codified. As an example, the "Siedlungsverband Ruhr" was founded in 1911 and exists until today. It has extensive powers in planning the settlement of the region around the Ruhr. The services and the standards had been raised after the Second World War to a quality of public services never known before. But with increasing costs on the one hand and decreasing

J. Dušek (✉)
College of European and Regional Studies, Žižkova tř. 6, České Budějovice 370 01, Czech Republic
e-mail: jiridusek@centrum.cz

A. Dias et al. (eds.), *Modeling Innovation Sustainability and Technologies*, Springer Proceedings in Business and Economics,
https://doi.org/10.1007/978-3-319-67101-7_12

inhabitants combined with decreasing tax income, municipalities have increasing problems in maintaining the quantity and quality of public services (Froecker and Hokkeler 2008). One way to cut costs and increase efficiency is to cooperate with neighbouring municipalities.

Since the 1950s, local governments in many European countries (Austria, Sweden, Finland, Germany, Switzerland, France...) have been cooperating and nowadays, the encouragement of cooperation among local authorities in the provision of local public goods remains on the political agendas of many central and local governments (Hulst and van Montfort 2007). There are several reasons for this widespread and persistent phenomenon—see details in e.g. Hertzog et al. (2008) or Frère et al. (2011):

- First, larger spatial units are expected to be more competitive in a globalized world.
- Second, as governments try to reduce the cost of providing public goods, the achievement of economies of scale in the provision of local public services is a strong incentive to cooperate.
- Third, fiscal cooperation allows jurisdictions to internalize spending spillovers: the benefits of public expenditure (infrastructure, road building, cultural facilities...) often spread across the boundaries of the supplying jurisdiction and affect the welfare of the citizens in neighbouring localities.
- Fourth, tax competition between municipalities has been observed, especially between urban municipalities, and tax cooperation is often seen as a useful corrective device for municipal levels of tax and spending that otherwise might tend to be too low.

One can speak of intermunicipal cooperation when two or more local governments work together to provide a public service. All gains and losses are shared between the participants of the cooperation. These cooperation spread from only coordinated behaviour up to founding a joint venture that settles the task for both participants. As the territorial consolidation often fails because of political resistance intermunicipal cooperation is a way to keep public services efficient and effective without territorial consolidation (Local Government and Public Service Reform Initiative, 2013). Intermunicipal cooperation can be divided into two categories: In case of a joint agreement municipals work together to operate a certain plant or share the provision of a service. The other category is called service agreement, where one town provides the service for the other town (Office of the State Comptroller, Division of Local Government and School Accountability, 2007). In both categories all types of cooperation can be found.

Among other things, cooperation is also a significant feature of current institutional theories of regional development where stress is put on mutual cooperation of a whole number of local and regional players (this cooperation being based on trust). It is exactly the level of trust which, according to Granovetter (1985), is a key characteristic of economy. Trust is being created by repeated (successful) interactions. There are networks of contacts with varying levels of quality and varying levels of trust. Every entity is connected to one of the networks, the quality of which significantly influences the entity's possibilities. A low level of trust leads to additional and/or increased costs, e.g. for safeguarding that delivered goods will be paid for, debt collection etc. Regional research inspired by this approach tries to analyse the role of personal relationships, mutual trust between partners and to study contacts as a form of social integrity, as well as to define their importance from the point of view of causes leading to differences between regions. At the same time, it should be noted that the presence of networks themselves is not a positive and beneficial phenomenon (there are also parasitic networks or even mafias). Prospects and chances of individual participants (individuals, municipalities, companies) depend not only on the intensity of their connection to networks, but also on the contents which individual networks provide (Wokoun et al. 2008).

Looking for sense in intermunicipal cooperation in the conditions of the current Czech Republic requires above all a change of attitude from passive application of the legal possibility to cooperate to an active application in the overall context of Czech Republic's public finances. Obviously, victory of democracy in the basic conflict with effectiveness has had its economic impact—insufficiently equipped small municipalities are not able to effectively satisfy public needs of their residents (reduced quality of public goods, neglect of long-term development, increased demands for subsidies, increase in municipality debt etc.). If we want to avoid the anti-democratic solution in the form of reducing the number of municipalities, let's turn to the other possibility, i.e. intermunicipal cooperation. Active voluntary cooperation is one of the possible ways towards integration in the area of territorial cooperation and towards better effectiveness in the area of public finances. Based on economic stimuli, larger entities are created, the boundaries of which are defined by the effectiveness of activities in the given area (Kotvalová 2001).

2 Material and Methods

The methodology of this contribution is in compliance with methods usually used in scientific research; it is based on the use of the latest theoretical knowledge gained from specialised literature, specialised research and studies, newspapers and materials published by individual participants in regional development. Also, the methodology is based on looking for and assessment of mutual relationships which contribute to the clarification of the problems solved and to a deduction and formulation of adequate conclusions which can be derived from such an analysis.

3 Results and Discussion

Theoretical and practical findings acquired during my work clearly confirm the significant role of intermunicipal cooperation in regional development. Municipalities are important actors in regional development—they are able to activate local and regional resources and thus can bring about synergic effects.

Before the actual analysis of intermunicipal cooperation, it was necessary to develop a representative data base characterizing the development of intermunicipal cooperation in the South Bohemian Region (Czech Republic). This proved to be a significant problem for several reasons. Not all forms of intermunicipal cooperation are recorded statistically and thus cannot be analysed (e.g. twin municipalities in case of international cooperation, public-private-partnerships, joint legal persons of municipalities and other entities etc.). However, despite the serious fragmentation and the high cost, it was possible to collect this data.

From the point of view of significance of individual forms of cooperation, clearly the most significant form in the South Bohemian Region is cooperation at the level of microregions and LAGs (local action groups); these two forms of cooperation represent about 60% of all forms of cooperation in the region (see Fig. 1). Other forms of cooperation are less significant since they are not as effective

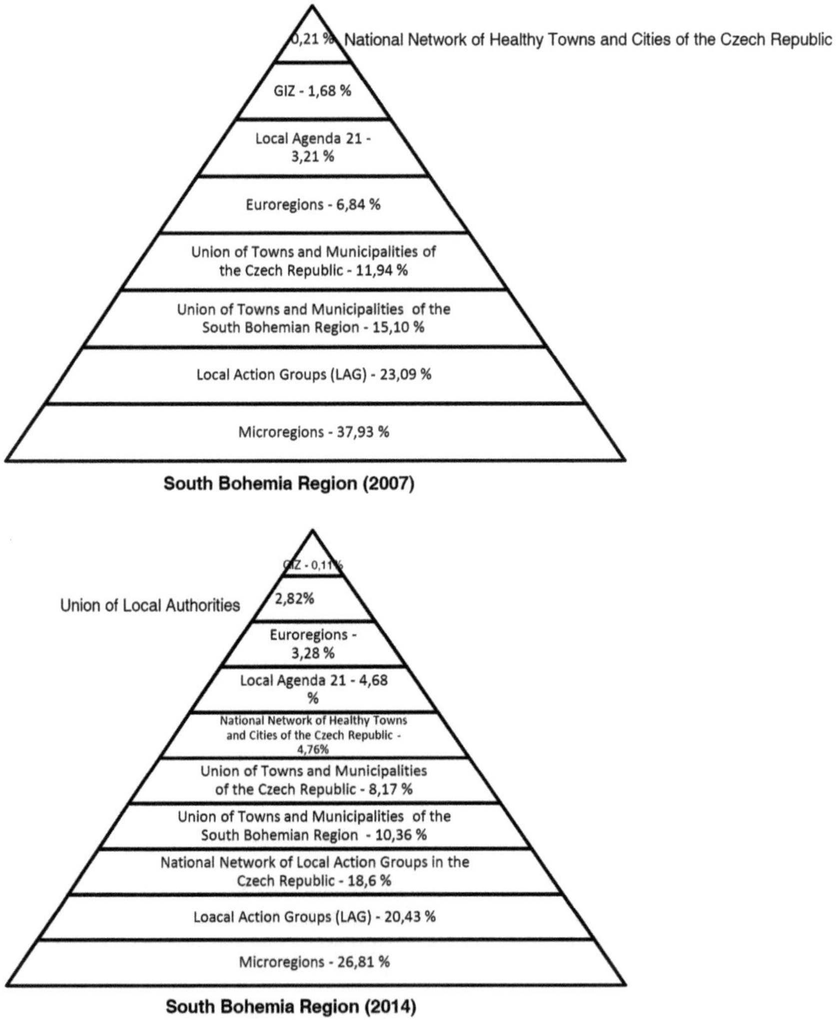

Fig. 1 Share of individual forms of intermunicipal cooperation in the South Bohemian Region

as cooperation in LAGs and among microregions, or because municipalities cannot participate in cooperation due to their geographical position.

In order for it being possible to compare the overall intensity of individual forms of cooperation in the South Bohemian Region, the so-called coefficient of cooperation (C_C) was created using the method of quantification. The maximum theoretically possible value of the coefficient of cooperation in the South Bohemian Region is 13, the minimum value is 0. The higher the C_C values, the more intensive the involvement of a town or municipality in intermunicipal cooperation in the South Bohemian Region (see Figs. 2 and 3). If necessary, the indicator can be modified to accommodate specific conditions of a different region (specific in the South Bohemian Region is the form of national structure of cooperation represented by the Union of Towns and Municipalities of the South Bohemian Region).

$$C_c = \sum_{i=MCR}^{GIZ} P_i$$

where MCR—microregion, GIZ—cross border impulscenters.

Although individual forms of cooperation differ in their importance and significance for the development of a region, all of the individual indicators were given equal weight. The reason for this decision was the fact that municipalities first become involved in microregions and/or local action groups and only later engage

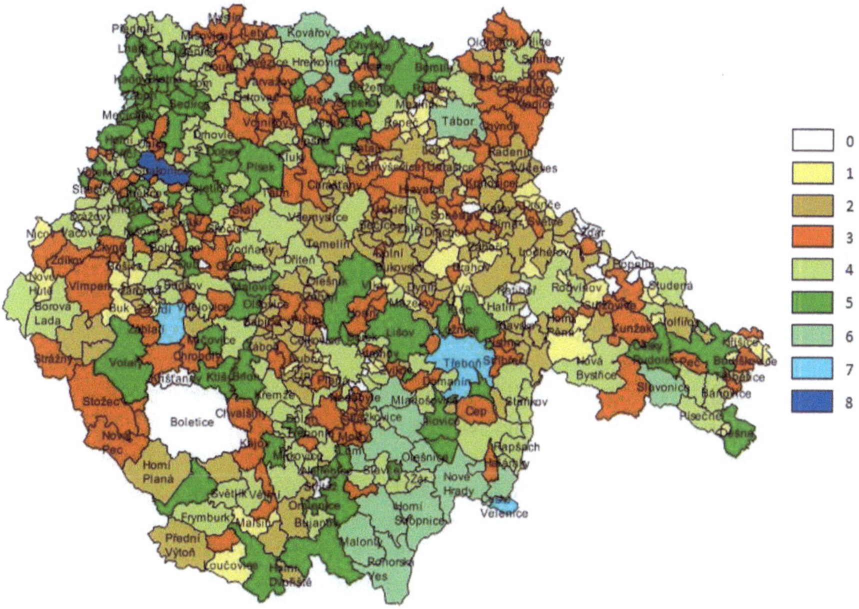

Fig. 2 Coefficient of cooperation (CC) in the South Bohemian Region in 2007

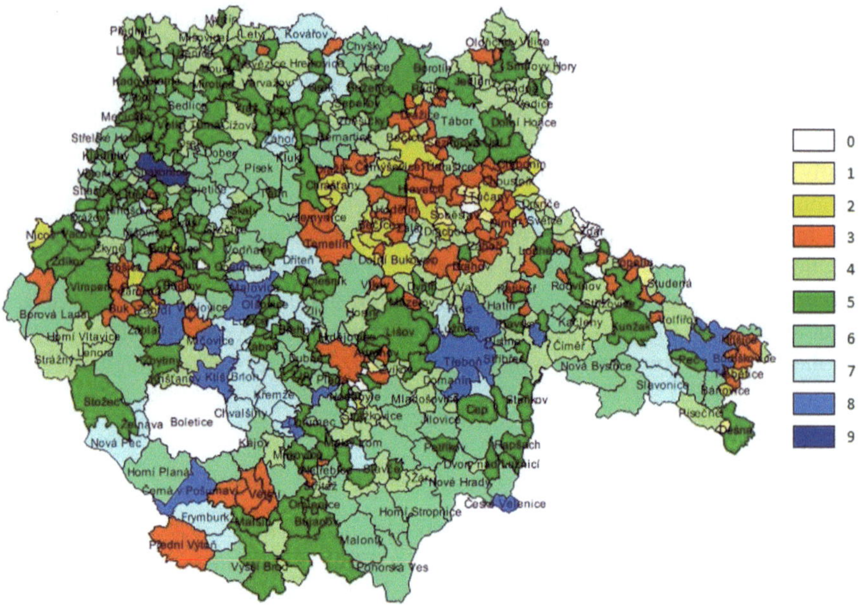

Fig. 3 Coefficient of cooperation (CC) in the South Bohemian Region in 2014

in further "above-standard" forms of cooperation. The scales thus reflect more the relevance of data base than the significance of individual features. With the defined equal weight, the coefficient of cooperation has better informative value—it is obvious at first sight how many times a town or municipality has been involved in any form of cooperation in a region.

4 Conclusion

An advantage of the coefficient of cooperation (C_C) is easy comparability of towns, municipalities and other pre-defined areas from the point of view of intensity of cooperation; a disadvantage are difficulties connected with the set-up and validation of basic data base, from which the coefficient is calculated.

By means of the coefficient of cooperation, it is possible to identify problematic areas of the region and categorize towns and municipalities of the South Bohemian Region according to their intermunicipal cooperation and suggest a specific strategy for their further development. The significance of the coefficient of cooperation thus moves from a theoretical level to a level of practical application by the public administration of the Czech Republic.

Cooperation of municipalities is an important factor of regional development and its significance should not be underestimated. The state and public administration should try to support cooperation between municipalities—financially, administratively and legislatively, as well as by consultancy and other forms of support. The results of questionnaire surveys have indicated that the main problem of intermunicipal cooperation was lack of finances and the right method of financing the cooperation. Legislation often inhibits inclusion of other partners in intermunicipal cooperation (e.g. business organisations, non-profit organisations and other institutions). For instance, no business entity can figure in a voluntary union of municipalities and it is very difficult to reach an absolute majority of entities outside of public administration in the case of an LAG.

The cooperation of local and regional entities can become an even more important factor of regional development in the future thanks to activation and better use of local resources, which will increase the competitive strength of a given region against other regions. The results of this work can be of value not only to individual participants in regional development in the South Bohemian Region (towns and municipalities, microregions, local action groups, South Bohemian Region), but also other institutions as a basis for assessment of the intensity and significance of cooperation among municipalities of the South Bohemian Region. The created model of assessment using the C_C coefficient can also be applied to other regions and administrative districts in the Czech Republic, which would be very interesting for reasons of the possible comparison of the results acquired.

References

Frère Q, Leprince M, Paty S (2011) The impact of inter-municipal cooperation on local public spending: empirical evidence using French data [online]. Groupe d'Analyse et de Théorie Economique Lyon, Lyon. https://www.gate.cnrs.fr/IMG/pdf/Paty2011.pdf. Accessed 18 Oct 2015

Froecker HJ, Hokkeler, M (2008) Interkommunale Zusammenarbeit. Friedrich-Ebert-Stiftung, Bonn, 84 pp. ISBN 978-3-89892-824-3

Granovetter M (1985) Economic action and social structure: the problem of embeddedness. Am J Sociol 91(3):481–510. ISSN 0002-9602

Hertzog R et al (2008) Toolkit manual: inter-municipal cooperation. Council of Europe, Strasbourg, 105 pp

Hulst R, van Montfort A (2007) Inter-municipal cooperation in Europe. Springer, Dordrecht, 238 pp. ISBN 978-1-4020-5378-8

Intermunicipal Cooperation and Consolidation Exploring Opportunities for Savings and Improved Service Delivery [online] (2007) New York State – Office of the State Comptroller, Division of Local Government and School Accountability, New York. http://www.municipal-cooperation.org/index.php?title=When_is_IMC_relevant%3F. Accessed 13 Oct 2015

Kotvalová J (2001) Meziobecní spolupráce. Veřejná správa 12(6):12, 21. ISSN 1213-6581

When is IMC relevant? [online] (2013) Local government and public service reform initiative, Budapest. http://www.municipal-cooperation.org/index.php?title=When_is_IMC_relevant%3F. Accessed 13 Oct 2015

Wokoun R, et al (2008) Regionální rozvoj (Východisko regionálního rozvoje, regionální politika, teorie, strategie a programování). Linde Praha, Praha, 480 pp. ISBN 978-80-7201-699-0

Jiří Dušek, Ph.D., vice-rector for science, research and development, is an assistant professor at the College of European and Regional Studies in České Budějovice (Czech Republic). The focus of his research and pedagogical activities is on public economy, public finances and regional development. He is author and co-author of more than 50 scientific papers/monographs, member of the editorial boards of several journals and has been an invited lecturer at many European universities and institutions.

Part III
Modeling Innovation and Sustainability of the Future Tourism Destinations

The multidimensional nature of tourism phenomena leads to discussion about the role of tourism in spatial development. The tourism sector plays an important role in economics and together with other sectors promotes wellbeing and enhances the development of places. According to the World Tourism Organization UNWTO, world tourism currently contributes to 30% of the world services exports, creates 9% of the GDP and over 8% of the jobs. In the most well developed countries, improvements on tourism bring significant economic benefits. This Part III focuses on places that do not belong among well-known tourist destinations but uniqueness and a modern approach implemented by tourism development seems to be promising.

Innovation processes in tourism are mainly aimed at increasing productivity, profitability and quality, thus improving the overall competitiveness of the tourism economy. Innovation is to be considered as a major driving force for structural changes in the tourism industry and transforming the tourism sector itself as well as tourism business models. Creativity, new ideas and innovation usher in smart solutions, create added value of tourism products and destinations and increase competitiveness of tourist destinations. Part III is aimed on innovations and modern trends in tourism and impact of the tourism sector and its activities on economics and development. The tourism sector is becoming very creative, dynamic and innovative. New tourism destinations are appearing as mushrooms after rainfall also because of new and modern approaches to tourism development. Tourists are becoming very active, hungry for new experiences, ready to be shifted from 'traditional' tourism destinations to 'new', yet to be experienced, cities, regions or countries. This trend is creating a big space for the development of new destinations or new basis of tourism. The following section brings unique research papers, exploring new, sometimes untouched places of our beautiful planet. The following part contains three original research papers based on case studies from four different countries, including rising star Poland but also still unexplored Slovakia and the unique untouched island in the mythical Baikal Lake in Russia.

Innovation Process in Mountain Destinations: Does Sustainability Matter? The High Tatras Case Study

Zuzana Gajdošíková, Tomáš Gajdošík, and Vanda Maráková

Abstract Nowadays innovations are considered as one of the key sources of performance and competitiveness among tourism destinations. As tourism development is primarily based on their natural sources, the innovation process should respect the criteria of sustainable development. The aim of this chapter is to analyse the implementation of innovations in mountain destinations and examine their effect on sustainable tourism development. The research adopts in-depth analysis of the Central European destination High Tatras, concerning primary and secondary data from a survey and interviews with destination managers, tourism stakeholders, local inhabitants and annual reports. The novelty of this analysis lies in examining the connection of implementing different categories of innovations within different stages of the destination life cycle. The chapter concludes that various categories of innovation with different effects on sustainability are implemented in different stages of the destination life cycle.

Keywords Destination life cycle • Innovation • Tourism destination • Sustainability

1 Introduction

Constantly changing environment, forced by demanding customer requirements, demographic changes, increasing competition, technology development and globalisation, put a pressure on tourism businesses and destinations to adapt their offer. In order to maintain a market position, tourism businesses and destinations need to implement innovations.

The growing interest in economic, social and environmental sustainability, as one of the most important sources of competitiveness of a destination, is particularly

Z. Gajdošíková (✉) • T. Gajdošík • V. Maráková
Department of Tourism and Hospitality, Faculty of Economics, Matej Bel University in Banská Bystrica, Tajovského 10, 975 90 Banská Bystrica, Slovakia
e-mail: zuzana.gajdosikova@umb.sk

A. Dias et al. (eds.), *Modeling Innovation Sustainability and Technologies*, Springer Proceedings in Business and Economics,
https://doi.org/10.1007/978-3-319-67101-7_13

important in the case of traditional mountain destinations, where tourism is often the main vehicle of development.

In a mountain destination, tourism is an important economic activity and important source of employment, because the development of other economic activities in such area is rather limited. Mountain destinations, in comparison with other types of tourism destinations, have certain particularities that need to be taken into account when implementing innovations. Innovations should focus on protecting the environment as a source of tourism development (environmental principles), ensuring the livelihood of the local population (economic approach) and improving their quality of life (social principle).

2 Theoretical Background

Tourism development in a destination is influenced by innovations focused not only on products and activities, but also on new technologies, management, marketing and cooperation of destination's stakeholders. According to the Strategy Europe 2020 the aim of the innovations is to create new products and services that ensure the economic development, create new employment opportunities and strengthens the competitiveness of tourism businesses and destinations in the tourism market.

The term innovation comes from Latin and means "restoration". It was introduced in the economic theory for the first time by Austrian economist J. A. Schumpeter in 1930s. In the last decade the importance of innovations was reflected also in tourism. The importance of innovations in the tourism sector was highlighted mainly by Hjalager (2002), Peters and Weiermair (2002), Jafari et al. (2003), Keller (2003), Decelle (2004), Weiermair (2004), Pechlaner et al. (2005), Pompl and Buder (2006), Walder (2007), Pikkemaat (2007), Hall and Williams (2008) and Aldebert et al. (2010).

As innovations in tourism are mainly a territorial phenomenon (Longhi and Keeble 2000), it is worth to implement them in a tourism destinations (Kämpf and Weber 2005). As the coordinator of tourism development in a destination is a destination management organisation, core business or a municipality, innovations can be implemented in several ways. Zach and Fesenmaier (2009) studied the innovations carried out by destination management organisations, while Flagestad and Hope (2012) pointed out the importance of the core business (operator of a mountain transport facilities) and its role in implementing innovations.

There are several types of innovations in tourism sector. Jafari et al. (2003) distinguish product, process, managerial, marketing and logistic innovations. OECD (2005) adds market and ad hoc innovations, while Mayer (2009) focuses also on organisational, institutional and customer innovations. Another point of view is presented by Castellacci (2008) who deals with demand driven innovations and innovations pulled by technologies and state interventions. Moreover Camisón and Monfrot-Mir (2012) aimed their research at innovations pulled by science and research and market driven innovations. However, the most significant classification of innovations in tourism is presented by Hjalager (2010) who creates the typology of innovations focused on product, process, management, marketing and institutions.

Product or service innovations refer to changes directly observed by the customer and regarded as something new. It can be a completely new product or service or new only for a particular business or destination. Process innovations are aimed at escalating efficiency, increasing productivity and flow (Hjalager 2010). Managerial innovations deal with new approaches to internal collaboration, motivating and empowering staff, building careers and compensating employees with extra pay and benefits (Ottenbacher et al. 2005). Marketing innovations are aimed at new marketing approaches targeting changes in communication between service providers and customers and building their positive relationship (Hankinton 2004). Institutional innovation is a new comprehensive collaborative or organisational structure or legal framework that efficiently enhances business in certain fields of tourism. Networks and alliances in tourism are considered essential for fostering the innovations in this sector (Lynch and Morrison 2007).

Recently, there has been a growing interest on tourism sustainability as one of the most important sources of competitiveness of a destination (Crouch 2007, Mihalič 2000, Poot 2000, Ritchie and Crouch 2003). The World Commission on Environment and Development (WCED) has defined sustainable development as: "development that meets the needs of the present without compromising the ability of future generations to meet their own needs" (WCED 1987). The World Tourism Organisation (UNWTO) expands this concept, stating that "sustainability includes quality of life for host communities, visitor satisfaction, takes care of the use of natural and social resources addressing the general objective of achieving a level of harmony among the various types of stakeholders involved in the tourism sector and/or interested in the way tourism in their area is managed and developed" (UNWTO 1996).

The natural environment is a prerequisite for development of tourism in the mountain destination. Kučerová (1999) and Keller (2012) points out that tourism in mountain destinations has limited potential for growth, because it depends on the natural resources used. Therefore the sustainability issue of mountain destinations was the object of the research of many authors. Todd and Williams (1996), Schendler (2003), Eydal (2004), Lewis (2005) examined the social and environmental nature of sustainable development in mountain destinations, while Hudson (2000) and Castle (2004) focused mainly on the quality of life of local inhabitants. Moore (2005) dealt in his research with social equity in mountain destinations and Flagestad and Hope (2001) were examining the democratic decisions about sustainable development. Matto and Scott (2008) created eight principles of sustainable development in mountain destinations and the sustainability in mountain destinations was also emphasised by Bürki and Abegg (2003) and Scott et al. (2006). Moreover as the destination evolves, it goes through different stages of destination life cycle (Butler 1980; Butler et al. 2006) and the tourism development has a diverse impact on the sustainability issues (Buhalis 2000).

Taking into account the above mentioned, the characteristics of innovations in mountain destinations can be proposed. Innovation in mountain destinations is a

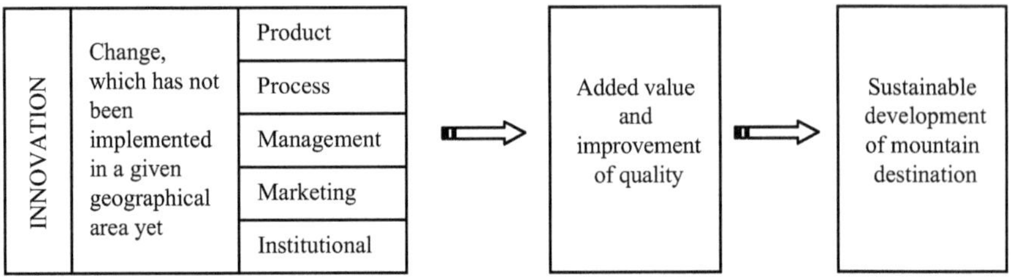

Fig. 1 Innovations in mountain destinations

process aimed at achieving added value and improvement of the quality of products and services for visitors, which should be carried out in accordance with the principles of sustainable development. Therefore the innovation in a mountain destination is the change, which has not been implemented in a given geographical area yet. It should be in line with the implementation of sustainable tourism management; focused on increasing the economic benefits for local residents and visitors and reducing the negative effects, increasing the contribution of tourism to the environment and reducing its negative impacts (Fig. 1).

Implementing innovations according to the sustainable development requires destination managers and local authorities to evaluate and monitor both the positive and the negative impacts (economic, environmental and social) of tourism in their destination and then regulate it to achieve a positive balance between these impacts.

3 Methodology and Results

Mountain destinations in Europe, due to their unique nature offer, play an important role in tourism development since the mid-eighteenth century. Progress in transport, development of technology and changes in consumer behaviour create a need to continually innovate products, processes, management, marketing and institutions. The innovation process has to be carried out systematically; it should not be aimed at new technologies and trends without clearly defined goals. Therefore it is desirable that innovations respect primary and secondary offer of a mountain destination, its market share and visitors' segment.

The geographical area of the Carpathians in Central and Easter Europe is so far under researched, however, due to its price competitiveness, special offerings and ease of accessibility, it is viewed as a future playground for central European tourism markets (Demiroglu et al. 2015; Vanat 2016). Tourism development in the Central and Easter Europe is mainly associated with skiing and hiking. The most important area is the High Tatras, as the traditional destination located in the highest mountains in the region (Fig. 2).

Tourism development in the High Tatras has started in the beginning of nineteenth century. In the twentieth century the first ski lessons for spa guests staying in hotels took place here. Since 2000s there has been a shift from the traditional

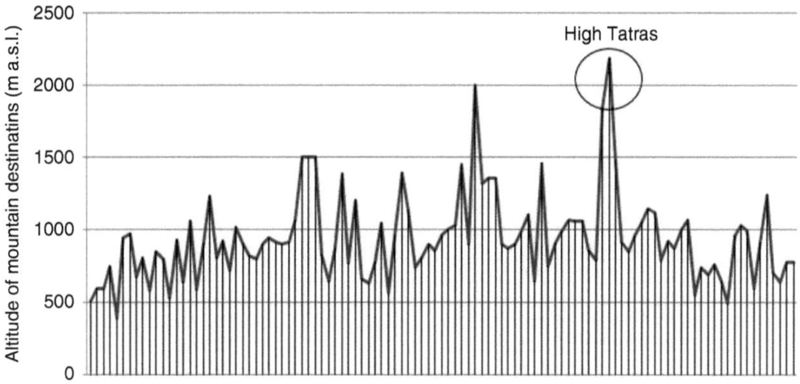

Fig. 2 Altitude of mountain destinations in the Western Carpathians

product focused on hiking and climatic spa to product more oriented on fun and entertainment. In 2009 the major investment company started to operate in the area of the High Tatras, which was the stimulus for rejuvenation and implementing the innovations.

Therefore the aim of the chapter is to analyse the implementation of innovations in the mountain destination High Tatras and to examine their effect on sustainable tourism development. Special attention is drawn to the link of implementing different types of innovations in diverse stages of destination life cycle.

The paper adopts the in-depth analysis of the destination carried out for 3 years, concerning primary and secondary data. Primary data were obtained by a survey among 91 tourism stakeholders and 1021 local inhabitants. Secondary data came from the annual reports, destination brochures, materials from destination management organisations, web sites of tourism stakeholders and destination management system. Moreover interviews with the stakeholders and destinations managers took part in order to clarify the results obtained.

3.1 Implemented Product and Marketing Innovations

Product innovations in mountain destinations are mainly focused on extending the winter season or creating the conditions for summer season in order to build a 365-day visitor experience. Mountain destinations are mostly influenced by the climatic changes; therefore it is worth to implement innovations that ensure the snow reliability.

As the traditional ski resorts in the Carpathians were developed from mountain villages (up to 1200 m a.s.l.) they need to implement innovations focusing on new technologies extending the winter season. According to Demiroglu et al. (2015), the ski resorts in the High Tatras would be naturally less snow reliable as the number of operational days with snow depth over 30 cm would be reduced from 66 to 55 by

Year	Innovation	Innovation type
2009	Parking place (capacity for 300 cars)	product
2009	Extension of ski trails by 6 km	product
2009	Accommodation packages including ski pass	product
2009	Marketing communication in Polish market	marketing
2009	Cooperation with Russian tour operators	marketing
2010	Educational paths for children	product
2010	New price politics	marketing
2010	Creating the accounts on social networks	marketing
2010	New web page of destination	marketing
2011	New 8-seat chairlift	product
2011	Tracks for mountain biking	product
2011	Ice rink	product
2012	Ski slopes regulation	product
2012	New seats on chairlift	product
2012	Visitor card	product
2013	New 15-seat cabin lift	product
2013	Extension of artificial snow systems	product
2013	Loyalty program Gopass	product
2013	Publishing of lifestyle magazine	marketing
2013	Online sell of Gopass card	marketing
2014	Improvement and expansion of accommodation capacities	product
2014	New après ski facilities	product

Fig. 3 Implemented product and marketing innovations in the High Tatras

2030. Therefore in the new introduction stage of destination life cycle (after the rejuvenation in 2009) it was necessary to implement product and marketing innovations (Fig. 3).

As a first step, the major operator of mountain transport facilities started to implement technological product innovations. In 2009 new parking places and extensions of ski trails were made. Consequently, with the growing number of visitors since 2011, new cable cars and renovation of existing ones were introduced. Moreover new snow making systems as well as the grooming of ski trails were introduced. The quality of accommodation infrastructure was improved and new après-ski facilities started their operation. In order to create conditions for summer season, new mountain bike trails were built.

As the tourism infrastructure was innovated in order to provide higher quality, the non-technological innovations started to be introduced and the growth stage of destination life cycle began. Product innovations focused on new accommodation packages and loyalty programs were introduced. These non-technological product innovations were implemented together with the marketing innovations. Innovations in marketing communication focusing on the neighbouring countries were made. Following the trends in information technologies the communication via social media, as well as the redesign of web page, were inevitable.

All the presented product and marketing innovations were introduced by the core business Tatry Mountain Resort, Inc. It was due to the fact that small tourism business (accommodation and catering facilities) were not able to cooperate and act under one coherent strategy. In order to reach new markets and increase the efficiency of marketing communication the networking and cooperation began to strengthen in the region and thus the institutional innovations started to be implemented.

3.2 Implemented Institutional Innovations

Institutional innovations in the High Tatras were focused on reengineering the organisational structure in the destination, consisting of private and public destination stakeholders. The first tourism organisation in the High Tatras was created in 1994 and was called Tourism Association of the High Tatras. It represented the interest of accommodation and catering facilities. Although this association created the baseline for cooperation, the networking and cooperation of destination stakeholders in the region was very low. In 2012 a new destination management organisation (DMO Region High Tatras) was created, which can be seen as the institutional innovation (Fig. 4).

The mission of the DMO is to increase the competitiveness and attractiveness of the destination High Tatras on domestic and international tourism market and to create a recognizable brand and image of the destination. This DMO represents the interest of three municipalities, the Tourism Association of the High Tatras, the operators of mountain transport facilities (TMR, Inc.) and aqua park. Tourism businesses (e.g. accommodation and catering facilities, tour operators) are represented by the Tourism Association. From the private sector, the most powerful stakeholder is the operator of mountain transport facilities—the company TMR, Inc. Besides the ownership of all cable cars and lifts in the region, it operates also six hotels, nine catering facilities, several sport and recreational facilities and shops.

The situation can be presented also by the retrospective network analysis (Fig. 5). The cooperation of stakeholders in 1995 was centred within the Tourism Association of the High Tatras. The intensity of cooperation, measured by weighted clustering coefficient of network analysis, was in 1995 very low (0.014). In 2015 the intensity was higher due to the implemented institutional innovation that

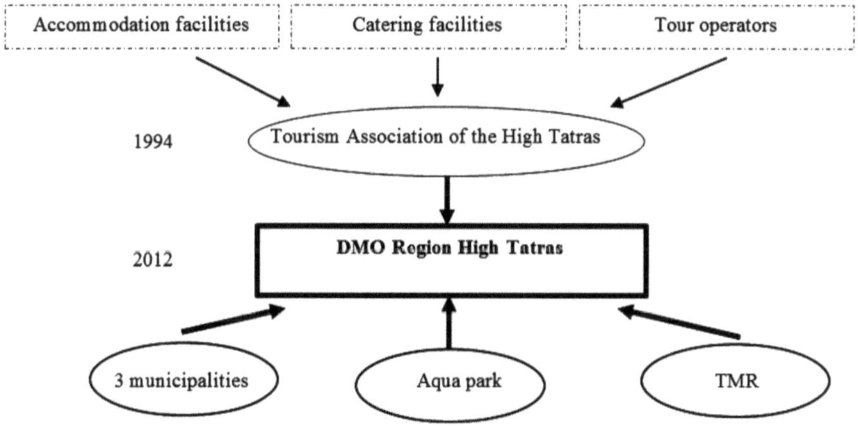

Fig. 4 Organisational structure of tourism stakeholders in the High Tatras

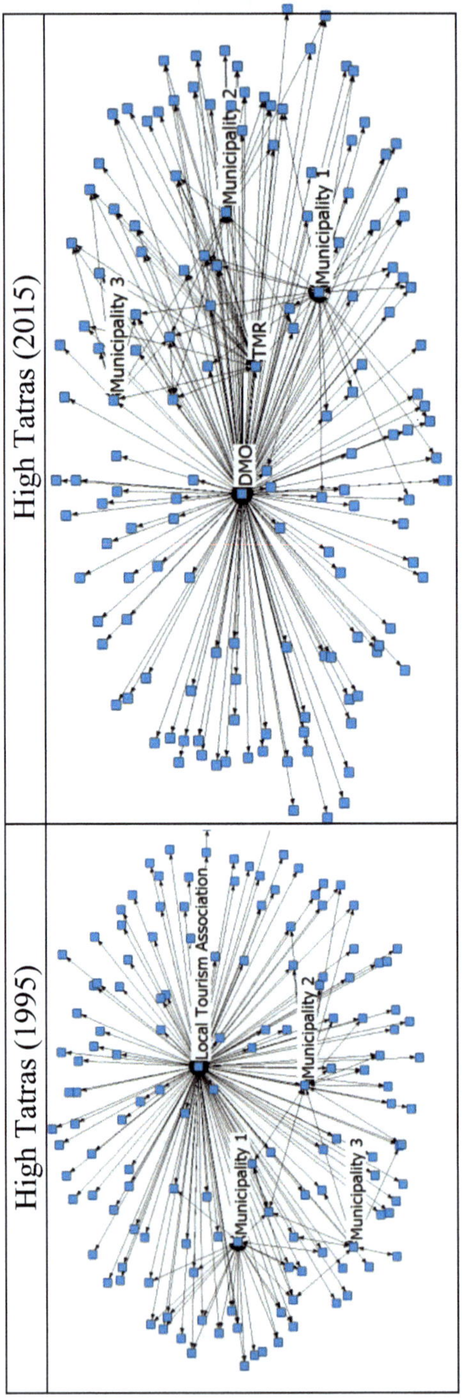

Fig. 5 Institutional change in the High Tatras

Outcomes of creating the DMO as an institutional innovation	Share of answers in %
Creating a complex tourism product	74,36
Creating the unified image and brand of a destination	64,10
Reaching new visitors' markets	61,54
Better exchange of information	51,28
Integrated marketing communication	38,46
Exchange of experience	28,21
Minimizing the costs	25,64
Better access to financial resources	23,08
Inclusion to booking system of a destination	7,69

Fig. 6 Outcomes of institutional innovations in the High Tatras

strengthened the product development and integrated marketing communication, leading to intensity of cooperation of 0.027.

The cooperation within newly established DMO created the positive outcomes for the whole destination. According to the primary research within 91 stakeholders, it can be concluded, that the institutional innovation led to creation of more competitive product reaching new market segments and created easier conditions for obtaining financial resources (Fig. 6).

Among the most important outcomes of the institutional innovation is, according to tourism stakeholders, the creation of a complex tourism product and creation of unified image and brand of the destination. Therefore it can be stated that the institutional innovation created the baseline for implementing professional destination management and marketing in the destination.

3.3 *Upcoming Process and Managerial Innovations*

After the institutional innovation, the process and managerial innovations started to be implemented in the region. However, the majority of these innovations are only intentions that would be implemented in the following years. The process innovations should be focused on efficiency, productivity and flow. They are connected mainly with implementing information technologies. In the High Tatras destination management system (DMS) Feratel Deskline started to be used. Although the system is now used only for booking the accomodation and presenting the organised events in the destination, its wider implementation will change the internal processes in the destination management organisation. The following processes will be innovated:

- destination product development,
- destination marketing,
- booking of services (accommodation, cultural events),
- management of visitors' cards,
- customer relationship management (CRM),
- management of destination's stakeholders.

Perspective	Example of a goal
Financial	- Increasing the economic effects of tourism in the tourism destination - Ensuring financial resources for tourism development
Visitor	- Reaching new visitors' markets - Increasing the share of regular visitors
Marketing	- Creation of competitive destination product - Usage of modern technologies in destination (smart destination)
Process	- Optimizing process in internal collaboration - Evaluating process aiming at visitors' satisfaction
Innovation	- Continuous monitoring the market rends - Creative differentiation in reaching new target groups
Employee	- Employee motivation

Fig. 7 Perspectives of Balanced Scorecard in tourism destination

Moreover, in terms of process innovations it would be worth to implement Balanced Scorecard method to evaluate and control the processes in the destination. According to specific features of tourism destinations, it is necessary to adjust the perspective of the BSC (Fig. 7) to the strategic aims of destination management organisation.

These perspectives are adapted to the specificities of tourism product and activities of destination management organisations. BSC optimize the processes performed in DMOs and control them using the selected criteria (key performance indicators). This improves the efficiency of utilizing public funds and optimises the performed activities.

BSC perspectives in tourism destination should be transformed to organisational goals, and for clarity drawn in the strategic map where the objectives will be linked based on the principle of cause–effect (Fig. 8).

Individual goals must be measurable, time-bound with allocated responsibilities. Determining managerial perspectives and goals of the organisation helps to streamline the strategic planning, management and marketing activities. The various perspectives adapted to the specificities of a destination can lead to a better overview of the management activities and determine the long-term objectives.

The managerial innovations in the High Tatras should be focused more on internal cooperation within the destination. Nowadays there are four other DMOs operating in the region with rather overlapping tasks and competences.

3.4 The Impact of Innovations on Sustainability

As the innovations in mountain destinations should be introduced according to sustainable principles, the impact of innovations on sustainability was examined in the destination. The first introduced type of innovations were product innovations. As it was mentioned, their aim in the High Tatras was to extend the winter season by ensuring the snow guarantee, and to raise the quality of visitor experience by

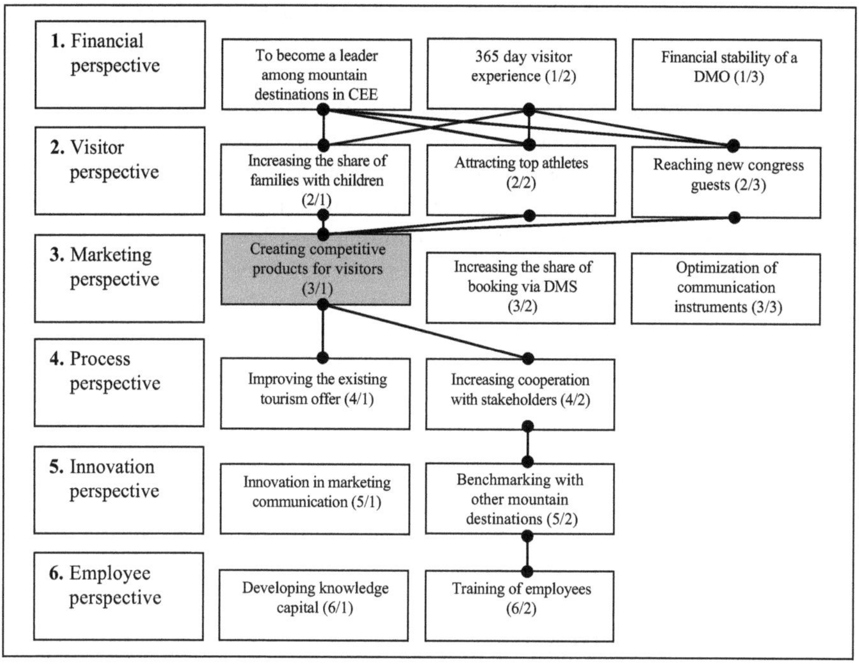

Fig. 8 Proposal of strategic map of BSC for the destination High Tatras

Quality level of skier	Speed of the skier (km/h)	Standard size of area (m^2)
Beginner	do 12,3	226
Average skier	12,3 – 21,5	226-520
Good skier	21,6 - 25,9	1174
Excellent skier	26,0 - 33,6	2000

Fig. 9 Standard sized area for different quality levels of skiers

shortening waiting times and thus increasing transport capacity and building transport facilities according to the latest energy efficient technologies, improving transport accessibility and parking capacity. As the new infrastructure is built, there is also a need to enlarge ski trails. This situation should take into account the impact on carrying capacity, environment, further destination development, as well as standard sized area for different quality levels of skiers (Fig. 9).

Nowadays the transport capacity in the High Tatras is very high, and in order to reach the comfort of Austrian mountain destinations, it would be required to build 126 km new ski slopes, which would have enormous negative effect on sustainable development.

Climate changes and snow uncertainty cause that technological product innovations in mountain destinations are mostly focused on artificial snow. Building of new or enlarging of existing snow making machines belongs to the most frequent product innovation. However, snow making has a significant negative impact on the environment. On the one hand it consumes a lot of water from retention lakes and there is also a necessity to incorporate the wiring and valves into the ground to produce artificial snow. In addition to large quantities of water the production of artificial snow consumes also electricity. On the other hand it contributes to extending the winter season, which has the impact on the economic sustainability.

The appropriate innovations in mountain destinations in lower altitudes should be aimed at the construction of ski park for children or ski school and related technical support (ropes, conveyor belts, children's ski lifts, etc.) This innovation does not require a high altitude, vertical drop or great snow cover. The impact on the environment is not so substantial.

Moreover the product innovations aiming at creating the favorable conditions for summer season include building bike trails. The positive aspect of this innovation is that it is a product for summer season, which is according to low altitude and climate changes considered as appropriate for mountain destinations. On the other hand the impact on soil erosion, appearance of the landscape and other impacts on the environment should be taken into account. Therefore the balance between environmental and economic sustainability of technological product innovations should be found.

Concerning the non-technological product innovations the positive impact on sustainability have those products that combine accommodation packages with free using of public transport in a destination. Although in the High Tatras in the summer season the public transport is included within the visitor card, the winter season does not have such product. However, this kind of innovations should be reasonably spread over seasons, and thus promoting the all-year-round utilization of tourism infrastructure. Such innovations are very common in foreign destinations, where transportation by mountain transport facilities, railways and buses is calculated within the price of accommodation. The product packages motivating visitors to use public transportation leads significantly to reducing the negative impact on the environment. Therefore these innovations should be introduced.

In the case of marketing innovations, the most frequently implemented are those connected with the internet, social media and smart-phones, which have no direct negative impact on the environment. It is necessary to take into account the number of visitors, the carrying capacity and the visitor management and thus use the information technology to influence the consumer behaviour according to the principles of sustainable development.

The impact of institutional, process and managerial innovations on sustainable development differs according to the nature of one specific implemented innovation and is focused more on economic and social pillar of sustainability. Hence, a negative influence on economic sustainable development is observed in 2009 when one strong business (TMR, Inc.) enter the market. The ownership of all core products in destination (transport facilities, important accommodation and hospitality facilities, and additional services) has a disputable influence on business environment and thus to sustainable destination development. Therefore the reengineering of the organisational structure in the High Tatras by creating the DMO has positive effect on sustainability. In destination management the role of

Tourism impacts		Share of respondents v %
Economic	Price inflation	44
	Increasing of employment	42
	Rising the standard of living	16
Environmental	Excessive construction	43
	Destruction of the environment	42
	Protection of nature	10
	High number of visitors	6
Social	Rising the awareness of destination	36
	Higher offer of leisure activities	28
	Privacy invasion	13
	Higher criminality	12

Fig. 10 Perception of tourism impacts by local inhabitants in the High Tatras

municipality and destination management organisation is crucial, because they can control the behaviour of a strong private business. This institutional innovation improved the relationships, cooperation and brought new opportunities for tourism development. According to the economic and social principle of sustainable development, it had a positive impact.

Based on the implemented innovations a survey among local residents was held focusing on the perception of economic, environmental and social impacts of tourism development (Fig. 10).

Among the economic effects of tourism, almost the half of the local population sees the price inflation. In the case of the environmental effects of tourism, the locals interfere with excessive construction connected with technological product innovations, which has also a huge effect on the disruption of the environment. Remarkable is the fact that the increased number of visitors interferes with only 6% of the population.

Residents of the High Tatras positively evaluated the raise of awareness of the destination and a wider range of leisure activities (more sport and cultural services). Nevertheless, 13% of citizens feel that tourism invades their privacy and 12% of the population met with higher crime as a result of the tourism development.

A relationship of innovation types and sustainable development is detected. According to the case study in High Tatras the authors observe that the biggest negative impact on sustainable development is generated by product innovations, followed by marketing, institutional, process and managerial innovation (Fig. 11).

Moreover, the case study in the High Tatras highlighted a development and implementation of innovations according to the stages of destination life cycle. The assumption, that innovations should be implemented in the stage of decline to ensure the rejuvenation, is out of date. Based on our experiences and definition of innovations in mountain destinations, it is concluded that different types of innovation should be implemented in different phases of destination life cycle. Moreover it is demonstrated that the implemented innovations depend on the life cycle phase and have different impact on sustainability.

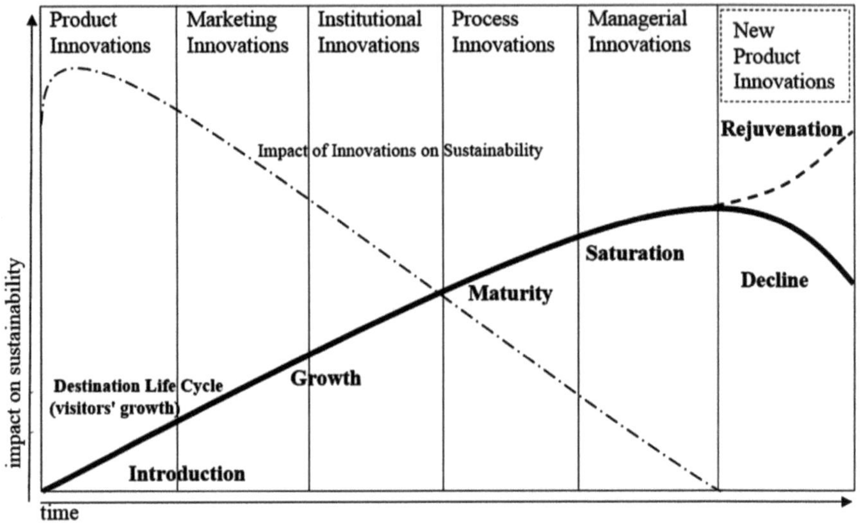

Fig. 11 Implemented types of innovations concerning destination life cycle

4 Conclusions

The aim of the innovation process at the beginning of "new era" of the High Tatras tourism development, considered as rejuvenation stage pushed by the new investor in 2009, was the rebuilding and modernisation of the mountain transport facilities, extension of ski slopes, ensuring the snow certainty with new artificial snow systems, improving the transport availability and parking possibilities. After completion of tourism infrastructure according to new trends, the accommodation, hospitality and après ski facilities were innovated. Afterwards, marketing, mainly the marketing communication, innovations were implemented. The increase in the number of visitors and destination popularity led to institutional, process and managerial innovations aimed at establishing new destination management organisation.

Mountain destinations are located in natural environment, which has the character of protected areas. Moreover, tourism in this area is mostly one of the most important income source for local inhabitants. All the implemented innovations should respect the principles of sustainable development in order to preserve the sources of tourism development for current and future generations. Various impacts of implemented innovations on sustainable development are observed in the paper, but it should be noted, that during the stages of destination life cycle, the negative impact of innovations on mountain destination sustainability is decreasing. Of course, the development of destination and the implementation of innovations cannot remain stable. To prevent the decline phase it is important to improve the quality of existing products and services and to respond to the internal and external changes, opportunities, challenges or new market trends. It is recommended to implement Balanced Scorecard method to evaluate and control processes in destination and to take into account possible strategies for reducing the impact of climate change on mountain destinations (e.g. technology: maintenance of slope, creating slopes at higher altitude and new snowmaking technologies; innovations in management and marketing tools: production of alternative products which are not dependent on the large snow cover). There is also a possibility of using renewable energy in mountain destinations, because they are in a unique position to capture especially wind and solar energy.

The development of tourism as an important sector of economy in mountain destinations should not destroy the natural resources. In the effort to gain a competitive advantage it needs to focus more on the sustainable development in its own best interest. Tourism planning and strategies need to be verified and evaluated not only in terms of their economic impact, but must also be assessed from the environmental and social point of view, which is mainly the role of municipality and legislative. Therefore the answer to the question in the title of the chapter "Does sustainability matter?" is positive. All innovations in mountain destinations should respect the principles of sustainable development.

Acknowledgements The research was supported by the research project VEGA 1/0809/17 Reengineering of destination management organisations and good destination governance conformed to principles of sustainable development.

References

Aldebert B, Dang R, Longhi CH (2010) Innovation in the tourism industry: the case study of tourism. Tour Manag 31(5):1204–1121. https://doi.org/10.1016/j.tourman.2010.08.010

Buhalis D (2000) Marketing the competitive destination of the future. Tour Manag 21(1):97–116. https://doi.org/10.1016/S0261-5177(99)00095-3

Bürki RH, Abegg B (2003) Climate change-impacts on the tourism industry in mountain areas. Paper presented at the First International Conference on Climate Change and Tourism Djerba. http://www.tec-conseil.com. Accessed 25 Apr 2016

Butler RW (1980) The concept of a tourism area cycle of evolution: implication for management of resources. Can Geogr 24(1):5–12. https://doi.org/10.1111/j.1541-0064.1980.tb00970.x

Butler RW et al (2006) The tourism area life cycle: applications and modifications. Channel View Publications, Oxford

Camisón C, Monfrot-Mir VM (2012) Measuring innovation in tourism from the Schumpeterian and the dynamic-capabilities perspectives. Tour Manag 33(3):776–789. https://doi.org/10.1016/j.tourman.2011.08.012

Castellaci F (2008) Innovation and the competitiveness of industries: comparing the mainstream and the evolutionary approaches. Technol Forecast Soc Chang 75(7):984–1006. https://doi.org/10.1016/j.techfore.2007.09.002

Castle K (2004) Vibrant villages or ghost towns? Ski Area Manag 43(3):50

Crouch GI (2007) Modeling destination competitiveness: a survey and analysis of the impact of competitiveness attributes. Cooperative Research Centre for Sustainable Tourism, Queensland

Decelle X (2004) A conceptual and dynamic approach to innovation in tourism. http://www.oecd.org/dataoecd/55/28/34267921.pdf. Accessed 25 Apr 2016

Demiroglu OC, Kučerová J, Ozcelebi O (2015) Snow reliability and climate elasticity: case of Slovak ski resort. Tour Rev 70(1):1–15. https://doi.org/10.1108/TR-01-2014-0003

Europe (2020) A strategy for smart, sustainable and inclusive growth. www.europa.eu
Eydal G (2004) The development of a sustainability management system for ski areas. School of Resource and Environmental Management, Burnaby. www.summit.sfu.ca. Accessed 25 Apr 2016
Flagestad A, Hope CA (2001) Strategic success in winter sports destinations: a sustainable value creation perspective. Tour Manag 22(5):445–461. https://doi.org/10.1016/S0261-5177(01)00010-3
Flagestad A, Hope ChA (2012) Identification and prioritisation of stakeholders in a winter sports destination for measuring organisational performance. A Norwegian perspective. Paper presented at Conference on Changing Paradigms of Sustainable Mountain Tourism, Brixen, 2012
Hall C, Williams M (2008) Tourism and innovation. Routledge, Oxon
Hankinton G (2004) Relational network brands. J Vacat Mark 10(2):109–121. https://doi.org/10.1177/135676670401000202
Hjalager AM (2002) Repairing innovation defectiveness in tourism. Tour Manag 23(5):465–474. https://doi.org/10.1016/S0261-5177(02)00013-4
Hjalager AM (2010) A review of innovation research in tourism. Tour Manag 31(1):1–12. https://doi.org/10.1016/j.tourman.2009.08.012
Hudson S (2000) Snow business: a study of the international ski industry. Wellington House, London
Jafari et al (2003) Encyclopaedia of tourism. Routledge, London
Kämpf R, Weber K (2005) Erfolgsfaktoren im alpinen Tourismus. Staatssekretariat für Wirtschaft (Seco), Bern
Keller PF (2003) Conclusions of the conference on innovation and growth in tourism in Lugano. http://www.oecd.org/cfe/tourism/34267885.pdf. Accessed 25 Apr 2016
Keller PF (2012) Changing paradigm in sustainable mountain tourism: a critical analysis from a global perspective. Paper presented at Conference on Changing Paradigms of Sustainable Mountain Tourism, Brixen
Kučerová J (1999) Trvalo udržateľný rozvoj cestovného ruchu. Univerzita Mateja Bela, Banská Bystrica
Lewis J (2005) Sustainable alpine tourism: the British ski industry's role in developing sustainability in the French Alps. University of London, London
Longhi C, Keeble D (2000) High technology clusters and evolutionary trends in the 1990s. In: Keeble D, Wilkinson F (eds) High technology clusters, networking and collective learning in Europe. Ashgate, Aldershot, pp 21–56
Lynch P, Morrison A (2007) The role of networks. In: Michael E (ed) Micro clusters and networks: the growth of tourism. Elsevier, Oxford, pp 43–62
Matto TD, Scott D (2008) Sustainable ski resort principles. In: Gössling S et al (eds) Sustainable tourism futures. Perspectives on systems, restructuring and innovations. Routledge, New York
Mayer M (2009) Innovation as a success factor in tourism: empirical evidence from western Austrian cable-car companies. Erkunde 63(1):123–139. https://doi.org/10.3112/erdkunde.2009.02.02
Mihalič T (2000) Environmental management of a tourist destination. A factor of tourism competitiveness. Tour Manag 21(1):65–78. https://doi.org/10.1016/S0261-5177(99)00096-5
Moore SR (2005) Corporate social responsibility and stakeholder engagement: a case study of affordable housing in whistler. School of Resource and Environmental Management, Burnaby
OECD (2005) Oslo manual: guidelines for collecting and interpreting innovation data. OECD, Paris
Ottenbacher M, Shaw V, Lockwood A (2005) Innovation performance in chain and independent hotels. In: Peters M, Pikkemaat B (eds) Innovation in hospitality and tourism. The Haworth Hospitality Press, New York
Pechlaner H, Fischer E, Hammann E (2005) Leadership and innovation processes-development of products and services based on core competencies. In: Peters M, Pikkemaat B (eds) Innovation in hospitality and tourism. The Haworth Hospitality Press, New York

Peters M, Weiermair K (2002) Innovationen und Innovationsverhalten im Tourismus. In: Bieger T, Laesser C (eds) Schweizer Jahrbuch für Tourismus 2001/2002. St. Gallen, pp 157–178
Pikkemaat B (2007) The scope for innovation research in tourism. Paper presented at the Conference of International Association for Tourism Economics, Palma de Mallorca
Pompl W, Buder C (2006) Notwendigkeit, Probleme und Besonderheiten von Innovationen bei touristischen Dienstleistugen. In: Pikkemaat B et al (eds) Innovationen im Tourismus. Erich Schmidt Verlag, Berlin
Poot J (2000) Reflections on local and economy-wide effects of territorial competition. In: Batey P, Friedrich P (eds) Regional competition. Springer, Heidelberg, pp 205–231
Ritchie JRB, Crouch GI (2003) The competitive destination: a sustainability perspective. CABI International, Wallingford
Schendler A (2003) Applying the principles of industrial ecology to the guest-service sector. J Ind Ecol 11(1):127–138. https://doi.org/10.1162/108819803766729221
Scott D et al (2006) Climate change and the sustainability of ski-based tourism in eastern North America: a reassessment. J Sustain Tour 14(4):376–398. https://doi.org/10.2167/jost550.0
Todd SE, Williams PW (1996) From white to green: a proposed environmental management system framework for ski areas. J Sustain Tour 4(3):147–173. https://doi.org/10.1080/09669589608667265
UNWTO (1996) Agenda 21 for the travel and tourism industry: towards environmentally sustainable development. UNWTO, WTTC and Earth Council, London
Vanat L (2016) International report on snow and mountain tourism. http://www.vanat.ch/RM-world-report-2016-vanat.pdf. Accessed 25 Apr 2016
Walder B (2007) Tourismus – Management von Innovationen. Tectum Verlag, Marburg
WCED (1987) Our common future. [online]. United Nations. http://conspect.nl/pdf/Our_Common_Future-Brundtland_Report_1987.pdf. Accessed 25 Apr 2016
Weiermair K (2004) Product improvement or innovation: what is the key to success in tourism? https://www1.oecd.org/cfe/tourism/34267947.pdf. Accessed 25 Apr 2016
Zach F, Fesenmaier DR (2009) Innovation in tourism: the case of destination marketing organizations. E-Rev Tour Res 7(1):27–36

Zuzana Gajdošíková is research assistant at the Department of Tourism and Hospitality, Faculty of Economics, Matej Bel University. Her research focuses on the development of mountain destinations, principles of sustainable tourism development and innovations in tourism destinations. To date she has published on these topics in well-known tourism journals such as *Tourism Management Perspectives, Quaestiones Geographicae, Economic Review of Tourism, Czech Journal of Tourism, Journal of Competitiveness, Journal of Tourism, Hospitality and Commerce or Forum Scientiae Oeconomia.*

Tomáš Gajdošík is a research assistant at the Department of Tourism and Hospitality, Faculty of Economics, Matej Bel University, Banská Bystrica. In his research he focuses on destination management, governance and leadership, as well as tourism information systems and smart tourism destinations. He has published in *Tourism Management Perspectives, Quaestiones Geographicae, Czech Journal of Tourism, Economic Review of Tourism, Journal of Competitiveness* and others.

Vanda Maráková is associate professor and Vice-dean for International Relations of Faculty of Economics, Matej Bel University. In her research, she focuses on destination management and marketing, destination image and corporate social responsibility. She deals also with efficiency of marketing communication tools. She has published and contributed to several scientific monographs, conference proceedings and professional journals such as *Tourism Management Perspectives, E&M Economics and Management, Quaestiones Geographicae, Economic annals-XXI, Economic Review of Tourism, Forum Scientiae Oeconomia* and others.

Modeling Innovation and Sustainability in Tourism via Competitive Advantage and Collaboration: Building Smart Tourism Destination on Olkhon Island in Baikal Lake

Kamila Borseková, Anna Vaňová, and Katarína Vitálišová

Abstract The main aim of the chapter is to define the process of identification, creation, building and exploitation of the local competitive advantage based on internal resources and cooperation of local private and public subjects, networking and participation via creation of unique tourism local production system and its influence on sustainability and ecological awareness. On the example of being a pioneer in developing a tourism industry in a hidden corner of the world without any tourism experience, this chapter tells an inspirational story of building a tourism destination from inauspicious beginnings, almost no trust and zero support from local, regional and national authorities through the slow building of a tourism industry until a sustainable, competitive and environmentally aware tourist destination of the twenty-first century.

Keywords Competitive advantage • Internal resources • Uniqueness • Location • Cooperation

1 Introduction

Nowadays, in time of many opportunities to travel and to relax, the place, which wants to be successful on the market, has to contain some unique feature or ability. Just the unique place without added value is not enough for a demanding customer.

The Olkhon Island is called the jewel of Siberia, because of its beauty, unique climate with rare species of plants and animals, cultural, natural and historical heritage. However, just this preposition was not enough to make from the island the tourism centre visited by tourists from all over the world. The impulse for writing this chapter is own empirical experience within almost half-year research stay in Siberia related with solving of seventh FP project focused on functioning of local

K. Borseková (✉) • A. Vaňová • K. Vitálišová
Faculty of Economics, Matej Bel University, Banská Bystrica, Slovakia
e-mail: kamila.borsekova@umb.sk

A. Dias et al. (eds.), *Modeling Innovation Sustainability and Technologies*,
Springer Proceedings in Business and Economics,
https://doi.org/10.1007/978-3-319-67101-7_14

production systems. Nikita's Homestead in the Olkhon Island is one of the best examples in the field of naturally created local production systems operating in tourism in Russia.

The paper presents the results of the project APVV SK-BG-2013-0018 Functioning of the local production systems in the times of crisis, the project KEGA 007UMB – 4/2015 Marketing in regional and local development and the project of seventh Framework program FOLPSEC.

2 Competitive Advantage and Innovation in Smart Tourism Destination

The tourism destination (territory) composes of various entities as natural sources, cultural and historical sources, entrepreneurs and organizations of public and private sector provided services and goods consuming in the tourism industry (Patúš 2012). They include several specific features. According to UNWTO (2002), it is a physical space in which a visitor spends at least one overnight. It includes tourism products such as support services and attractions, and tourism resources within one day's return travel time. It has physical and administrative boundaries defining its management, images and perceptions defining its market competitiveness. Its effective development assumes to incorporate all relevant actors from the public, private and non-profit sector. All their activities should be controlled and managed from the quality, time and content point of view (Borseková and Petríková 2015). The key preposition to build successful tourism destination is an identification of competitive advantage and sustain it via innovation.

The common unit linking cooperation, innovation and territory is a cluster, and when a dominant orientation of all activities within the territory are in field of tourism, it is a tourism cluster or a tourism local production system. Tourism production system includes all economic activities that contribute to production and distribution of tourism products and services, i.e. products and services that generate tourist experiences; the social groups, cultural features, and physical elements that are incorporated into tourism products and services; and agencies for regulating the commercial behaviour and social externalities associated with such production and distribution (Britton 1991, p. 455–456). The tourism production system characterizes a high concentration of small and medium sized companies and their specialization in the tourism industry. The topic of tourism cluster is researched by many experts (inter alia Beni 2003; Ferreira 2003; Capone 2004). They agree that the tourism cluster is associated with the limited geographical area with interconnected active partners (e.g. companies, institutions, service's providers, suppliers, policy makers, universities, competitors etc.) interconnected in tourism activities. The cooperation among partners, excellent management of the network, and the coordination of the production chain can generate the original product—a tourist destination as a competitive advantage (Beni 2003; Ferreira 2003). The role of tourism cluster is to connect SMEs in cooperation resulted in establishing the unique tourism offer. By Soteriades (2012) it can be one of the best available tools in fostering tourism development. Clusters are vital for development of regional economy, increasing the productivity, performance, innovative capacity and local businesses' critical mass (Novelli et al. 2006).

The uniqueness and competitiveness of tourism offer in territories is created by combination of competitive advantage and innovation. The definition of competitive advantage in the tourism destination explains the approach of regional competitiveness and regional competitive advantage. Kitson et al. (2004, p. 993) considered, at the regional level, concepts of regional competitiveness and regional competitive advantage as very similar and in some aspects even as synonymous. Authors dealing with competitiveness at regional level (inter alia Martin 2002; Kitson et al. 2004; Gardiner 2003; Skokan 2004) agree that emphasize is on citizens in terms of quality of life and growth of living standards, or growth of regional product and regional employment. The concept of regional competitiveness has its origins in several theories—institutional economics, evolutionary economics and the theory of learning regions as well as in Porter's approach of strategic planning and the theory of competitive advantage. Understanding of the spatial dimension of competitiveness adds approach of a new economic geography (for detail information see for example Hudec 2007).

In the literature, it can be found several theoretical approaches to the competitive advantage on spatial level (Borseková et al. 2012). As two basic approaches can be considered market-oriented approach (Porter 1999; Kotler et al. 2005; Vaňová 2006; O'Leary 2000, etc.) and approach to competitive advantage based on resources (Grant 1991; Ulrich and Lake 1991; Powel 1992; Hall 1993; Pfeffer 1994; Barney 1997; West et al. 2010, etc.).

Despite several differences in these approaches, generally authors agree that competitive advantage is a certain higher value compared to the competition. Several authors, inter alia Ansoff (1965), Solomon et al. (2006) or Vaňová (2006) associate it with the term uniqueness, others agree that it is related to profit (Porter 1999; Barney 2002; Besanko et al. 2000 and others).

First, market orientated approach, focuses mainly on external factors—opportunities, threats and sectoral competition. In this approach, market and environment identify subjects that are non-effective or offer products for consumers not willing to pay required price.

Second approach, competitive advantage based on resources focuses on the resources of the territory and the competitive advantage is created through internal resources. Unique resources help to create a unique position in the market.

A compromise between these two approaches is an approach based on value networks (Barney 1997; Kotler and Armstrong 1992; Porter 1999; Solomon et al. 2006) combining modern approaches to spatial development, like marketing places (Kotler 1982; Hanuláková 2004; Vaňová 2006), strategic marketing planning (Vaňová 2006) and smart specialization. According to this approach, building a sustainable competitive advantage is based on the positioning of subjects in value networks (Hollensen 2010; Mcphee and Wheeler 2006). This approach to competitive advantage involves a balanced mix of modern and innovative approaches to spatial development based on an efficient and sustainable exploitation of internal resources, using opportunities and eliminating threats originating from the external environment by respecting the demand of the market. It can be explained as a smart approach aimed at building a sustainable competitive advantage for a spatial unit, with the goal of assuring not only smart growth but also smart development. Innovation is the crucial essence in this approach as innovation can act as a vital tool for building a competitive advantage and as a driver of development (Borseková et al. 2017). To sustain the competitive advantage in territory, it is necessary to innovate the territorial offer based on uniqueness. In other words, to sustain a competitive advantage means constantly beating the competition through "shifting the target" or through constant improvement and innovation.

The innovation should be seen as a source of higher competitive ability. It means also an incorporation of new knowledge about supply and demand into the production. In the tourism, there are several definitions of innovation, but, generally the following types of innovation in tourism sector appears to be useful—product or service innovation; process innovation; organisational and managerial innovation. Product or service innovations refer to changes directly observed by the customer and regarded as new. It can be a completely new product or service or new for a particular enterprise or destination. Process innovations are aimed at escalating efficiency, increasing productivity and flow (Hjalager 2010). Organisational innovations deal with new approaches to internal collaboration, motivating and empowering staff, building careers and compensating workers with extra pay and benefits (Ottenbacher and Gnoth 2005). Management innovations are aimed at new management and marketing approaches targeting changes in communication between service providers, customers, and building their positive relationship (Hankinton 2004).

We can conclude that the key to tourist destinations competitiveness and sustainable competitive advantage as well as attraction new or repeat customers, is innovation. Figure 1 illustrates the interconnections between innovation, competitive advantage, tourism and spatial development.

In the next part of the paper, we present a case study of Olkhon Island in Baikal Lake, building its competitive advantage in a smart way based on innovation, partnership and ecological awareness.

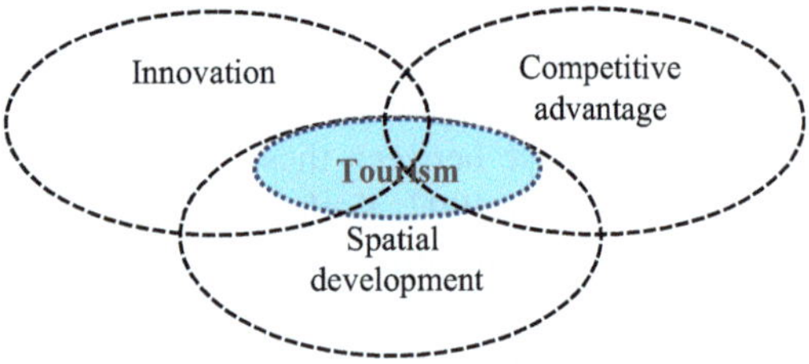

Fig. 1 Interconnection among innovation, competitive advantage, tourism and spatial development (Source: Borsekovó et al. 2017, p. 41)

3 Building of Smart Tourism Destination on Olkhon Island in Baikal Lake (Case Study)

The case study presents the results of own empirical research realised in Russia, Siberia. Empirical research and data collection were realised in 2014, in the form of structured and unstructured interviews with employees of Nikitas Homestad—the biggest tourist base in Olkhon Island in Baikal Lake and additional questionnaire with one of co-owners of Nikitas Homestead, which is a main contact point for Olkhon destination. The primary data are supplemented by the secondary data, e.g. website of tourism operators, tripadvisor.com, lonelyplanet.com, etc.

Olkhon Island is located in Lake Baikal, situated in south-east Siberia in Russian Federation and from 1995 enlisted on UNESCO World Heritage List. Lake Baikal is the oldest (25 million years) and deepest (1700 m) lake in the world. It contains 20% of the world's total freshwater reserve. Known as the 'Galapagos of Russia', its age and isolation have produced one of the world's richest and most unusual freshwater ecosystem, which is of exceptional value to evolutionary science (UNESCO, for more information see http://whc.unesco.org/en/list/754).

Olkhon Island, according to many resources, is the most amazing place in Lake Baikal area. Olkhon is the biggest island inside the Baikal Lake, only one inhabited, with an area of 730 km^2 and the fourth largest lake-bound island in the world. The island is 72 km long and 15 km wide with population around 1500 people, consists mostly from aboriginal inhabitants—Buryats. The most of island population (around 1200 people) live in the capital Khuzir, which is the main village of island. There are four other settlements in the island; most of inhabitants are fishers, farmers and cattle-ranchers. Due to an increasing number of tourists from all over the world, many residents work in this sector as well, and tourism has become an important part of the economy in Olkhon.

Olkhon is located approximately 8 hours drive (300 km) from the city Irkutsk, which is the important transport hub. Irkutsk is located on the Trans-Siberian route. From Irkutsk, you can reach two important destinations—by using Trans-Siberian route to Vladivostok or Ulan-Ude route through Ulaanbaatar to Beijing. Although Olkhon is located only 300 km from Irkutsk, the route shall take 8 hours, as from the half way there is dirty road only.

As a potential competitive advantage of the island can be identified the landscape, special climate conditions and relating fauna and flora on the island. Scientists still debate whether "Olkhon" translates as "little forest" or "dry" as both names fit well. The amount of precipitation is extremely low here, about 200 mm per year, which corresponds to biotope of semi-desert.

The second potential competitive advantage is Olkhon richness in archaeological landmarks. Olkhon is a geographical, historical and sacred centre of Baikal, the heart of many legends and fairy tales and it is believed to be the home of many Baikal spirits. Legends say that Khan Gutababai came here and was sent by high spirits from the Heavens. He was the head of all khans. His son Shubunkua still lives here as a white eagle. The indigenous Buryats, adherents of shamanism,

Fig. 2 Richness of Olkhon Island

believe the island to be a spiritual place; one of the groups of deities' adhered to in Buryati yellow shamanism is called the "thirteen lords of Olkhon". On the western coast, close to Khuzhir, is Baikal's most famous landmark, the Shamanka, or Shaman's Rock located on the Burkhan cape. Natives believe that Burkhan, a modern religious cult figure of the Altai peoples, lives in the cave in this rock. Burkhan Cape (Shaman Rock)—known as one of the palaces of Heaven's gods, tengrii, and one of nine Asia's most sacred places. This is the most famous of Olkhon's capes. Here, many testimonies of ancient people's lives were found. The strongest of heavenly tengriies (gods) chose Shaman Rock to be his home, and people were not allowed to go there. Many years ago, people used to cover hooves of their horses so as not to disturb the Great Spirit and master of the rock. Now people are not supposed to think negatively or behave badly here. Figure 2 shows the natural and cultural richness of Olkhon Island.

Nikita Bencharov, owner of Nikita's Homestead was in the role of strategic planner and manager of tourism development of an entire area. Gradually, he implemented his vision and strategy by necessary steps that significantly influenced the territorial development on the Olkhon Island in positive way.

Nikita's Homestead is located only 5 min walk from Burkhan cape and Shaman Rock and also runs the tourist information office outside its premises. Nikita's Homestead is family-run tourist base, offering wide range of tourist services, including accommodation, traditional homemade food and beverages, sport services, culture events, excursion and trips, bike rental, horse riding, tourist information services, souvenirs etc.

Nikita Bencharov was former very good and successful Russian representative in table tennis, and he is first person who started to work with tourists in Khuzhir. He is a pioneer of tourist industry and he used to say that Olkhon is the point where people can get rid of problems. When he moved to Khuzhir in 1989, nobody thought that it's a good opportunity to run tourist business there. After 25 years, it seems that his tourist base is the most famous among foreign tourists heading to Olkhon or even to Lake Baikal. One of the reasons is that Nikita's place has own, well processed web site including accommodation reservation system in four languages—Russian, English, German and French. Natalia, his wife, moved to Olkhon 18 years ago from Moscow. She worked at the Mars CIS Company as a financial manager. They both are very good in English, which is not usual in tourist destinations in Russia. Their tourist base looks like Siberian wooden estate with log cabins for guests, brick stoves, Russian banya and large dining room. They host foreigners as well as Russian tourists. They also invite foreign volunteers, who teach languages at local school, provide them with free accommodation and meals.

There are many internet resources with information about Olkhon Island. The most of them are mentioning also Nikita's Homestead, usually as the option number one in Olkhon Island, including resources as wikitravel.com,[1] lonelyplanet.com,[2] tripadvisor.com,[3] hostelworld.com,[4] waytorussia.net,[5] lakebaikaltravel.com,[6] grandbaikal.com,[7] baikalnature.com,[8] baikalex.com,[9] etc. We can conclude that all these web resources highly recommend this place to stay and reviews of tourists are mostly positive or very positive.

As an example, we can mention probably the most famous Lonely Planet tour guide. It includes seven facilities in Olkhon Island, Nikita's is recommended as a top choice. Trip advisor presents 51 facilities on Baikal Lake and Nikita's has the most reviews—overall 215 reviewers from all over the world, from more than 30 countries from Europe, Asia, Australia and New Zeeland, North, South and Middle America. Figure 3 shows ratings on TripAdvisor for Nikita's Homestead. It is the second best evaluated facility in the Baikal lake and has the certificate of excellence from Trip advisor.

Olkhon is visited by about 50,000 people every year (mostly in summer), most of them (80%) are just campers, the rest approx. 10,000 stay at the tour basis, guesthouses, yurt camps. Overall, there are 35 facilities in Olkhon Island that offer accommodation or services for tourists. The most of them are guesthouses or homesteads (usadba). Three facilities have character of hotel, few camp places and tourist cottages. Prices start at 600 rubbles per night (approx. 13 euros), the

---

[1] http://wikitravel.org/en/Olkhon

[2] http://www.lonelyplanet.com/russia/siberia/olkhon-island/hotels

[3] http://www.tripadvisor.com/Tourism-g304562-Olkhon_Irkutsk_Oblast_Siberian_District-Vacations.html

[4] http://www.hostelworld.com/hosteldetails.php/Nikita-s/Olkhon-Island/28961

[5] http://www.waytorussia.net/Baikal/Destinations/Olkhon.html

[6] http://www.lakebaikaltravel.com/trip_planner/tour_baikal/159-olkhon-island-guide.html

[7] http://www.grandbaikal.com/qa/1057.html

[8] http://www.baikalnature.com/hotels/57872

[9] http://www.baikalex.com/travel/olkhon.html

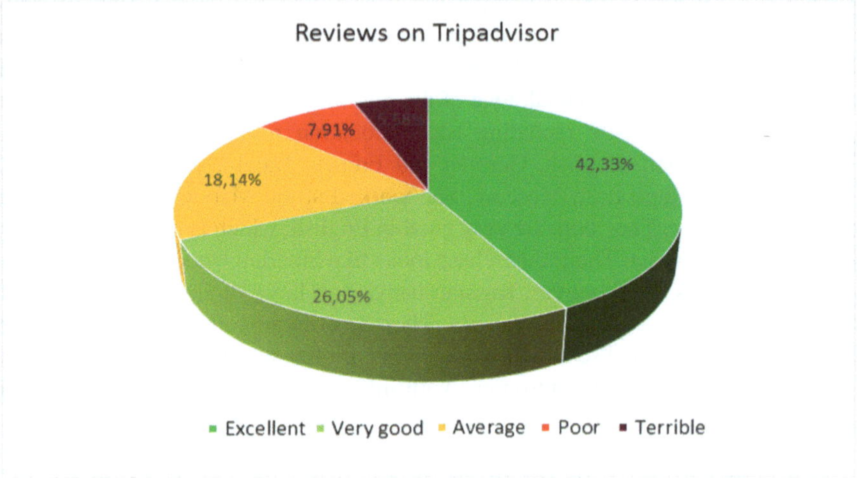

Fig. 3 Reviews of Nikita's Homestead on TripAdvisor (Source: https://www.tripadvisor.sk/Hotel_Review-g304562-d1151987-Reviews-Nikita_s_HomesteadOlkhon_Irkutsk_Oblast_Siberian_District.html)

most common price is around 1000 rubbles (21 euros), included meals in some cases. The most expensive facility in Olkhon is Baikal View Hotel where the price for one night is 3600 rubbles (77 euros) including breakfast. Prices in Nikita's Homestead are between 1000 (21 euros) rubbles and 1500 rubbles per night per person and including full board or half board—differences are influenced by the season. Price for tourist accommodated in neighbourhood is 950 rubbles (20 euros), including full board served in Nikita's Homestead. According to information from Natalia Bencharova, they are accommodating approx. 2500 tourist yearly (which is almost two times more that the whole island population). With the overall capacity 113 persons + tens of other accommodation possibilities in neighbourhood, Nikita's Homestead is the biggest tourist facility in the Olkhon Island and with the share of approximately 25% of overall accommodates tourists is a leader on the market. Booking in the high season (July and August) is needed to be done very much in advance, sometimes even year before. During the summer, the demand highly exceeds supply. Due to this reason owners start to cooperate with their neighbours in Khuzir village and in case of need they are able to accommodate tourists in neighbour houses and offer all services within their tourist base. Nikita's Homestead cooperates with 18 subjects in neighbourhood who provide accommodation for tourists and 22 local households that are food suppliers (milk, fish, meat, vegetables). Suppliers initiate this cooperation as they come to Nikita's Homestead and offer their services. According to owners of Nikita's Homestead, the cooperation with local suppliers helps them to cover the demand from tourists more effectively and local suppliers are thankful for the opportunity to work and have stable income at least during the summer season. During the peak season (July, August) there is 70 employees, the rest of the year requires 12 stable employees.

They also provide a tourist information office. Besides, it was established a local museum and cooperation with two bike rentals. On requirement, it is possible to provide horseback riding as well. They also cooperate with Buryat village with aim to show the way of life of aborigines people—Buryats and their religion—shamanism. Thanks to the boom of tourism and cooperation, the Olkhon Island was reconnected to the electricity grid in 2005 and mobile phones now work in Khuzhir.[10]

Owners of Nikita's Homestead highlight the level and importance of cooperation. All their business activities are based on building smooth and partner relationships in their team, with guests and locals. Their rule is that the guest is not just a person who paid money and left. The good relations resulted in various kinds of activities, e.g. humanitarian projects,[11] promotion of island Olkhon and Nikita Homestead, and cooperation in expeditions. In 1997 Nikita Homestead hosted Jacques-Yves Cousteau team in winter time; they shot a documentary about Baikal Lake. It helped (amongst others) the following on several occasions with their projects: ZDF television company (Germany), shootings for reality-show "Whisper of the stars", Channel 1 (Germany), with their set of documentary movies about Baikal, by Klaus Bernaz awarded in Germany as "The best documentary of the year". It provided accommodation for movies and magazine teams: panoramic shooting for movies Silk *(2007, Director—François Girard)* and Serko *(2006, Director—Joël Farges)*, photo sessions for ELLE magazine. In March 2010 Jim Denevan made his art work at Lake Baikal ice. In 2014 the German TV station ZDF recorded two series of document here. It also hosted different professional symposiums for physicists, psychologists, doctors, sociologists and others.

Besides being pioneer in tourism industry, owners of Nikita's Homestead started to be active in the development of culture and ecological awareness. In 2001, with the initiation of owners of Nikita's Homestead, the Olkhon children's organization 'Berkut—A New Baikal Generation' was established. As a part of the organization's 'New Generation' initiative, several artistic and environmental projects were implemented. Organized clean-ups on the island have been taking place for many years now, alongside related regional competitions and hiking trips. In 2004, a grant was awarded to fund "Clean Olkhon", a project that focused on clearing up the rubbish dumps surrounding the village of Khuzhir. A significant number of local people got involved in this project, especially youngsters and teachers from the village's high school. Together, they took part in the litter pick on the island, whilst also helping to organize games, competitions and hiking trips, all with the goal of drawing attention to the wider problem of waste management on the island. Later, the group "New Generation" became a partner in the "Clean Baikal" initiative, a project that was successful in installing litterbins across the island. On the ferry that connects Olkhon with the mainland, a ridgepole was installed where the winners of the children's poster competition "We want to live pure" had their works exhibited.

[10]http://www.lonelyplanet.com/russia/siberia/olkhon-island#ixzz34Sb02GZW

[11]http://askbaikal.com/2011/05/love-story-olkhon-island-and-nikitanatalia-bencharovs/

The aim being to draw tourists' attention to the problem of garbage collection in remote areas.

The project succeeded in removing the rubbish dumps, whilst a separate collection of waste glass was also successfully organized. During first year, around ten lorry-loads of refuse left the island, whilst seven lorry-loads of glass bottles were recycled. The project was developed and supported by the local administration. Until these days, this cooperation persists in the form of activities with the national park and volunteers to carry out clean-up operations on the island. By considering the enormous importance of Lake Baikal as the biggest reservoir of drinking water in the world, these activities are extremely important and smart.

Nikita's Homestead was over time active in organizing cultural events for guests and local community. Over the years many artistic events including theatres, concerts and art exhibitions were organized. In addition, many trips for local children to travel abroad, for example to China, France, Germany, Poland and Mongolia were organized as well. In 2012, in partnership with the district administration, Nikita's Homestead opened up a new music academy in Khuzhir. In the last few years, the music scholars have won numerous awards, including the regional piano duet competition "Olkhon Lyra" twice. Children from the academy also took second and third place in the recent 2014 regional competition for young composers.[12]

4 The Impact of Local Production System with Dominant Subject on Building a Smart and Sustainable Tourism Destination

In the previous case study, we analysed unique and smart local production system—Nikita's Homestead. Local production system should be defined as territorial union or network of different subjects or actors, whose focus their effort on interconnected activities and have a common goal or set of goals. Cooperation and interaction among subjects within an LPS play a critical role in the development and competitiveness of an LPS as a whole (Garcia 2006; La Rovere 2003). As the most common type of LPS should be regarded a cluster. There are several different definitions of clusters, but the most influential is undoubtedly that of Michael Porter. Porter (1998) defines a cluster as a geographic concentration of interconnected companies, specialized suppliers, service providers, associated institutions and firms in related industries. The subjects networked with each other and operating in close proximity are seen to have a competitive advantage. Following Fig. 4 shows Nikita's Homestead as a local production system.

We analysed internal and external environment, market and competition and we can define competitive advantage of Nikita's Homestead. The local competitive advantage is based on the very unique location, close to one of the nine most sacred

[12]http://www.olkhon.info/en/news/social-initiatives.html

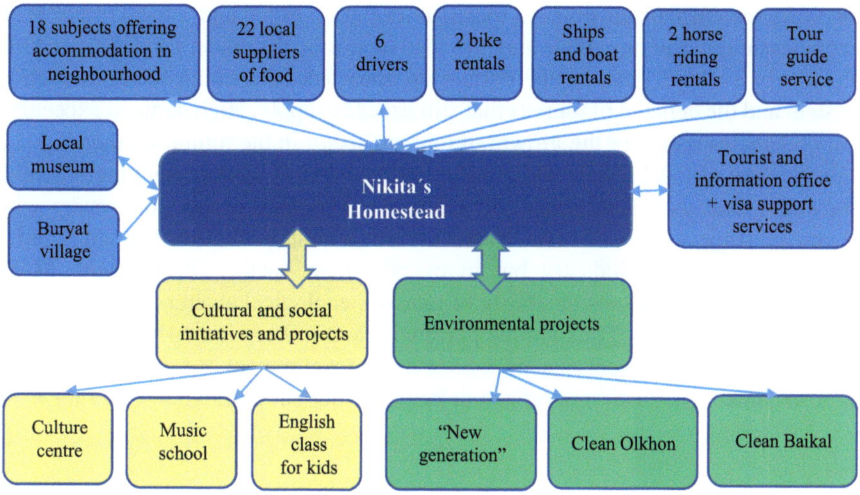

Fig. 4 Local production system Nikita's Homestead

places in Asia, and high level of developed cooperation. Even Nikita's Homestead covering only small local area in the Baikal region, we can say, based on the theoretical and empirical analysis, that Nikita's Homestead has not only local but also regional competitive advantage. The factors that influence shift from local to regional competitive advantage can be divided into three categories: natural, obtained and market.

The natural factors are especially a unique location, a unique natural environment of Olkhon Island, an interesting history of the Olkhon Island and religious, a cultural and spiritual richness of the place. The obtained competitive advantage factors are a great image of the Nikita's Homestead among the tourists from all over the world confirmed also by the most prestigious tour guides and tourist information web sites, well developed cooperation with local suppliers that resulted into formation of the unique local production system.

The market factors are still increasing number of tourists coming to Olkhon, demand that exceeds supply, new trends in travelling including spiritual trips and finding inner peace of people. At last, but not the least, we have to highlight smart approach to building the tourism industry in Olkhon Island with high level of ecological awareness and sense for sustainability and preservation this unique heritage for future generations.

5 Conclusions

According to the theoretical knowledge and empirical research results, we can define Nikita's Homestead as an effective and smart functioning local production system. Local production system includes several local suppliers and subjects and two own established subjects—a tourist information office and a local museum. It includes unique Buryat village and provides support in the matter of visa procedure. Positively evaluated can be the bottom-up approach of creating local production system and the demand of tourists that still exceeds supply. High demand exceeding supply creates space for the growth of tourism sector in the future. Partnership and involvement of local people and relevant stakeholders is an important ingredient in the smart and sustainable tourist development. The local production system with dominant subject—Nikita's Homestead became an important development impulse not only in tourism industry but in overall smart spatial development. Much emphasis is placed on sustainability of tourism, ecological awareness and respecting roots and culture of this unique place with aim to keep it unaffected for future generations. In the current time of mass tourism development, we should consider it as a sustainable and smart approach in building tourism destination of a future.

References

Ansoff HI (1965) Checklist for competitive and comptetence profiles. McGraw-Hill, New York

Barney JB (1997) Gaining and sustaining competitive advantage. Addison-Wesley, Ontario

Barney JB (2002) Gaining and sustaining competitive advantage. In: Rumelt R (ed) (2003) What in the world is competitive advantage? 5 p

Beni M (2003) Globalizaçâo do Turismo; Megaicndências do Sector e a Realidade Bra.sileira (Tourism globalisation: industry megatrends and Ihe Brazilian realily). Editora Aleph, Sao Paulo

Besanko D, Dranove D, Shanley M. (2000) Economics of strategy. In: Rumelt R (ed) (2003) What in the world is competitive advantage? 5 p

Borseková K, Petríková K (2015) Tourism local production systems in the Slovak Republic. In: Novoselev AS, Selivertsov VE (eds) Local production systems: analysis and forecasting of regional economic development. IEIE SB RAS, Russia, p 362

Borseková K, Petríková K, Vaňová A (2012) The methodology of use and building competitive advantage on the regional level. J Secur Sustain Issues 2(1):41–50

Borseková K, Vaňová A, Vitálišová K (2017) Smart specialization for smart spatial development: innovative strategies for building competitive advantages in tourism in Slovakia. Socio Econ Plan Sci 58(2017):39–50

Britton S (1991) Tourism, capital and place: towards a critical geography of tourism. Environ Plan D: Soc Space 9:452–475

Capone F (2004) Regional competitiveness in tourism local systems, 44° European Congress of the ERSA. Regions and Fiscal Federalism, Universidad de Porto

Ferreira M (2003) Estrategia e Planeamento Regional do Turismo (Strategy and regional planning of tourism). Investigaçâo em Turismo – Livro de Acias, pp 67–82

Garcia R (2006) External economies and competitive advantages in local productive system: the Marshall, Krugman and Porter's views. Ensaios FEE 27(2):301–324

Gardiner B (2003) Regional competitiveness indicators for Europe – audit, database construction and analysis. 21 p

Grant RM (1991) The resource-based theory of competitive advantage: implications for strategy formulation. Calif Manag Rev 33(3):114–135

Hall R (1993) A framework linking intangible resources and capabilities to sustainable competitive advantage. Strateg Manag J 14(8):607–618

Hankinton G (2004) Relational network brands. J Vacat Mark 10(2):109–121

Hanuláková E (2004) Marketing places. Ekonóm, Bratislava
Hjalager AM (2010) A review of innovation research in tourism. Tour Manag 31(2010):1–12
Hollensen S (2010) Marketing management: a relationship approach. PEL, Edinburgh
Hudec O (2007) Regionálne inovačné systémy – Strategické plánovanie a prognózovanie. Ekonomická fakulta TU, Košice, p 200
Kitson M, Martin R, Tyler P (2004) Regional competitiveness: an elusive yet key concept? Reg Stud 38(9):991–999
Kotler P (1982) Marketing for non-profit organizations. Prentice-Hall, Englewood Cliffs, NJ
Kotler P et al (2005) Moderní marketing. Grada Publishing, Praha, p 1041
Kotler P, Armstrong G (1992) Marketing. SPN, Bratislava
La Rovere RL (2003) Competitive strategies in systems of micro, small and medium companies: the importance of the information management. In: Small company: cooperation and local development. Relume Dumará, Rio de Janeiro, RJ
Martin R (2002) A study on the factors of regional competitiveness. The European Commision
Mcphee W, Wheeler D (2006) Making the case for the added-value chain. Strateg Leadership 34 (4):39–46
Novelli M, Schmity B, Spencer T (2006) Networks, clusters and innovation in tourism: a UK experience. Tour Manag 27(6):1141–1152
O'Leary N (2000) Unique selling proposition. Adweek Western Edition
Ottenbacher M, Gnoth J (2005) How to develop successful hospitality innovation. Cornell Hotels Restaur Admin Q 46(2):205–222
Patúš P (2012) Cieľové miesto ako systém. In: Gučík M et al (eds) Manažment cieľového miesta cestovného ruchu. Ekonomická fakulta UMB, Banská Bystrica, pp 58–68
Pfeffer J (1994) Competitive advantage through people: unleashing the power of the workforce. Harvard Business School Press, Boston
Porter M (1998) Competitive advantage of nations, 2nd edn. Macmillan Business, London
Porter ME (1999) Competitive advantage of nations. Free Press, New York
Powel TC (1992) Organizational alignment as competitive advantage. Strateg Manag J 13 (2):119–134
Skokan K (2004) Konkurenceschopnost, inovace a klastry v regionálním rozvoji. Repronis, Ostrava, p 160
Solomon RM, Marshall GW, Stuart EW (2006) Marketing: real people, real choices. Computer Press, Brno
Soteriades M (2012) Tourism destination marketing: approaches improving effectiveness and efficiency. J Hosp Tour Technol 3(2):1–23
Ulrich D, Lake D (1991) Organizational capability: creating competitive advantage. Acad Manag 5(1):77–92
UNWTO think-tank (2002) http://destination.unwto.org/content/conceptual-framework-0. Accessed 2 May 2016
Vaňová A (2006) Strategic marketing planning of territorial development. Faculty of Economics UMB, Banská Bystrica
West D, Ford J, Ibrahim E (2010) Strategic marketing: creating competitive advantage. Oxford University Press, New York, p 585

Kamila Borseková is a Head of the Research and Innovation Centre at the Faculty of Economics and Coordinator of Research at Matej Bel University in Banská Bystrica, Slovakia. She is currently the project manager in several international projects and a member of several research teams in domestic and international projects. She is author, or co-author of more than 50 scientific papers, chapters and studies. She is active in organization of scientific events (e.g. in cooperation with Regional Science Academy, organization of successful international workshop and Advanced Brainstorm Carrefour—Smart People in Smart Cities) as well as active attendance on scientific events (inter alia in London, Amsterdam, Rome, Lisbon, Porto, St. Petersburg). She was awarded by the Rector for excellent results in solving international research projects.

Anna Vaňová is a Vice-dean for Development and assoc. prof. at the Department of Public Economy and Regional Development, Faculty of Economics, Matej Bel University in Banská Bystrica, Slovakia. Her teaching and research activities are supplemented by international projects for American Express, Open Society Fund Bratislava, the British Know How Fund, the International Institute of Marketing, communications and business etc. She is active in expert consultancy, lecturing, reviewing, and evaluation activities.

Katarína Vitálišová is a Head of Department of Public Economics and Regional Development at the Faculty of Economics, Matej Bel University in Banská Bystrica, Slovakia. She is an author or co-author of more than 45 scientific papers. She actively participates in national and international scientific events (e.g. London, Rome, Malta, Bucharest, Porto, Bordeaux etc.) She has participated in several international and domestic projects.

Events and Places: What Strategies for Cities and Regions Marketing Choices? Remarks on Event Sector Development in the Post-Industrial City of Łódź (Poland)

Agnieszka Rzeńca and Mariusz E. Sokołowicz

Abstract Over the last decade the term "event marketing" has become both popular and ambiguous. It is used to describe small community initiatives gathering dozens of people as well as cyclic mass events with multimillion-dollar budgets. There is a variety of reasons why many people and organizations are using events as a marketing tool. For enterprises they are one of the instruments of corporate branding as well as ways of building relations with stakeholders. Public organizations perceive events as an instrument of building positive image among the public, as well as a way to convince people of certain social and political ideas. From the participants' (users) point of view, events are more and more alternatives for spending free time or building a culture of participation in social life.

Thus, a complex and multifaceted relation between an event and the place where it is held can be a very interesting subject to analyze. To what extent a given territory is a source of the event's specificity? How much do organizing events (especially mega-events) and event marketing contribute positively to local economic development? What are the costs and benefits of organizing events not only for its organizers but also for the territory and the community involved? Is it more effective to organize huge mega-events or rather smaller projects prove to be more beneficial and efficient from the place-marketing point of view? These questions arise more and more often among local policy makers, MICE sector representatives, local activists, and so on. This chapter, by presenting the development potential of event hosting in the post-industrial city of Łódź, is an attempt to answer some of them in this particular local context.

Keywords Local economic development • Marketing events • Tourism marketing mix

A. Rzeńca (✉) • M.E. Sokołowicz
Faculty of Economics and Sociology, University of Łódź, Łódź, Poland
e-mail: agnieszka.rzenca@uni.lodz.pl; mariusz.sokolowicz@uni.lodz.pl

A. Dias et al. (eds.), *Modeling Innovation Sustainability and Technologies*, Springer Proceedings in Business and Economics,
https://doi.org/10.1007/978-3-319-67101-7_15

1 Event Marketing as an Element of Place Marketing

Event marketing can be defined as an attempt to coordinate communication around a planned or self-created event gathering target groups in a particular place and time as a result of information transmitted (Eckerstein 2002: 13). It is also perceived as a promotional tool aiming at organizing various events, in order to achieve the organization's objectives vis-à-vis its environment: clients, decision-makers, collaborators, opinion leaders or other stakeholders (Kolber 2016). According to this, we can follow Getz (2008: 404) and argue that planned events are spatial-temporal phenomena and each of them is unique because of interactions among the setting, people, and management systems—including design elements and the program.

Despite the fact that the so called MICE sector (meetings, incentives, conferences, and exhibitions) has almost become a separate branch of the economy, the category of events is much more broader and covers different types of occurrences, such as: 1. Cultural events (festivals, carnivals, celebrations, ceremonies and galas), 2. Sports tournaments (World Cups, Olympic Games, recreational meetings), 3. Business occasions (the aforementioned MICE industry), 4. Art and entertainment events (concerts, festivals, galas, ceremonies), 5. Scientific and educational events (conferences, seminars, lectures and speeches), 6. Private activities (family gatherings, parties and social events), 7. Religious meetings (e.g. World Youth Day), and, last but not least, 8. Military events (performances, battle reconstructions, etc.) (see Getz 2008: 404).

From the marketing-mix point of view, event marketing is usually classified as a part of promotion (Fig. 1). Other marketing tools that go under this section are: advertising, sales promotion, personal sales, direct sales, public relations, and sponsoring. Thus, event marketing is not a substitute for any of the other components but rather a complementing element (Eckerstein 2002: 15). Although events are being created and organized according to the "promotion-mix rules" and use a wide range of promotional tools, they also refer to other marketing-mix elements, including product design and management (event as product marketing), pricing strategies (e.g. free events vs. admission fee events), people, processes, and, last but not least—the place and physical evidence.

Fig. 1 Event marketing in relation to marketing-mix and promotion-mix. Source: Own study

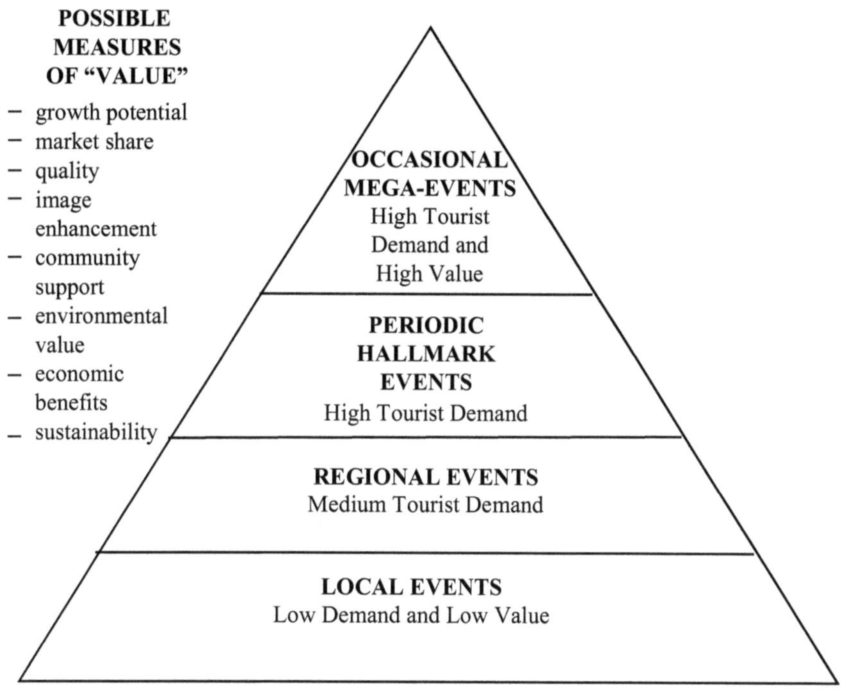

Fig. 2 The portfolio approach to event tourism strategy-making and evaluation. Source: Getz D. (2008) Event tourism: Definition, evolution, and research. Tourism Management 29(3): 407

From this perspective, we can classify events from the point of view of their size and scale, which is presented in Fig. 2. It presents a variety of event options, dependently on their scale and scope.

It seems to be obvious that one can observe a relatively large number of local and regional events, organized in many cities and towns around the world, whose impact upon both the place where they are organized and the audience is relatively moderate. They generate low or medium tourist demand, as well as "value", perceived in terms of such elements as pure economic benefits, but also in terms of growth potential or market share, impact on the community, environment, or sustainability. On the other hand, hallmark events or mega-events, usually occasional from the point of view of a particular location, generate high tourist demand and high value.

What can be interesting from the point of view of the places that host events, it is the level of their embeddedness. It is usually claimed that, in contrast to the location of a geographical destination, the location of events is not necessarily fixed (Leenders et al. 2015: 759). It concerns mainly large events (especially mega-events which, by their nature, are organized each time in a different location). However, also medium or small scale local or regional affairs can be the "wandering" events (Fig. 3).

	Large events	**Small events**
Locally specific (high level of embeddedness)	Rio Carnival Oktoberfest, Munich Mardi Gras, New Orleans	Burning Man, Nevada Armada de Rouen Urban festivals and celebrations of historical incidents Neighborhood days
Locally non-specific (low level of embeddedness)	Olympic Games World Cups World Youth Days EXPO fairs	G-8 summit Grand Slam TEDx events

Fig. 3 Combination of the scale and territorial embeddedness of events—examples. Source: Own study

Figure 3 shows how, both large scale and small scale events may be strongly or weakly "attached" to the place where they are organized. One cannot imagine such globally recognizable feasts and festivals like Rio Carnival or Oktoberfest being organized in New York or Budapest. Their transfer to another place in the world would probably deprive them of the element making them attractive on a global scale. At the same time, such mega events as Olympic Games, EXPO fairs, World Youth Days or Rolling Stones concerts would lose their attractiveness exactly when held every time at the same place.

On the other hand, relatively small or medium size events (e.g. Burning Man festival or Armada de Rouen in France), including many small neighborhood initiatives, do not even have any ambition to grow as the organizers aim to integrate mainly local communities. Bearing this in mind, the question whether large scale events or small scale affairs are more suitable for individual cities and regions remains open. It depends on a number of determinants being both "local" (budgetary constraints, civic engagement, scale of the city, geopolitical location, etc.) and "global" (domestic and international competition of other events, existing awareness of the city or town even is planned, ability to propose something new and attracting visitors, or even coincidence, which may decide on the growing recognition of a new event).

2 Which "Event Strategy" for Not Major Towns and Cities?

The history of mega-events dates back to the ancient times (Olympic Games). However, the époque of mega-events organized intentionally to promote cities hosting them, started in modern times with the nineteenth century World Fairs, known today as Expos. Such mega-events are large cultural, commercial, sporting events, having a dramatic character, mass popular appeal and international significance (Roche 2000: 1).

The expansion of these events was due to the convergence of tourism, sport and business in the late twentieth century, which is expressed through sponsorship rights, exclusive broadcasting rights, advertising space for sponsors and merchandizing. On the other hand, nations, regions and cities are increasingly interested in the marketing and image-related benefits of hosting large sports events (Skavronska 2015: 139).

Today, more and more cities and regions assume, that organizing or hosting existing mega-event is a good polity for boosting local economic development, improving place branding, and transforming urban spaces. In consequence, in the past three decades, mega-events have reached a size that has made them transformative ventures for entire cities, regions, and sometimes whole countries. Such events now routinely command more than US$10 billion in capital investment, take up several hundred acres of land, and require the infrastructure to accommodate, move, and keep secure hundreds of thousands of visitors and tens of thousands of athletes, officials, and journalists (Müller 2015: 6). However, as Müller (2015: 6–7) notices, also negative impacts of mega-events on cities and regions are well documented and occur in almost every case.[1]

This "mega-event" syndrome leads very often to overestimation of positive effects of mega-events, subordination of any public policies to the event priorities, inequitable distribution of resources or even suspension of regular rule of law (Müller 2015: 7). This twofold potential of mega-events leads to the question whether every city is a suitable place for organizing them. Interestingly, not only new insights are generated on what an event can do for a destination or regional appeal, but also research has started to answer the question on the importance of a location for the festival (Leenders et al. 2015: 756).

Among the most often indicated benefits of organizing large scale events one can mention an increase in employment, development of tourism, multiplier effects resulting from new investment, improved infrastructure and, consequently, economic restructuring. Also such arguments as the possibility of making iconic urban projects, improving the image of the place, or gaining new skills and competences, appear. On the other hand, there are many indications as to the disadvantages, such as unjustifiably high costs of organizing events, crowding-out effects caused by tourists, forcible relocation of people and businesses, or transfer of revenues outside of the territory. However, as Roche (2000: 10–11) points out, there is also a potential way of building some degree of mega-events localism. As most of these

[1]For example, after the 1994 World Cup in the United States, host cities experienced a net economic loss rather than the predicted gain (Baade and Matheson 2004). The Olympic Summer Games in Athens cost at least 3.4% of the gross domestic product (GDP) of Greece at the time and left a legacy of underused sports facilities and environmental destruction (Gold 2007). In Rio de Janeiro, preparations for the 2014 Football World Cup and the 2016 Olympic Games exacerbated socio-spatial polarization, as authorities evicted and resettled tens of thousands of residents (de Paula 2014).

Table 1 Large scale events vs. local and regional events—potential advantages and disadvantages

Advantages	Disadvantages
Large scale events	
– Potentially big revenues – Occasion and possibility of improving infrastructure – New workplaces – Skills development – Better international branding – Building pride of the city within citizens – Accelerating local business by spin-offs	– High costs of organization – Underused facilities after the event – Lack of territorial specificity of the event – Exclusion or expulsion of local community – Collapse of local businesses – Negative environmental footprint – Public takes risk for private benefits
Local and regional events	
– Relatively low costs of organization – Possibility of implementing "small innovation" solutions – Building community engagement – Better fix to local needs and aspirations – Building territorial specificity – Engaging local business	– Weaker recognition effect – Weaker chances for internationalization of a brand – Risk of the dispersion of efforts – Coordination costs – No guarantee of community involvement

Source: Own study

events are organized in multi-annual cycles, they are localized in space and time. In consequence, they have also distinctive "urban level" characteristics, providing a dedicated architecture, habitation areas, public functions and services, which produces an opportunity of exposing particularities of the place (Table 1).

In case of local and regional events, among the advantages one can mention relatively lower organization costs, bigger potential of territorial embeddedness and better fix to local needs, or better prospects for the local community and local businesses engagement. Potential disadvantages, on the other hand, are effects of the event scale, resulting in weaker recognition effect or weaker internationalization potential, as well as the risk of the dispersion of efforts or high coordination costs in case many local initiatives appear at the same time.

The final decision about organizing mega-events or hallmark events in a given city or a town, or replacing them with the strategy of activating local communities and economies by a set of local or regional affairs, should be always a consequence of comprehensive and reliable cost-benefit analysis. This analysis may not disregard local socio-economic conditions, as well as territorial specificity of the place. To ensure this, a reliable set of quantitative and qualitative data should be collected and processed using various research methods. Although this seems obvious, the collection of such data, due to the heterogeneity of available statistics or unavailability of data, makes this assumption a real challenge.

3 Can Post-Industrial City Be an Event City? Example of Łódź (Poland)

Łódź is one of the largest cities in Poland, third ranked by population, situated in the centre of Poland and Europe. Although its history dates back to the Middle Ages, its today's specificity is the outcome of economic processes of the nineteenth century, especially the development of craftsmen and workers' settlements. Unusual rate of growth of the textile industry that started in the 1820s within one hundred years transformed an agricultural town inhabited by merely several hundred people in the late eighteenth century into an industrial metropolis with several hundred thousand residents (ca. 500 k in 1915) (Dzieciuchowicz 2009: 113–115).

Spatial development processes that accompanied the dynamics of population increases have resulted in a specific arrangement of urban tissue with its physiognomical and architectonical divergences, where areas permeate and get mixed and where buildings that perform different functions neighbor one another: residential buildings—houses for rent and factory workers' houses (locally referred to a "famulas"), residences—villas and palaces of industrial tycoons, and industrial premises (weaving and spinning mills, warehouses, etc.). It was also the time, when harmonious urban assumptions in the spirit of Classicism were developed for craftsmen and industrial settlements, such as Nowe Miasto (New Town) or Osada Łódka (Łódka Settlement), together with factory-residential-housing settlements erected in the second half of the nineteenth century by, e.g., the Grohmans, I.K. Poznański's industrial quarter, or K. Scheibler's Księży Młyn (Priest's Mill) (Janiak 2016: 63).

The last decade of the twentieth century brought rapid social, politic, and economic transformations to the city. Markets for the so far produced textiles were lost, competition from cheap Asian manufacturers intensified producing deep changes in social and economic structures. However, despite physical destruction of a substantial portion of factory complexes and painful changes to the labor market, Łódź preserved the specific climate of a (post)industrial city. The threat of depopulation of the city centre and the loss of local identity created the need to look for new paths of social and economic development. One of these direction is an ambition to promote Łódź as the CENTRE OF THE CREATIVE INDUSTRIES claiming that "creative industries based on enterprise and creativity of its residents is what Łódź should be famous for in the future" (Łódź Brand Management Strategy for the Years 2010–2016 2009). The community of Łódź decided to select four general sub-products: economy, culture, education and tourism as elements responsible for territorial identity. In a more detailed way, economic offer of the city is mainly represented by trade and services, cultural activities connected with film, fashion, textile, and clothing industry, as well as by infrastructure for international fairs. Cultural sub-product has such specific features as the pre-war tradition of artistic avant-garde and modernity, supported by post-industrial heritage and design activities. Tourism in Łódź, according to the Authors, should concentrate on city-break type of guests, post-industrial infrastructure and sentiments of pre-war inhabitants and their descendants. Finally, educational sub-product, except the universal offer, has its specific value represented by art universities (Łódź Film School, Academy of Fine Arts) and a broad offer of design teaching programs (Sokołowicz 2013).

Łódź, which until recently was considered a dull and sad city associated predominantly with simple and poor workers' culture and specific but not very well identified traditions or smoke from factory chimneys nowadays attracts with spectacular regeneration effort and adaptation of monumental industrial architecture to new functions as well as with efficient municipal transport arrangement in the city centre. Industrial facilities used as office spaces, hotels, apartments or shopping and entertainment centers are being increasingly more seamlessly intertwined with the image of the city. These spaces represent a powerful potential for hosting various events.

On top of that, the city still carries traces of its multicultural past. In urban landscape it can be seen in the diversity of religious and lay architecture but also in regional cuisine, vocabulary and local design or avant-garde applied art.

Łódź as the former capital of film and Polish cinema and a vibrant academic center of artistic universities (Strzemiński Academy of Fine Arts, Leon Schiller National Film, Television and Theater School, Academy of Music) has also got a wide offer of a variety of cultural events representing many fields of art. The most prominent among them include: Łódź Design Festival, Light Move Festival, Łódź 4 Cultures Festival, and Łódź Ballet Festival. Sports facilities, among which we should mention three stadiums and Atlas Arena the biggest sports and entertainment venue in Poland, host top world class events.

Community living in this amazing and specific city experiences subsequent stages of evolution towards building the awareness of its potential; from ignorance to full recognition and identification of broadly understood cultural values and industrial heritage. New public spaces emerge to perform service functions (EC1, Nowe Centrum Miasta [New City Centre]) while those already unknown or underappreciated are intensely promoted, like, e.g., Off-Piotrkowska "the place where culture, art and creative business develop" considered one of the wonders of Poland. Recently, efforts are being undertaken to host in Łódź one of the major mega-events: EXPO 2022.

4 Łódź: From Industry to Event "Industry"?

Recent years witnessed dynamic changes in events organized in Łódź also when it comes to their diversity. However, we lack comprehensive and systematic studies that would discuss the growth of this sector of economy in all its aspects, that is why presented data are fragmentary. Using available statistical data coming from two sources: Poland Convention Bureau of Polish Tourist Organisation and Central Statistical Office of Poland (GUS) we conducted quantitative analysis. Different classifications of events have made the analysis more difficult. In its statistics GUS focuses exclusively on mass events (for which authorization is required). Nevertheless, using both of them has resulted in a complete picture of the situation

Table 2 Events classified in research programs conducted in Poland

Poland Convention Bureau of the Polish Tourist Organisation	Central Statistical Office
– Conferences/conventions – Corporate events – Motivational events fairs	– Art and entertainment events (e.g., concerts, performances and theater productions, shows, screenings, festivals, cabarets) – Sports events – Interdisciplinary events (combination of sports events with art and entertainment)

Source: Own study based on data from Poland Convention Bureau (2014, 2015, 2016) and Central Statistical Office of Poland (2015)

Table 3 Number of events in Polish cities

	2013		2014		2015	
City	No. of events	Share %	No. of events	Share %	No. of events	Share %
Gdańsk	4422	24.4	1676	8.9	3242	10
Kraków	3413	18.9	3986	21.2	5196	15
Poznań	1986	11	1706	9	2979	9
Wrocław	1393	7.7	2791	14.8	2729	8
Sopot	1062	5.9	–	–	–	–
Warszawa	644	3.6	3586	19.1	13,322	39
Kielce	632	3.5	659	3.5	687	2
Łódź	580	3.2	1241	6.6	975	3
Katowice	331	1.8	1576	8.4	1997	6
Bydgoszcz	444	2.5	779	4.1	1061	3

Source: Own study based on data from Poland Convention Bureau (2014, 2015, 2016)

with differences among the main cities (Table 2). Authors' own studies based on existing materials (range of offered events, calendar, reports, etc.) allowed to grasp and describe the specificity and nature of events in Łódź.

The above data clearly suggest that Łódź belongs to top ten key event cities, although the share of "Łódź events" in the domestic market is minor. Big fluctuations in the number of organized events are characteristic but negative phenomenon, especially when in the majority of cities included in the research we can observe their systematic increase (Table 3).

Central Statistical Office of Poland keeps records of mass events (Table 2). When it comes to mass, open events Łódź ranks in the top in Poland (it ranked ninth in 2013 and sixth in 2014) (Fig. 4).

Characteristically, in all the cities sports indoor events dominate the market (over 50% of all events). On the one hand, it proves how popular these events are and how easy it is to monitor the number of their participants. The fact that these events are organized everywhere in a similar format prevents us from identifying the main ones. The only differentiating factor is the importance of sports event and focusing on a particular discipline. Łódź specializes primarily in volleyball-related events organized in sports and entertainment venue Atlas Arena.

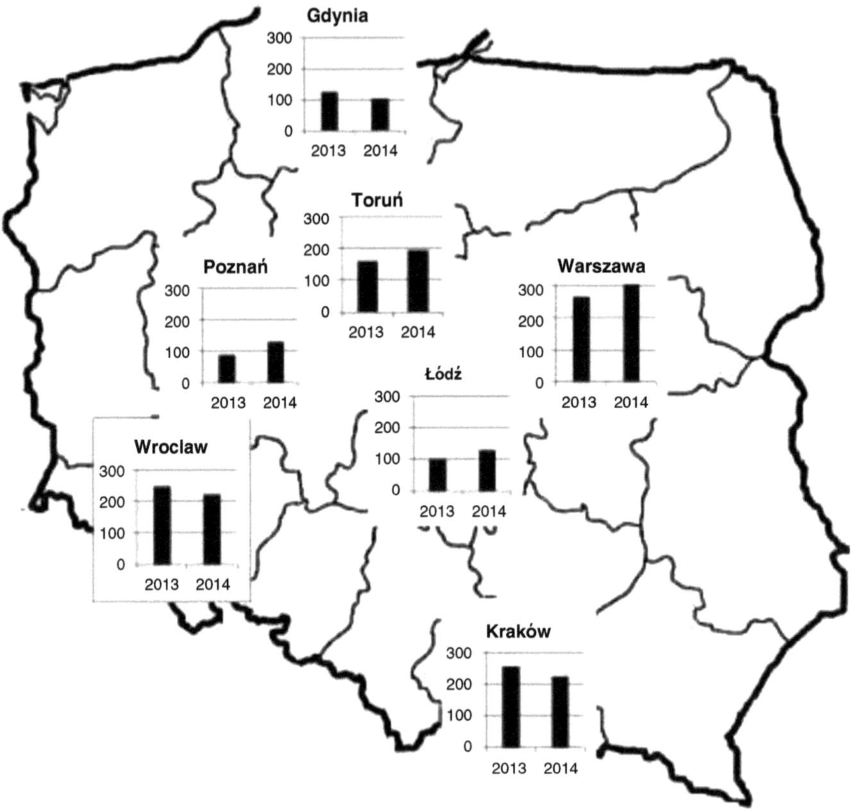

Fig. 4 Number of mass events organized in selected cities in Poland. Source: Own study based on data from Central Statistical Office of Poland (online database)

Analyzing the relationship between collected statistical evidence we may conclude that the convergence of data obtained from two different sources of statistical data is relatively strong as Pearson correlation coefficient for them is 0.6. Close relationships between various categories of organized events demonstrate the potential of the city and its well exploited opportunities to meet demand in the "event industry". It means, Łódź is one of the key cities that attract events of different nature with development predispositions in this area.

Statistical data, however, do not convey the climate of the city and events that shape its style. How the city is perceived is closely related with temporary events and images in public space (Dymnicka 2013: 146). Łódź is famous for events that have entered the calendar of key events and whose origins and potential derive from its endogenous resources, such as human capital, tradition and identity, strong institutions of culture, academic center, material structures, etc. (Table 4).

Industrial spaces, previously hermetic and closed up, through events re-enter the urban tissue and acquire a new profile, they become open and public (universally accessible). They become integral elements of the city and transform its

Table 4 Selected events connected with Łódź industrial heritage

Event connected with industry and industrial tradition	"Universal" event, which acquires specificity from being organized in postindustrial space
– *Łódź Design Festival*—makes references to the tradition of industrial design reflected nowadays in operations of the creative sector – *Łódź Triennale*—the oldest (1972), the largest and the most prestigious exhibition in the world that promotes modern art of fabrics – *International Contest for Young Clothes Designers Golden Thread*—makes references to the tradition of textile industry, clothes design, fashion, etc.	– *Light. Move. Festival.*—shows of light installations, illuminations in urban space and in post-industrial establishments – *Factory Owner Marathon*—sports event (marathon for various age groups) organized in post-industrial space (nowadays Special Economic Zone—Łódź Center), factory-residential and housing complex – *Manufaktura's Birthday*—festival in regenerated space and in industrial buildings today used for shopping, services and entertainment functions

Source: Own study

physiognomy and morphology. They are places where meetings are held and where activities are developed and concentrated also in the field of creative industries.

Among events that build up the city image we can find some inspired with experiences of other cities and/or modeled after events from the neighborhood. They develop dynamically in Łódź, which is confirmed by the scope of their impact and repeatability. Many festivals are totally new initiatives that have been "approved" by the city and are gaining in importance. The already mentioned Light Move Festival can be an example of a universal event that could be held anywhere but which acquires a new dimension in post-industrial spaces of Łódź. Within a short time, the festival has become the showcase of the city, an initiative that has fostered city identity, highlighted and enhanced advantages of the city and built relationships and emotions not only amongst its residents. By its "embeddedness" in the local community it has become its integral element. The same can be said about the International Festival of Comics and Games organized since 1991 as the biggest event of its type in Poland and in Central and Eastern Europe. City authorities, however, seek big international projects, in which Łódź could be present in a wider dimension and get recognition in Europe and in the world. Currently (2016), they have launched application process to host International Expo in 2022, i.e., the so called small Expo, which will surely stimulate the discussion about using mega-event as a driving engine for post-industrial city development.

5 Conclusion

Leisure time activities are getting commercialized and they are being transformed into orchestrated performances organized to entertain and please the participants but also to bring profit to the organizers. Hence, a playful approach with ethnic and cultural roots is increasingly often a mass phenomenon. Playful forms, once attributed only to the elite, such as, e.g., golf, polo or tennis remain such only in terms of finance not culture (Mokras-Grabowska 2015: 11). In the age of place marketing development for cities, events that stimulate emotions are the key. The counterbalance may come from place authenticity, sophisticated format of the event based on endogenous potential, such as the space, premises, human creativity, or tradition. Nevertheless, defining a city and building its potential based on architecture, even original and authentic, is not enough in dynamic times. There is a need for new resources, "energy" of impulses that could give new dimension to tradition and history, highlighting its relevance for the development. Thus, huge role should be played by public institutions, which co-develop cultural offer of the city, groups of performers, urban activists, and culture industry strongly connected with the creative sector.

In the case of Łódź, similarly to the textile industry, which in the past exerted an impact upon the distribution of other municipal functions, "event industry" may today importantly stimulate economic and social changes. From the point of view of post-industrial cities and their future development, sectors of the "event industry" should be profiled against the following criteria: target group, location and concentration of event space, potential range of impact, embeddedness of the event or residents engagement/involvement. Putting aside single, incidental, ad-hoc events and thinking in terms of brand creation we may identify stages of "event lifecycle" starting from "incubation", through early development, growth stage, necessary restructuring or modernization up to the decline stage. When we have many events and competition, these elements must be monitored.

Our research demonstrated that "event industry" sector is a new and little examined area from the perspective of city marketing. Statistics kept for a short time are incomplete while the multiplicity and divergence of events and their initiators produce the risk of dispersion of efforts and hamper synchronization and effective management. The key challenge is how to skillfully "manage event portfolio", which should be embedded in the specificity of the place. On top of that, a serious challenge for each city and region actively involved in marketing is to strike a proper balance between small local and regional events and mega-events. There is no single universal answer to the question.

References

Baade RA, Matheson V (2004) The quest for the cup: assessing the economic impact of the World Cup. Reg Stud 38(4):343–354

Central Statistical Office of Poland (2015) Kultura w 2014 r. Central Statistical Office, Warsaw. http://stat.gov.pl/download/gfx/portalinformacyjny/pl/defaultaktualnosci/5493/2/12/1/kultura_w_2014.pdf. Accessed 02 May 2016

de Paula M (2014) The 2014 World Cup in Brazil: its legacy and challenges. Heinrich-Böll-Stiftung, São Paulo

Dymnicka M (2013) Przestrzeń publiczna a przemiany miasta. Wydawnictwo Naukowe SCHOLAR, Warsaw

Dzieciuchowicz J (2009) Ludność Łodzi do 1918 roku. In: Liszewski S (ed) Łódź. Monografia miasta. Łódzkie Towarzystwo Naukowe, Łódź, pp 116–168

Eckerstein A (2002) Evaluation of event marketing. Important indicators to consider when evaluating event marketing. International Management Master Thesis No. 2002:25

Getz (2008) Event tourism: definition, evolution, and research. Tour Manag 29(3):403–428

Gold MM (2007) Athens 2004. In: Gold JR, Gold MM (eds) Olympic cities: city agendas, planning and the world's games. Routledge, Abingdon, pp 265–285

Janiak M (2016) The last fascinating and unusual and yet undiscovered city – Łódż. Renowacje i Zabytki 1(57):47–63

Kolber D (2016) Event marketing – organizacja imprez czy coś więcej? http://www.epr.pl/event-marketing-organizacja-imprez-czy-cos-wiecej,event-pr,1226,1.html. Accessed 27 Apr 2016

Leenders MAAM, Go FM, Bhansing PV (2015) The importance of the location in hosting a festival: a mapping approach. J Hosp Market Manag 24(7):754–769

Łódź Brand Management Strategy for the Years 2010–2016 (2009) http://www.kreatywna.lodz.pl/data/dataPublicator/lodz_strategia_zarzadzania_marka_eng_1.pdf. Accessed 16 Apr 2013

Mokras-Grabowska J (2015) Czas wolny w dobie postmodernizmu. Folia Turistica 34:11–30

Müller M (2015) The mega-event syndrome: why so much goes wrong in mega-event planning and what to do about it. J Am Plann Assoc 81(1):6–17

Poland Convention Bureau (2014) Poland meetings and events industry report. Poland Convention Bureau/Polish Tourist Organization, Warsaw

Poland Convention Bureau (2015) Poland meetings and events industry report. Poland Convention Bureau/Polish Tourist Organization, Warsaw

Poland Convention Bureau (2016) Poland meetings and events industry report. Poland Convention Bureau/Polish Tourist Organization, Warsaw

Roche M (2000) Mega-events and modernity: Olympics and expos in the growth of global culture. Routledge, London

Skavronska I (2015) Organization of sports mega-events as a tool of regional and urban transformation. In: Kociuba D, Sokołowicz ME (eds) Place-based approach to regional policy Polish, Slovakian and Ukrainian Youth Perspective. Selection of Proceedings. Monographs of ERSA Poland. Biblioteka, Łódź, pp 133–157

Sokołowicz ME (2013) Is the community of Łódź informed about living in creative city? City of Łódź' branding strategy and its perception. Studia Ekonomiczne. Zeszyty Naukowe Uniwersytetu Ekonomicznego w Katowicach 149:256–267

Agnieszka Rzeńca, Ph.D.—specialized in economics and environmental protection. Her scientific interest concerns environmental economics, especially in the context of sustainable development issues. Agnieszka Rzeńca research concentrates on contemporary challenges for urban development, such as resource-efficient urban management, revitalization, ecosystem-based management, eco-innovation, and environmental aspects of quality of life in the urban areas. Her expert activity concerns advising on strategic management in local self-government, resulting in numerous work and expertise for territorial units and functional areas.

Mariusz E. Sokołowicz, Ph.D.—born in Łódź, Associate Professor in the Institute of Spatial Economy at the Faculty of Economics and Sociology, University of Łódź. Author and editor of over 80 publications as well as expert reports, devoted to the issues of strategic management in local self-government and public sector institutions. Among his research interest one should mention place marketing, urban economics, and institutional economics. He also gained experience as a local government officer, responsible for revitalization of the city centre. The President of Board of Piotrkowska Street Foundation—an non-governmental organisation actively promoting city centre revitalization. He deepens his knowledge and experience during numerous internships abroad, i.e. to France, Georgia, Germany, Scotland, Portugal, Russia, Ukraine, Slovakia, and Bulgaria.

The Necessity for a Local Level of Gastronomic Tourism Standardization: The Case of Torino's City Branding

Henk J. de Vries, Frank M. Go, and Sophie A. Alpe

Abstract This study investigates the role and opportunities of gastronomy within the cultural re-branding of a city. Entrepreneurs and tourism authorities can emphasize the uniqueness of local cuisine but the common offer by different companies also requires alignment and standardization in order to define what they have in common, to assure product and service quality, and to communicate the service offer to potential customers. This creates a tension between standardization and authenticity: both concepts seem to contradict each other but these are needed in combination. This paradox is explored using the case of an alternative food network in the Italian city Torino. Our findings show the potential of this combined approach of authenticity and local standardization.

Keywords Tourism entrepreneurship • Cultural branding • Local gastronomy • Standardization

1 Introduction

Academic attention for urban tourism[1] pays little attention to the broader context in which tourism creates benefits as well as challenges for urban host communities. Apart from the benefits for companies as well as tourists, tourism brings problems

---

Earlier versions of this paper were presented at three conferences:
Standards: Boosting European Competitiveness of the European Academy for Standardisation EURAS (Brussels, Belgium, 2013); Changing Asian Consumption Patterns and Emerging Dilemmas: A New Research Agenda for the Next Practice in Tourism and Hospitality (Guangzhou, China, 2013); Modelling Innovation Sustainability and Technologies (Carcavelos, Portugal, 2015).

[1]E.g. the workshop entitled "Integrating City Tourism(s) into the Urban Research Agenda, in L' Aquila, Italy, 2015, organized as a follow up to the 3rd Global Summit on City Tourism organized by the United Nations' World Tourism Organization entitled "New Paradigms in City Tourism Development".

H.J. de Vries (✉) • F.M. Go • S.A. Alpe
Rotterdam School of Management, Erasmus University, Rotterdam, The Netherlands
e-mail: hvries@rsm.nl

A. Dias et al. (eds.), *Modeling Innovation Sustainability and Technologies*,
Springer Proceedings in Business and Economics,
https://doi.org/10.1007/978-3-319-67101-7_16

such as cultural commodification and gentrification. Such issues have led to growing discontent with tourism in many cities in Europe and beyond, despite the business advantages tourism brings. Paradoxically, sustainable differential advantages in the worldwide economy are often embedded in local assets, including knowledge, creativity, innovation and corporate-community collaborative initiatives. Another paradox is the emerging conviction that sustainable economic growth depends increasingly on the Social Progress Imperative for creating a shared language, gender equality and common goals. Public-private partnerships create opportunities to maximize social progress and minimize discontent. So the challenge is how to overcome such paradoxes and to preserve or even strengthen local authenticity while also benefiting from the efficiency and effectiveness of modern business approaches, and to do this in a cooperation between local stakeholders. Zeng et al. (2012) show that the paradox of authenticity and standardization can be bridged at company level, this paper explores how to do that at the level of a local community of companies and other stakeholders. We focus on developing common standards for locally produced authentic food to support gastronomic tourism, and take the case of the Italian city of Torino.

The paper is organized as follows. First, the literature review discusses the fields that need to get interconnected: community-based tourism, city branding, authenticity and standardization, and then highlights the combination of authenticity and standardization in alternative food networks. Next, the research methodology is presented. Then we present the empirical study on gastronomic tourism in Torino. In the last section we discuss our findings.

2 Literature

2.1 *Community-Based Tourism*

The past 50 years show expansion of transnational corporations. Olins (1999) suggests that TNCs and government entities have traded identities, i.e., they "are taking on each other's role". This is a difficult context for those who seek production and consumption at a small scale, and authenticity. Authentic geo-brands often must "break away from the hegemonic grasp of tour operators and the oligopoly of wealthy elites at the national level" (Timothy 2002, p. 150). This has spawned a community-based tourism approach, which advocates a form of tourism that seeks to leverage the contribution of a variety of tourist goods and services, including the engagement of residents toward community goals thereby warranting public support (Murphy 1985, p. 37). Community-based tourism uses cultural heritage assets to respond to the curiosity motives of recreational travellers. They can experience recreation in different ways, encompassing active physical and/or passive elements. Passive recreation tends to be dependent on the media (e.g. films, television, magazines) and life-entertainment (e.g. concerts, theatre and stage productions).

Active recreation increasingly involves travel for leisure purposes (e.g., learning a language, sport and related leisure activities, gastronomy).

A community is an independent group of people that share values and institutions (Brieger 2005) tied by mutual trust based on intense personal interactions (Hayami 1997, p. 241). However, the multitude of stakeholders and stakes may hinder common decision making in community-based tourism development (Orbasli 2000). Insiders (residents and local entrepreneurs and the public sector) and outsiders (international investors and business interests as well as tourists) may compete for scarce local resources to meet their own interests (Go and Trunfio 2011). The challenge is how to align these interests to preserve the authentic qualities over time. To achieve this, new forms of clustering and networked partnership are needed (Vaňová et al. 2010) such as Regional Innovation Ecosystems (Markula and Kune 2015). These broaden the role of enterprises, public sector, citizens and universities towards participative innovation processes and experiences, to yield multiple gains and insights in return. Place branding requires involvement of stakeholders, both as service providers in the branding process and as "brand ambassadors." Participation of local people is especially important because companies deliver the services and local residents impact the goodwill of the place brand, contributing to competitive positioning and stakeholder satisfaction (Freire 2009).

2.2 Rebranding Post-Industrial Cities

Tourists experience place encounters i.e., situations shaped by cultural and historical features, social and infrastructural contexts as well as the values and representations of political entities and projected images (Fan 2006; Govers et al. 2007). For instance, many cities with an industrial past are currently engaged in a process of repositioning their cultural identity and want to be perceived as creative by internal and external stakeholders (Vanolo 2008). The identities of post-industrial cities are often monolithic in nature: rooted in the success of a specific industry that tends to yield homogeneity in economic, political and social arenas (Vanolo 2008; Cannavale and Canestrino 2009). Rebranding such a post-industrial city is more difficult than its corporate variant (e.g. an airline brand). This challenges the local authorities to align the contributions of service providers to balance the assembly of components like authenticity and hospitality that contribute to create a new profile building on local values and intangible heritage.

Because the aim of an effective place brand is to help circulate its "products" within the market place, it is necessary that it remains within the relational context of functional similarity—standardization (Aronczyk 2008, p. 53). Therefore, the elements of the tourism supply need to be in line with the needs of consumers, with place branding efforts aimed at matching the demand and supply sides of the market (Go 2005). In most local areas the tourism sector consists of a heterogeneous supply of small and medium-sized enterprises, including transportation, attractions, restaurants and accommodation. However, the tourist demand is also heterogeneous as tourists have different incomes, tastes, interests and attitudes. Accordingly, to satisfy their needs, tourists are in search for a 'product constellation' (Go 2005), i.e. a group of relevant suppliers capable of bundling their resources in a wider network towards the joint creation of a value proposition which responds to the (mass-)customized needs of a particular tourist profile. In this sense, territorial units need to draw upon common associations of stakeholders (Fan 2006).

2.3 Authenticity

Due to the alleged alienation consumers suffer in modern society they are increasingly in search of authentic places or travel experiences (Zeng et al. 2012). Consequently, authenticity has become a powerful marketing and branding tool in tourism. Authenticity should be rooted in a concept that is not solely the outcome of a contemporaneous production line, but rather aligned to history and tradition (Taylor 2001). A territorial unit's identity is critical to the delivery of authentic services and raises a dilemma with regard to the direction of causality of place branding—does the identity of a city allow this city to claim a competitive position? Or does the place branding process influence the city to shape its competitive place brand identity? (Fan 2006). Go and Govers (2011) and Hankinson (2004) claim that organic (authentic) images for cities with a history of industrial decline require more than marketing communications, so that the assembly of a value proposition reflects the identity of the destination. Moreover, Askegaard and Kjeldgaard (2007) suggest that local authenticity does not always require traditional and historical connotations. They claim that place branding practices that draw upon local cultural capital offer potential to construct a local culture, which does not only refer to nostalgic attributes, but also to the future and modernity.

2.4 Standardization

In tourism, transnational corporations (TNCs) have levered standardization to establish and sustain consistent service quality and brand image, in an attempt to outperform competitors unable to do so (Go and Ritchie 1990). The uniformity brought about by McDonaldization (Ritzer 2012) is not appreciated by everybody, many mobile, knowledgeable and value-oriented international tourists would prefer more appealing alternatives (Weiermair and Steinhauser 2003). Martinez Badillo (2008) suggests to use standardization for common branding of the specific local touristic offer. This would be a combination of the national and global level of standardization—local to describe unique characteristics, international to specify an internet platform accessable all over the world to present this touristic offer to tourists.

In his seminal book *Standardization—A New Discipline*, Verman (1973) distinguishes five levels of standardization: the individual, company, association, national and international level. In this list, the local level is missing. Currently, the emphasis is on the international level—'standards facilitate co-ordination and co-operation on a global scale' (Brunsson and Jacobsson 2000, p. 1). Most national standard bodies advocate the importance of global standards. For example, the German standardization strategy emphasizes: 'German standardization has a consistent international agenda. (...) Germany is at the forefront in bringing future-oriented topics into standardization on a worldwide scale' (DIN 2016). Finally, Verman (1973) already signalled that associations had shifted their geo-organizational perspective from the national to the global level. The American Society for Testing and Materials, for instance, repositioned itself as ASTM International. And NGOs challenge governments and industries to set their standards in a way that encompasses the global level (Peters et al. 2009).

Though the local level was not included in Verman's (1973) dimensions of standardization these were common in Europe during the Middle Ages when local guilds set their own standards related to craftsmanship (Epstein 2008). Currently a movement can be observed which resembles the Middle Ages, wherein smaller territorial units retain an important role and render the concept of localization increasingly relevant (Mateo and Seisdedos 2010; Go and Trunfio 2011). This paper explores to what extent territorial units might apply a local level of standardization to enhance their attractiveness by developing identities and market these in a distinctive and sustainable manner (McKinsey Global Institute 2011) for competitive advantage and simultaneously attempt to mitigate human feelings of anxiety (Giddens 1991, pp. 38–39).

2.5 Alternative Food Networks

A form of combination of authenticity and standardization can already be observed in Alternative Food Networks (AFNs). AFNs share four main characteristics: a short distance between producers and consumers, small-sized farms, the existence of localized purchasing venues, and a commitment to the environmental, economic and social dimensions of production (Venn et al. 2006; Renting et al. 2003). They emerge as a consequence of changing consumer behaviour, with increased emphasis upon diversity and distinctiveness of food. Locally produced food offers potential to enhance a destination's attractiveness. An appropriate combination of localization and globalization, including local, national, regional and global standards for food, food production and food consumption, can contribute to the convergent development of an urban area, economically, socially and environmentally, and also provide a theme for destination branding (Rand and Heath 2006).

AFNs attach information and quality cues to products, allowing consumers to make a value judgment in relation to their knowledge of the production method and place (Renting et al. 2003; Venn et al. 2006). Transparent information in the form of certificates may give consumers justified confidence in producers' claims (de Vries et al., 2010). It provides consumers with value-laden information, delivering them the spatial-temporal link necessary to discern the indexical authenticity of a product as described by Grayson and Martinec (2004). Consumers are therefore more involved in AFNs, while in conventional food networks they largely remain passive (Morgan and Murdoch 2000).

Extended AFNs make use of authenticity to deliver satisfactory consumer experiences (Renting et al. 2003). Zeng et al. (2012) show that authenticity can be combined with standardization to commoditize traditional produce and expand market opportunities. Standards and certifications in AFNs allow for the differentiation of given products from conventional equivalents, based on quality, processes and place of production (Higgins et al. 2008). Arguably, standards become even more stringent for AFNs, as delivery of superior quality and authenticity is crucial to the survival of the network, under conditions of increasing regulatory and ethical boundaries (Morgan and Murdoch 2000).

Standards and certifications for AFNs render benefits for both producers and consumers but are also a cause of concern for AFNs. Smaller producers lack resources to meet standards or may have difficulties to obtain certification (de Vries et al. 2009; Higgins et al. 2008). Standardization has the potential to deplete an expression of local identity, a key driver of authenticity (Zeng et al. 2012). This occurs, for example, when AFNs give in to the demands of powerful value chain partners who apply standards and certification to raise competitiveness aimed at serving shareholder interests as opposed to a sustainable business approach which seeks to be accountable to stakeholder interests (Higgins et al. 2008).

2.6 Conclusions

In tourism, globalization and the resulting homogeneity have put into motion a counter-movement of localization and authenticity. Localization benefits from the economies of scale enabled by standardization. Standardization and authenticity seem to be conflicting concepts but Zeng et al. (2012) argue that entrepreneurs may use a combination of authenticity and standardization to meet customer wishes for authenticity in a cost-effective way and expand their business. However, they do not address the level of local host communities. First attempts can be observed to have Alternative Food Networks that combine authenticity and standardization at a local level. However, research insufficiently relates this to the broader context of city branding. This paper aims to fill this gap.

3 Methodology

The complex and contemporary nature of the subject justify explorative research suing a case study design (Yin 1994; Kitchin and Tate 2000). The Italian city of Torino was selected as a case, because its regeneration program seeks to improve its urban image and the cultural sector's capacity to raise tourism expenditures. Torino is the capital of the Piedmont region known for its authentic food—this combination seems to be an opportunity. A gastronomic tourism approach then would seek to pair the authenticity to local standards aimed at the preservation of local food traditions and knowledge transfer which contributes to social progress within the region and allows to better communicate the city's offer to external people.

We need empirical data and use triangulation to obtain these: in-depth interviews as the primary source of information, combined with desk research. Secondary data collection took place one month prior to the interviews, allowing for a greater degree of insight, related to the quality and completeness thereby reducing risks (Malhotra and Birks 2007). We contacted 27 members of Torino's gastronomy sector via email for participation. Interviews could be arranged with 12 of them, enough to reflect the heterogeneity of Torino's gastronomic tourism system. Included were the views of administrative bodies representing the local, provincial and economic dimensions, as well as representatives of the gastronomy network. All participants were provided with the same questions. Questions were first developed, screened and approved in English, and subsequently translated in Italian to limit information loss during the interview process. The questionnaire was pre-tested among three stakeholders. Three interviews were face to face, the other ones by phone. Furthermore, interviews were recorded, allowing for reliable transcripts from which categories and data reduction processes could follow (Maimbo and Pervan 2005). Interview time varied between 26 and 56 min (average 43 min). Verbatim transcript length varied between 2254 and 5689 words (average 3662 words).

4 Case Study: Torino's Slow Food Approach

4.1 Torino: Historical Ingredients for Creative City Branding

Torino is located within the Piedmont province, in the northwest of Italy, and is the country's fourth largest city (Camera di Commercio di Torino 2011; Crisci and Heins 2005). The city witnessed various "eras" (Fig. 1) each of which left their specific legacy. For instance, the Roman (military) era is reflected in Torino's urban layout: a squared blueprint with two primary perpendicular roads (Visita Torino 2011a). The Risorgimento era left the city with a style of architecture, arts, urban landscapes and rhetoric of the foregone monarchy manifest in the royal residences

	Until ca. 1400	ca. 1400 – 1600	ca. 1600 – 1899	ca. 1900 – 1995	ca. 1995 - present
The city's evolution and challenges	***Taurini*** – Celtic population migrates and inhabits the Po river. **Torino** (*Augusta Taurinorum*) is part of the roman empire – important communication centre. **Barbaric** invasions. **Feudal** state during the Middle Ages.	**Torino** is a Savoy city - administrative, political and diplomatic capital of the Savoy reign. **French** invasions. **Restitution** to Savoy reign.	**Torino** is prominent in Savoy reign. **Barocco** era. **Torino's** political representatives fundamental contributors to Italy's unification. **Italian** capital 1861-1865. **Loss** of role of capital – profound crisis: diminished political importance, population decrease.	**Torino** starts exploiting distilleries, textile factories, foundries to become an industrial hub. **FIAT** is founded – Torino automotive capital. **First and Second World War** - destruction of the city. **Post-war**: reconstruction, growing immigration (south to north), expansion of automotive sector.	**Automotive** engineering and electronic crisis. **Unemployment. First** branding plan 1966. **Workforce** diversification. **Pursuit** of creative/cultural city ideal – 2006 Winter Olympics. **Culture**-oriented city: emphasis on quality, history, sustainability and creativity – cultural commoditization.
Brand identity	Pre-Savoy City	Military Savoy City	Risorgimento City	Industrial City	Modern - Cultural City
European context: evolution and challenges		**Agricultural** revolution – efficiency. **Population** growth – migration towards cities. **Renaissance** and banking revolution **Global** discovery. **Expansion** of European culture – colonization.	**European** global dominance. **Global** industrial revolution. **Industrial** systems applied to agriculture: food abundance – population growth. **Cities** grow in population. **Armament** of nations.	**First and Second World War** – disorganization, vast migrations, agricultural damages. **Political** alliances – end of communism. **Wealth** society – literacy, life expectancy growth, mass media, growing disparity between rich and poor.	**Globalization** – ICT, innovation, mass media **Resource limits** – aging Europe, ecosystems, sustainability **Transition** – well-being, governance, leadership, public engagement.

Fig. 1 Torino brand genealogy. Sources: Visita Torino (2011a, b); Gambarotta et al. (2004)

or the renowned *Via Po'* (Visita Torino 2011b; Comoli 2012). Torino's industrial era leaves numerous unoccupied factories throughout the territory.

In addition, the city can be characterized by the notion of "high quality" derived from the historical presence of the royal family and the city's political importance: examples are the refinement of high quality fabrics and tobacco, followed by the engineering and automotive sectors (Palmucci and Visconti 2012).

Torino's recent history (1900 onwards) is linked to the automotive sector as headquarters of FIAT are in the city. Upon production peaks (1960–1970) the factory employed a third of the city's work force, and Torino became a "one company" manufacturing city (Huxley 2010). The 1980s–1990s automotive crisis led the city into a deep socio-economic crisis. Between 1990 and 2002, FIAT's Turin based production decreased from 60% to 30% of total production and resulted in a social as well as an economic crisis for Turin, and a 20% increase in the unemployment rate (Colantino et al. 2014). Torino was forced to engage in a process of deliberation regarding its future to overcome its manufacturing image (Huxley 2010; Verri 2005). In response, public sector initiatives included the launch in 1996 of the city's first branding plan, a campaign for workforce diversification, the staging of the 2006 Winter Olympics, and the pursuit of a modern creative city. Torino's industry and its reputation have since diversified, specializing in innovative and high-tech knowledge sectors, providing alternative paths to low-skilled employment (Huxley 2010).

4.2 Slow Food

Though Torino is the capital of the Piedmont province known for its Slow Food, this food tradition has been distinct from the city's rebranding efforts. The Slow Food association was founded in 1986 by Carlo Petrini in Bra, Piedmont, as *Arcigola*. It is considered an innovative and creative response to globalized food systems (Simonetti 2012). The aim of the Slow Food association is to re-introduce and safeguard the pleasures of authenticity and traditions of food, promoting flavour uniqueness and quality (Simonetti 2012). Furthermore, the association aims to re-educate consumers to the notion of *gusto* (taste), preserving agricultural biodiversity and promoting a new model of alimentation, respectful of the environment and national identities.

Slow Food aims to re-generate and protect traditional food and wine produce by recognizing those that meet the qualities of "*buono, pulito e giusto*" (good, clean and fair). These are included in a list called "*L'Arca del Gusto*" (The Ark of Taste). Listed products must be traditionally produced (authentic), tasting well (re-education of taste) and be compatible with an economically and environmentally sustainable system (fair trade). Through this selective process, Slow Food singularizes food produce emphasizing distinctiveness (Lotti 2010). For Slow Food to endorse and support a given product, the latter must adhere to a series of standards. Here meat production serves as example to explain the selection criteria:

- The animal must belong to an autochthonous breed and be raised with respect to the environment and animal life.
- Meat product derivatives must reflect the local gastronomy history as well as *good taste*.
- The products and animals (in this case) are considered endangered (Lotti 2010).

Adherence to such criteria ensures that more general standards of quality, authenticity and sustainability are met (Slow Food West Michigan 2012). Subsequently, the methods of production can be established: This process is conducted differently for each product Slow Food supports. The SF association considers the EU-wide standards that regulate food productions and hygiene as a source which endangers conservation of agro-biodiversity (Lotti 2010). Therefore, the stringent production standards that each participant in a Slow Food presidium must meet are designed in a collaborative process between producers and the SF association on an individual basis, further emphasizing their products' uniqueness. These standards allow variation for each product in order to respect the notion of uniqueness, respectfulness of traditions, tastiness, etc.

4.3 Analysis

Food is a pivotal aspect of any destination, being one of the four systemic aspects that compose tourism experiences (Gnoth 2002). According to consumer studies and travellers' ratings, Torino's gastronomy sector already scored (8.63) above Italy's national average (8.38) (Salamon 2008). Torino might take better integration of the famous nearby Slow Food in its gastronomy sector as a spearhead. Further emphasis may attract even more tourists and contributes to the preservation of Piedmont's traditions and gastronomic products (Lotti 2010). This helps to present Torino with an opportunity to position itself as the place to be for the sustainable production and consumption of authentic food.

So far, Torino's branding efforts did not result in a set of attributes to construct the city brand nor weights that show their relative importance (e.g. weights assigned to the role of gastronomy). The city lacks an agreed-upon set of functional attributes of authenticity. Laying down such attributes in a set of standards would allow to assign labels to companies and their processes, products and services, and this would create opportunities within the global competitive market to raise the value of tourist experiences. It would facilitate Torino's process of transforming from a post-industrial city to a creative and high-quality city.

Such a transition is not easy. Our interviews reveal a lack of effective interaction amongst SMEs in Torino. This hinders common place branding (Cannavale and Canestrino 2009; Go and Govers 2011; Huxley 2010). However, the mass media and social media harbour the potential for communicating and enforcing a local brand identity. So the time may have come for stakeholders in Torino to overcome their culture of self-made business and establish a partnership for city branding.

Quality is inherent to Torino's reputation stretching to a number of sectors that operate or have operated in the city:

- High quality textile factories—a legacy left by the royal family;
- A well-developed technological innovation sector (more than its expertise in car manufacturing—Torino's polytechnic university ranks first in Italy);
- A chocolatier sector, offering a high quality gastronomy product.

Therefore we think the notion of quality should be taken into account in the city's re-branding efforts and this may establish the bridge to the inclusion of Slow Food in the positioning of the city. Of course, this choice must be endorsed by the internal stakeholders so that a cohesive and organic image can emerge (Hankinson 2004; Go and Govers 2011), as opposed to relying primarily upon marketing and communication efforts (Govers 2012) to drive branding efforts.

To demonstrate quality of products and services, standards are needed: standards to be able to compare local performance with performance elsewhere, and standards to assure that the quality level is maintained and improved. Normally these would be international standards but in case the products and services as such are typically local, local standardization can establish and promote the level of quality. This could provide a suitable option for Torino's entrepreneurs to avoid the loss of the distinguishable aspects of their products and the weakening of reputation of both their firm and the city (Paniccia 2007).

The development of such local quality standards can build on the spirit of excellence manifest in Torino's tacit knowledge capabilities that are at the core of "indigenous" production (Paniccia 2007). This tacit knowledge should be codified in local standards which allow for a degree of flexibility for contextual differences as well as a transparency in the choice of methods. Then performance standards are better than standards that prescribe solutions (CEN/CENELEC 2010; De Vries 1998). Such standards help network partners to define criteria for their processes, products and services and these allow them to monitor quality assurance.

Torino has potential not only for a retro-marketing approach, but more importantly to advance insight to what extent the city has adopted a shared language and common goals and style. It should balance between serving the demands of the international community and preserving and presenting the memories and place associations from the past that the host city residents hold dear. Locally developed standards would aid in the preservation of the traditional business-base, pivotal to the development of a sustainable destination (Maheshawari et al. 2011).

Torino does not have to start from scratch. The guidelines developed by the Slow Food association regarding e.g., autochthonous breeds, and valorization of Piedmont's local produce have already yielded positive results for its agricultural economies and rural communities (Paniccia 2007; Maheshawari et al. 2011). The standards used for Slow Food's processes have proven relevance for thousands of local gastronomy entrepreneurs, administrative bodies and consumers around the world.

In the development of local standards, the local community may make use of international standards and guides. In particular ISO/IEC Guide 76:2008 'Development of service standards—Recommendations for addressing consumer issues' (COPOLCO 2008) can be of help: though designed with international standards in mind, this guide can also be used for the systematic development of local standards—not only for the food but also for the restaurants.

The combination of the city branding process with the standardization process links public sector initiatives with private and academic resources. It requires a combination of marketing knowledge, standardization knowledge, strategic thinking and organizational changes leading to a common ground for coordinated cooperation. The stakeholders should integrate technological, market and organizational change within a city brand based on the functional attribute of quality that spans across various sectors (Paniccia 2007). In order to enable such boundary spanning knowledge processes, Torino must transform towards networked collaborative partnerships upon which to define inter-network functional attributes contributing to collaborative behaviour, upgrading competitiveness (Berg et al. 2008).

However, the city faces a problem. Our findings suggest that individualistic behaviour and lack of ability to think at a strategic level hinders a collective approach to branding, and impedes delivery of a coherent gastronomic tourist experience. In another study Mertens (2011) also used standardization to explore the feasibility of a partnership of bars and restaurants under a city brand umbrella and reported similar problems: lack of strategic thinking and lack of cooperation attitude, leading to sub-optimal results at both the city and the firm level.

To conclude, Torino reinvents itself as a culturally branded unit (European Commission 2007; Gambarotta et al. 2004). However, this new Torino brand identity can be further strengthened. The city leadership should rely on the community in order to manage the integration of technological, market and organizational change to accomplish a publicly engaged regeneration strategy for the city and the surrounding region. Communities of practice get underway internally by valorising new resources. The leadership must therefore be institutionalized so that local partnerships feel empowered to respond promptly and effectively to changing consumer demands and resource pressures, and to overcome individualistic behaviour of single players. Institutionalization enhances the chance of long-term continuity in ways that are valued and rewarded by the community.

5 Discussion and Conclusions

The case of Torino shows the potential of standardization to support place branding. Place branding presupposes commonality and its essential characteristics may be incorporated in standards. These can be performance standards specifying minimum quality levels for process, product or service characteristics, test methods, and standards to communicate these characteristics to external parties such as tourists. In developing such local standards, international standards may be used, but the typical local situation will necessitate either additions or modifications. Cooperation is imperative to develop such local standards. In the ancient guilds, such cooperation was institutionalized. Some local communities may already have organizations that are prepared to develop such standards. However, the combination of knowledge and skills needed to develop and implement these standards may require a knowledge-transfer framework, supported by a dedicated team based in a dedicated center. Both formal education and professional education would be well served to learn more about standards which are omnipresent in society while professionals who serve the tourism sector need to learn how standards impact their work, career and profession (de Vries 2014).

In the European context one may question whether the emergence of local standards would hinder the single European market without barriers to trade (Czaya 2007). This could in particular be the case if local products are being exported. However, this issue is manageable, using the European Union's schemes of geographical indications and traditional specialties, known as Protected Designation of Origin (PDO), Protected Geographical Indication (PGI), and Traditional Specialities Guaranteed (TSG). Moreover, in the case of localization, services are being delivered nearby, so there is no export and thus probably no conflict between local standards for the service offering and international or European service standards. Local standards would support the economic and cultural development of cities and in this way support the development of Europe in the direction of 'unity in diversity' (Arts et al. 2003). This would be a welcome addition to the ongoing unification, which increasingly causes irritation among European citizens and undermines trust in the common European adventure. In a similar way, the opportunities of combining authenticity and standardisation for the sake of sustainability, cultural heritage and prosperity apply all over the world, as already exemplified in the studies on Chinese restaurants (Zeng et al. 2012) and tourism in developing countries (Marinez Barbillo 2008).

This work contributes, first, to the standardization literature, which tends to ignore the local level of standardization and with a few exceptions also fails to relate to the concept of authenticity; second, to the tourism and hospitality literature by linking city branding to standardization and addressing the authenticity versus standardization issue. Our findings have managerial relevance for local entrepreneurs and tourism authorities who are capable of envisioning the potential of applying a common branding standard. Moreover, our case also shows that thinking in such a strategic way can be a challenge.

References

Aronczyk M (2008) "Living the brand": nationality, globality and the identity strategies of nation branding consultants. Int J Commun 2:41–65

Arts W, Hagenaars J, Halman L (eds) (2003) The cultural diversity of European unity. Findings, explanations and reflections from the European values study. Brill, Leiden

Askegaard S, Kjeldgaard D (2007) Here, there, and everywhere: place branding and gastronmical globalization in a macromarketing perspective. J Macromark 27(2):138–147

Brieger WR (2005) Definitions of Community. Johns Hopkins University, School of Public Health, Baltimore, MD. (PDF presentation)

Brunsson N, Jacobsson B (2000) The comtemporary expansion of standardization. In: Brunsson N, Jacobsson B et al (eds) A world of standards. Oxford University Press, Oxford, pp 1–17

Camera di Commercio di Torino (2011) Torino economia: Rapporto Sulla Provincia di Torino. Camera di Commercio di Torino, Torino

Cannavale C, Canestrino R (2009) International co-operations and entrepreneurship development: a cross-cultural perspective. Rev Int Comp Manag 10(2):254–265

CEN/CENELEC (2010) CEN/CENELEC guide 17 guidance for writing standards taking into account micro, small and medium-sized enterprises (SMEs) needs. CEN/CENELEC, Brussels

Colantino A, Burdett R, Rode P (2014) Transforming urban economies: policy lessons from European and Asian cities. Routledge, Oxon

Comoli V (2012) 50 Luoghi del Barocco a Torino. From Comune di Torino – Torino Turistica. http://www.comune.torino.it/servizionline/schede/torinoTuristica.php?

context=torinoTuristica&submitAction=homeIndice&id=4&idRoot=1&refLanguage=it. Accessed 10 July 2012

COPOLCO (2008) ISO/IEC guide 76 'development of service standards – recommendations for addressing consumer issues. International Organization for Standardization & International Electrotechnical Commission, Geneva

Crisci M, Heins F (2005) L'invecchiamento della popolazione nelle regioni dell'Unione Europea dei 15. Processi e risposte politiche. Istituto Ricerche Economico-Sociali del Piemonte, Torino

Czaya A (2007) Das Europäische Normungssystem aus der Perspektive der Neuen Institutionenökonomik. Peter Lang, Frankfurt am Main

De Vries HJ (1998) The classification of standards. Knowl Organ 25(3):79–89

De Vries HJ (2014) How to implement standardization education in a country? In: Jakobs K (ed) Modern trends surrounding information technology standards and standardization within organizations. IGI Global, Hershey, PA, pp 262–275

De Vries HJ, Blind K, Mangelsdorf A, Verheul H, van der Zwan J (2009) SME access to European standardization – enabling small and medium-sized enterprises to achieve greater benefit from standards and from involvement in standardization. CEN and CENELEC, Brussels

De Vries HJ, Feilzer AJ, Gundlach HCW, Simons CAJ (2010) Conformity Assessment. In: Hesser W, Feilzer AJ, de Vries HJ (eds) Standardisation in companies and markets, 3rd edn. Helmut Schmidt University Hamburg, Hamburg, pp 871–904

DIN (2016) The German standardization strategy. DIN, Berlin

Epstein SR (2008) Guilds, innovation and the European economy, 1400–1800. Cambridge University Press, Cambridge

European Commission (2007) State of European cities. European Commission, Brussels

European Commission (2015) Corrigendum to commission implementing regulation (EU) No 668/2014 of 13 June 2014 laying down rules for the application of regulation (EU) No 1151/2012 of the European parliament and of the council on quality schemes for agricultural products and foodstuffs. Official Journal of the European Union, L 39, 2015-02-14, p. 39

Fan Y (2006) Branding the nation: what is being branded? J Vacat Mark 12(1):5–14

Freire JR (2009) Local people' a critical dimension for place brands. J Brand Manag 16:420–438

Gambarotta B, Ortona S, Roccia R, Tesio G (2004) Torino: Il grande libro della citta'. Edizioni del Capricorno, Torino

Giddens A (1991) Modernity and self-identity: self and society in the late modern age. Polity, Cambridge

Gnoth J (2002) Leveraging export brands through a tourism destination brand. J Brand Manag 9 (4):262–280

Go FM (2005) Co-creative tourists: an idea whose time has come. In: Keller P, Bieger T (eds) Innovation in tourism – creating customer value. International Association of Scientific Experts in Tourism, Brainerd, MN, pp 77–89

Go FM, Govers R (2011) Introduction. In: Go FM, Govers R (eds) International place branding yearbook 2011: managing reputational risk. Palgrave Macmillan, UK, pp XXIV–XLIV

Go FM, Ritchie JRB (1990) Tourism and transnationalism. Tour Manag 11(4):287–290

Go FM, Trunfio M (2011) E-services governance in public and private sectors: a destination organization management perspective. In: D'Atri A, Ferrara M, George JF, Spagnoletti P (eds) Information technology and innovation trends in organizations. Springer, Berlin, pp 11–19

Govers R (2012) What is place branding. Rotterdam School of Management, Erasmus University, Rotterdam

Govers R, Go FM, Kumar K (2007) Promoting tourism destination image. J Travel Res 46 (15):15–23

Grayson K, Martinec R (2004) Consumer perceptions of inconicity and indexality and the influence on assessments of authentic market offerings. J Consum Res 31:296–312

Hankinson G (2004) The brand images of tourism destinations: a study of the saliency of organic images. J Prod Brand Manag 13(1):6–14

Hayami Y (1997) Development economics. From the poverty to the wealth of nations. Clarendon Press, Oxford

Higgins V, Dibden J, Cocklin C (2008) Buiding alternative agri-food networks: certification, embeddedness and agri-environmental governance. J Rural Stud 24:15–27

Huxley J (2010) Turin: meeting the new crisis with a proven package of responses. URBACT cities repond to the economic crisis. URBACT

Kitchin R, Tate NJ (2000) Conducting research in human geography: theory, methodology and practice. Prentice Hall, London

Lotti A (2010) The comodization of products and taste: slow food and the conservation of agrobiodiversity. Agric Hum Values 27:71–83

Maheshawari V, Vandewalle I, Bamber D (2011) Place branding's role in sustainable development. (Emerald, Ed.) J Place Manag Dev 4(2):198–213

Maimbo H, Pervan G (2005) Designing a case study protocol for application in IS research. The ninth Pacific conference on information systems. PACIS, University of Hong Kong

Malhotra N, Birks D (2007) Marketing research: an applied approach, 2nd edn. Financial Times Press, Harlow

Markula M, Kune H (2015) Orchestrating regional innovation ecosystems. In: Lappalainen P, Markkula M, Kune H (eds) Orchestrating regional innovation ecosystems. Aalto University in cooperation with Laurea University of Applied Sciences and Built Environment Innovations RYM Ltd, Aalto

Martinez Badillo JJ (2008) Standards implementation for e-tourism in developing countries – a research approach. In: Jakobs K, Söderström E (eds) Proceedings 13th EURAS Workshop on Standardisation. Aachener Beiträge zur Informatik, Band 40. Wissenschaftsverlag Mainz in Aachen, Aachen, pp 177–186

Mateo C, Seisdedos G (2010) Different branding strategies from the use of the territory of origin by commercial brands: the brand-territory matrix. J Town City Manag 1(2):175–185

McKinsey Global Institute (2011) Urban world: mapping the economic power of cities. McKinsey&Company

Mertens J (2011) Kennisdeling op basis van Social Capital en Modulariteit – Het delen van kennis in de context van Koninklijke Horeca Nederland, Regio Oost-Brabant. Master thesis, Rotterdam School of Management, Erasmus University. Rotterdam

Morgan K, Murdoch J (2000) Organic vs. conventional agriculture: knowledge, power and innovation in the food chain. Geoforum 31(2):159–173

Murphy PE (1985) Tourism a community approach. Routledge, New York

Olins W (1999) Trading identities: why countries and companies are taking on each other's roles. The Foreign Policy Centre, London

Orbasli A (2000) Tourists in historic towns urban conservation and heritage management. E&FN Spon, London

Palmucci L, Visconti M (2012) 100 luoghi dell'industria a Torino e in Piemonte. Comune di Torino – Torino Turistica, Torino. http://www.comune.torino.it/servizionline/schede/torinoTuristica.php?context=torinoTuristica&submitAction=homeIndice&id=16&idRoot=1&refLanguage=it. Accessed 10 July 2012

Paniccia I (2007) The recent evolution of Italian industrial districts and clusters: analytical issues and policy implications. Frammentazione e sviluppo locale: modelli interpretativi e scenari di politca economica. MIUR – PRIN, Rome

Peters A, Koechlin L, Förster T, Fenner Zinkernagel G (2009) Non-state actors as standards setters. Cambridge University Press, Cambridge

Rand GE, Heath E (2006) Towards a framework for food tourism as an element of destination marketing. Curr Issue Tour 9(3):206–234

Renting H, Marsden TK, Banks J (2003) Understanding alternative food networks: exploring the role of short supply chains in rural development. Environ Plan A 35:393–411

Ritzer GF (2012) The McDonaldization of Society, 8th edn. Sage, Thousand Oaks

Salamon E (2008) L'italia delle grandi citta'. Doxa. Doxa, Venezia

Simonetti L (2012) The ideology of Slow Food. J Eur Stud 42(2):168–189
Slow Food West Michigan (2012) Standards and procedures. Slow Food West Michigan. http://www.slowfoodwestmichigan.org/snp. Accessed 3 March 2013
Taylor JP (2001) Authenticity and sincerity in tourism. Ann Tour Res 28(1):7–26
Timothy D (2002) Tourism and community development issues, aspects of tourism, tourism and development, concepts and issues. Channel view publication, Clevedon, pp 149–164
Van den Berg L, van der Meer J, Otgaar A, Speller C (2008) Introduction. In: van der Meer J, Speller C, van den Berg L, Otgaar A (eds) Empowering metropolitan regions through new forms of cooperation. Ashgate, Aldershot
Vanolo A (2008) The image of the creative city: some reflections on urban branding in Turin. Cities 25:370–382
Vaňová A, Borseková K, Foret M (2010) Importance of partnership and cooperation for territorial development. Theoret Appl Econ 10(551):73–82
Venn L, Holloway L, Cox R, Dowler E, Tuomainen H (2006) Researching European “alternative” food networks: some methodological considerations. Area 38(3):248–258
Verman LC (1973) Standardisation – a new discipline. Archon book. The Shoe String Press, Hamdon, CT
Verri P (2005) Torino: a future of knowledge and creativity. ISoCaRP Congress, Torino
Visita Torino (2011a) Storia della Citta’ di Torino. Visita Torino, Torino. http://www.visitatorino.com/storia_torino.htm. Accessed 6 June 2012
Visita Torino (2011b) Vie e piazze di Torino – Via Po’. Visita Torino, Torino. http://www.visitatorino.com/via_po.htm. Accessed 10 July 2012
Weiermair K, Steinhauser C (2003) New tourism clusters in the field of sports and health; the case of alpine wellness. 12th International Tourism and Leisure Symposium, Barcelona
Yin R (1994) Case study research: design and methods, 2nd edn. Sage Publications, Beverly Hills, CA
Zeng G, Go FM, Vries HJ (2012) Dilemma of authenticity and standardisation: expansion strategies of restaurant groups in China. Int J Hospitality Manage 31(4):1090–1100

Henk J. de Vries is Associate Professor of Standardisation at the Rotterdam School of Management, Erasmus University, and Guest Researcher at Delft University of Technology, Faculty of Technology, Policy and Management. His research and teaching focus on standardisation from a business point of view. Henk is President of the European Academy for Standardisation EURAS. He is (co-)author of more than 350 publications in the field of standardisation. See http://www.rsm.nl/hdevries.

Frank Go passed away Spring 2017. He was professor emeritus of the Erasmus University in the Netherlands. Previously, he held the Bewetour Chair of Tourism Marketing at the Rotterdam School of Management and served at the business faculties of universities in Toronto, Calgary in Canada and Hong Kong; as visiting professor at the George Washington University, USA, Leuven, Belgium, Open University Business School, UK, Rikkyo University, Japan.

Research interests focused on brand identity and image, innovation, entrepreneurship and marketing strategies reaching into wider social areas such as globalization and sustainability.

He has authored and co-authored publications in Annals of Tourism Research, Tourism Management, Journal of Travel Research, Brand Management, Tourism Recreation Research, The Economist Intelligence Unit.

With Robert Govers he is co-author of Place Branding (2009) and co-editor of a series of edited book volumes on Place Branding In the Age of Innovation (2010), Reputational Risk (2011) and Managing Smart Growth and Sustainability (2012) all published by UK-based Palgrave MacMillan.

Sophie Alpe graduated Cum Laude from the Rotterdam School of Management, Erasmus University with a Master's degree in Marketing Management, focusing her studies on destination branding and event management. Her work experience has focused on both these fields. She currently works in the events management industry, but has had multiple experiences within destination management and marketing consulting.

Part IV
Scalable and Sustainable Information and Communication Technologies

Humans and computers have been in close dialog, since the dawn of the computing era. From the first attempts of natural communication using speech (including automatic speech recognition—ASR and text to speech synthesis—TTS) at the Bell Labs in the thirties, or of natural communication using graphical primitives at MIT during the sixties, we have seen the emergence of the WIMP (Windows, Icons, Menus, Pointer) paradigm of interaction in the early eighties, making it possible for all of us to interact today with computer-mediated communication apps and services via simple means. Today, we all adhere to the proliferation of touch-based WIMP interfaces, popular in mobile devices such as smartphones or tablets, whose worldwide penetration is increasing, not only in the consumer space but also in the industry and the businesses in general. In addition to WIMP-type of interaction, other more natural human–computer interaction modalities are gaining interest and adoption, notably speech, and soon we will witness the mainstream emergence of other modalities. Gaze (tracking both eyes positions), human body motion and hands gesture, all derived from visual and depth sensors (such as Kinect), thoughts evaluated from EEG—Electroencephalography signals, or emotions computed from processing raw data obtained from a plethora of biometric sensory devices (such as EMG—Electromyography, ECG—Electrocardiography, EDA—Electrodermal Activity, visual and depth sensors to capture facial expressions), are soon becoming the next step of human–computer interaction technology.

Much in the same way, computer-mediated communication services have evolved from simple text-based platforms, like Bulletin Board Systems, Internet Relay Chat and e-mail, extremely popular during the eighties, to more complex, multimedia-enabled services such as audio–video conferencing, instant messaging and, more recently, social media services. Such services, facilitate today the social interaction between humans, with the most popular service (Facebook), currently having 1860 million daily active users on average (December 2016), were over 1149 million are daily active in mobility, and where approximately 87.5% of such daily active users, are based in countries outside the US and Canada. This social media interaction revolution has made it virtually possible for a person to reach anyone, anywhere, anytime, with great ease. Such fast evolution has led to increased interaction among people of all ages.

Ageing is another trend that deserves analysis. Europe, in particular, is ageing fast as life expectancy is increasing and the birthrate is declining dramatically. In 2060, it is estimated that the number of seniors (citizens aged over 65 years) will reach up to 30% in the European Union. One needs to acknowledge that that demographic ageing is one of the most relevant events and conquests in the history of mankind. As an answer to this matter of affairs, quite evident in the European case but also prevalent in Japan, public R&D institutions from 22 European countries, have created the Ambient Assisted Living (AAL) Association, that owns and manages the Active Assisted Living Programme (http://www.aal-europe.eu). This programme funds collaborative R&D projects using innovative ICT, targeted at creating better conditions of life for the older adults and allowing to overcome the difficulties age brings to people, including promoting active aging at work, foresting ageing well in the community and in the society, aiming at reducing the impact of social exclusion observed with the elderly population and, finally, facilitating ageing well at home and in the family. At the same time the AAL programme intents to strengthen the industrial opportunities in Europe in the Health and AAL sectors. Today, enabling technologies such as, mobile computing devices (smartphones, tablets), Internet Of Things (IoT), multimodal human–computer interfaces adopting universal design principles (design for all) and using natural modalities, such as speech, body and hand gesture, touch, gaze, or even thoughts and emotions, are extensively adopted in AAL projects and solutions.

This part of the book addresses some of these challenges and highlights the importance of the adoption of ICT technologies by consumers, corporations and business, with a focus on long term adoption.

Impact of ICT Utilization on Innovations and on Labor Productivity: Micro-level Analysis for Poland

Lukasz Arendt and Wojciech Grabowski

Abstract This chapter studies the relationship between innovations, use of Information and Communication Technologies and labor productivity in the Polish enterprises. The research framework takes advantage of the Crepon–Duguet–Mairesse (CDM) model and the new firm paradigm, which focuses on the co-innovative productivity sources. Within this study, factors determining implementation of innovations (product, process-organizational and marketing) are identified and joint impact of innovations, ICT use and complementary factors on labor productivity, are evaluated.

Estimation of the simultaneous discrete choice model revealed that there is complementary relationship between ICT and all types of innovations. We also showed that human capital positively influences product and marketing innovations, while organizational readiness enhances product and process-organizational innovations. These two types of innovations have positive impact on labor productivity in the Polish firms. The results confirm the mediating role of co-innovative productivity sources - ICT alone do not enhance labor productivity, but we recorded productivity gains if ICT were complemented by other mediating factors.

Keywords Innovations • ICT • Productivity • ICT complementarities

L. Arendt (✉)
Faculty of Economics and Sociology, Department of Economic Policy, University of Lodz, Lodz, Poland
e-mail: larendt@uni.lodz.pl

W. Grabowski
Faculty of Economics and Sociology, Department of Econometric Models and Forecasts, University of Lodz, Lodz, Poland
e-mail: emfwog@uni.lodz.pl

A. Dias et al. (eds.), *Modeling Innovation Sustainability and Technologies*, Springer Proceedings in Business and Economics,
https://doi.org/10.1007/978-3-319-67101-7_17

1 Introduction

Information and Communication Technologies (ICT) are undoubtedly innovative technologies—since development of the first microchip, we have seen a continuous process of new inventions in the ICT industry: personal computers, world wide web, cloud computing, or recently big data and the Internet of Things. It has been argued, that the pace of ICT development is much faster in comparison with previous groundbreaking technologies, like steam engine or electricity,[1] and shall ultimately lead to improved economic performance (at micro and macro-level) measured by productivity. ICT is a general purpose technology (GPT)—it influences innovations and productivity in all sectors, not only in the ICT-producing sector. The changes that ICT brings (or may bring) to companies encompass delivering completely new products, reengineering business processes, restructuring organization's structure by flattening hierarchies and delegation of responsibilities (Hempell et al. 2006). These changes have been perceived as so profound, that the new concept—"new firm paradigm"—has been developed (Diaz-Chao et al. 2015; Arvanitis et al. 2013).

The growing importance of ICT, and extensive research within the framework of the new firm paradigm have led to booming interest in analyzing the joint link between ICT, innovations, and productivity. This new, combined research framework took advantage of two streams of research that were focusing either on the relationship between ICT and productivity, or on the link between innovations and productivity. The first stream of research was rooted in the analysis related to the productivity paradox, that finally brought positive results and revealed ICT is enhancing productivity and GDP growth. However, it appeared that ICT become productive only if is supported by complementary factors[2] (Brynjolfsson and Hitt 2003; Brynjolfsson 2005). The other stream of research, rooted in the Schumpeterian concept of innovations, has been dealing for a long time with the issue of the proper measurement of innovation. Traditionally, innovations were measured through inputs—analyses were usually focused on R&D expenditures. However, this method did not capture the effectiveness of innovative effort. The other solution was to measure innovation's output—in this approach patents were usually used as a proxy for innovations. Nevertheless, also this framework is not perfect, as patents capture only a part of innovative effort undertaken by companies (Griffith et al. 2006). Moreover, Cainelli et al. (2006) found, that R&D expenditures

[1]To illustrate the unprecedented pace of ICT development we may quote the so-called Moore's Law, according to which the number of transistors that could be placed on a microprocessor is doubling every 2 years (this prediction, which appeared to be true, was formulated by Gordon Moore in 1965). To better understand how fast this development is taking place, Brynjolfsson and Saunders (2010) presented the Moore's Law in perspective of the aviation industry—it is a story of a commercial flight from New York to Paris, which cost about 900 USD and lasted 7 h in 1978. If the aviation industry has been developing according to Moore's Law, the same flight in 2010 should cost about 1 penny and take not more than 1 s!

[2]In some research approaches innovations are treated as ICT complementarities.

and patents are not appropriate measures of innovation activities in the service sector. To overcome these issues, Crépon et al. (1998) have developed approach (known as CDM model) which combines indicators measuring inputs and outcomes of innovations and relation between them. The CDM model has been recently used extensively for analyzing relationship between R&D, innovations and productivity. Generally, it has been revealed that companies' growth (including productivity) is related more to the results of technological activities, than to inputs (measured by investments in physical or ICT capital), which previously was a dominant belief. In Europe, most of research studies in this area have been based on Community Innovation Survey (CIS)—as a result analyses are limited mainly to product and process innovations. However, more and more focus has been recently put on marketing innovations (Stojcic and Hashi 2014), and new innovative models and methods have been conceptualized. One of the newest approaches to innovations in the European Union, strongly promoted by the European Commission, is the paradigm of Open Innovation 2.0. It refers to the concept of melting pot in which universities together with industry, government and consumers come up and realise new innovative ideas (formally speaking, it is a quadruple-helix innovation approach) (Curley and Salmelin 2013). On top of that, CDM model has been extended in order to capture the impact of ICT on innovations and productivity within one framework (Nguyen Thi and Martin 2011). Innovations and ICT are usually perceived as complementary factors, thus decisions on investing in ICT and introducing innovative solutions within enterprises are not exogenous, and shall lead to the same result, which is increased productivity.

The research studies on the relationship between innovations, ICT and productivity have been focused primarily on highly-developed countries. Unfortunately, such analyzes in emerging economies are scarce (Aboal and Tacsir 2015), and undoubtedly there is a need for more detailed insight on this issue in countries which are innovations and ICT followers, but they are heading (to some extent) the development paths of technology leaders. This chapter elaborates on the interactions between innovations, use of Information and Communication Technology, and labor productivity in Poland—a post-communist country from the Central and Easter European region, that is usually classified as a developing economy. Poland has undergone remarkable transition from the centrally-planned to market economy since the beginning of 1990s, and joined European Union in 2004. In spite of dynamic structural changes and economic growth (even during the last world crisis) productivity, as well as innovation capability are still below average levels of developed economies.

This research study is based on firm-level data collected within a survey carried out in the Polish companies in the first half of 2015. We believe, that in order to assess the joint impact of innovations and ICT on company's labor productivity, different types of innovations, different factors enhancing productivity, as well as joint ICT-innovations activities should be taken into account. To capture these elements, we developed research approach that combines augmented form of the CDM model (Polder et al. 2009; Hall et al. 2012) and the "new firm paradigm" concept. We refer to the types of innovations defined in the Oslo Manual (2005), however in the econometric modelling we aggregated process and organizational innovations into one innovation category. We also paid a lot of attention to the role of ICT complementarities as for better understanding of ICT diffusion effects, taking into account the fact, that organizational change may be treated as an organizational innovation.

The goal of this study is to identify factors which determine introduction of different types of innovations, and to evaluate the joint impact of innovations, ICT use and complementary factors on labor productivity in Polish companies. The chapter is organized as follows. The next section presents literature review on interactions between innovations, Information and Communication Technologies and productivity. Next, a statistical overview of sources of GDP growth, labor productivity, innovation system and innovativeness in Poland is depicted. Description of data, modelling approach and the results of estimation is then discussed. The final section elaborates on concluding remarks.

2 Literature Review

At the beginning of the ICT-driven information revolution it was a common belief that ICT will ultimately enhance economic growth and productivity. Unfortunately, research studies until mid 1990s did not find positive relationship between ICT and productivity. The US data in the early 1970s (beginning of the ICT revolution) showed a downshift in TFP growth, which lead to the well-known statement expressed by Solow (1987) "*You can see the computer age everywhere but in the productivity statistics*" launching a bulk of research studies on the productivity paradox. At some point it became evident, that relationship between ICT and productivity is not straightforward, and usually ICT investments must be complemented by additional investments/changes in the area of human capital, organizational change or innovation, to reap the ICT-related benefits (Pilat 2006).

This refers to the two types of effects, which Information and Communication Technologies, through technical progress channel, generate in relation to GDP and productivity growth (Jung and Mercenier 2014). The so called first-order effect captures the impact of investments in ICT infrastructure (hardware and software) on the stock of capital. The growth of ICT-capital, ceteris paribus, shall positively influence the GDP growth. The second-order effect is a result of complementary changes that are induced by ICT investments—these complementarities affect Total Factor Productivity (TFP), and as a consequence—productivity and economic growth.

ICT-driven organizational change (workplace innovation) has become the most popular explanation of the productivity paradox.[3] The conceptual models explaining

[3]The other explanations emphasize the measurement issues, problems of lags, redistribution and dissipation of profits, and mismanagement of ICT (Brynjolfsson and Saunders 2010; Yang and Brynjolfsson 2001; David 2002).

interactions between ICT and company's productivity may be found in Freeman and Lazear (1995) or Milgrom and Roberts (1990). Freeman and Lazear (1995) argued that providing more decision-making power to employees in management and production processes (up to the socially optimal level) in exchange for higher share of rents, brings more profits to companies. Milgrom and Roberts, in their more influential paper, put more focus on synergy effects related to introduction of the whole system of complementary solutions leading to increasing productivity (Black and Lynch 2004). Since late 1990s many research studies have confirmed the positive relationship between ICT and productivity, firstly in highly-developed countries (Jorgenson and Stiroh 2000; Oliner and Sichel 2000; Stiroh 2002), and then in emerging economies (Van Ark and Piatkowski 2004; Dimelis and Papaioannou 2010; Jorgenson and Vu 2010).[4] However, it appeared, that research results are volatile to the level of analysis—Cardona et al. (2013) found that while macro and mezo-level studies showed differences between European countries and the US (favorable to US), micro-level data suggested no significant differences amongst countries, as far as ICT and productivity was taken into account.

The relationship between a company's innovativeness and productivity is also not straightforward. In the classical approach, Schumpeter (1994) perceived innovation as a tool which shall make firm more competitive. However, the aim of each company operating on the competitive market is to maximize profit—thus innovation does not necessarily mean higher productivity, at least at the firm level. Nevertheless, Schumpeter argued that innovation coincides with creative destruction, which enhances productivity growth.

Cainelli et al. (2006) discussed three types of mechanisms describing relationship between innovations and economic performance of a company. The first two mechanisms (encompassing three theoretical approaches which they labeled Schumpeterian I, Schumpeterian II, and Schmooklerian) assume one-way relationship between innovation and economic performance, while the third one is of dynamic nature, where this relationship operates in two ways—innovations and economic performance are reinforcing themselves.

Recently, a framework presented by Crépon et al. (1998)—the CDM model—has become the most popular method to test empirically interactions between innovations and productivity at the firm level in different national contexts[5] (Goya et al. 2013). The CDM model shifted the focus to output definition of innovation activities, and it has been argued to be a major improvement comparing to the extended Cobb-Douglas production function approach, which dominated previously (Mohnen and Hall 2013). In the CDM framework three stages of the

[4]For the comprehensive literature review on productivity paradox see (Pilat 2006; Cardona et al. 2013).

[5]Szczygielski and Grabowski (2014) analyzed interactions between innovation strategies and productivity in Polish companies. They found negative relation between innovation performance and productivity in the cluster of firms introducing marketing innovations, while in the case of firms adopting innovations, introducing process innovations and market-oriented innovations, positive relationship was reported.

innovation activities are analyzed: allocation of resources to R&D activities, delivery of the innovation outcomes, and translating innovation outcomes into firms' productivity. Thus, the R&D effort is transmitted to productivity through technological outputs—innovations. Research studies based on the CDM framework provided many examples of positive relationship between innovations and company's performance. Mohnen and Hall (2013) argued, that all types of innovations, as defined in the Oslo Manual, translate into better productivity performance, and emphasized existence of complementary relationship between these types of innovations, pointing out that it is difficult to describe the exact processes behind this complementarity phenomenon. However, in some cases positive link between innovations and productivity was valid only with respect to process innovations in some countries, while for other economies it applies only to product innovations (Hall et al. 2010; Bertschek et al. 2010; Griffith et al. 2006). It was also showed, that R&D generate indirect (spillover) effects, which is a similar observation to the one broadly discussed in the literature on the ICT-driven productivity and organizational change.

The classical CDM model has been modified to test some additional propositions. Huergo and Moreno (2011) adapted CDM approach in order to analyze the influence of persistence of decisions of R&D investments and production of innovations on productivity of Spanish companies. They proved that the state dependence in technological inputs and outputs influence current innovations, and thus the omission of this persistence may lead to overestimation of innovations impact on productivity in the short run. At the same time, it appeared that innovation activities have long-run effects on productivity. Unlike many others studies based on CDM approach, Huergo and Moreno (2011) results did not confirm positive impact of product innovations on productivity (TFP growth).[6]

Another stream of modifications in the classical CDM framework has been focused on accommodating the effects of ICT—in this approach ICT is incorporated into the model as an innovation input. The extended CDM framework includes the three-step analysis: assessment of the determinants of innovation input—R&D and ICT simultaneously; impact of R&D intensity and ICT use on innovation outcomes; and finally influence of innovation outputs on labor productivity. The work of Polder et al. (2009) is considered to be one of the first that combines, in the empirical study, two literature streams on R&D driven technological innovation and non-technological innovation complemented by ICT (Nguyen Thi and Martin 2011).

This and other developments gave a basis to studies on joint impact of innovations and ICT on productivity, and interactions between innovations and ICT. And it appeared to be a challenging task, as interactions between different elements of the analyzed system seem to be even more complicated. Kleis et al. (2012) distinguished three main channels through which ICT influences company's innovations.

[6]However, in most cases the relationship between product and process innovations, and labor productivity was analyzed. This shows, that the choice of productivity measure may influence the results.

In the first one ICT improve management of knowledge that is necessary to create innovations. In the second one, ICT is perceived as a tool enabling more efficient cooperation with external partners. In the third one, ICT directly contribute to innovation processes by playing a supportive role in generating new ideas, designing new products, and finally in delivering these products to the market. Quite similar approach may be found in Gago and Rubalcaba (2007), who argued, that ICT plays an important role as a driver of innovations in services, especially if innovation is based on a new ways of client-provider cooperation or more advanced use of business services.[7] In their approach ICT play the following roles as for enhancing innovations: it acts as a driver (in ICT related services and processes), a facilitator (e.g. usually in case of distant provision of services—outsourcing, offshoring), and an agent, in case of use of innovative ICT-oriented services. Thus, innovations and ICT are usually treated as complementary factors that interact with each other. ICT provides efficiency gains through reduction of transaction costs, improvement of business processes, facilitating coordination and increasing diversification (Koellinger 2005). It is a General Purpose Technology, so it provides platform for product and process innovations and creates spillover effects based on external R&D effort (Gretton et al. 2004). ICT also enables closer link between businesses, suppliers, customers, competitors and other partners (Spiezia 2011).

Hempell et al. (2006) used extended production function approach to analyze relationship between ICT and innovations (treating innovations and ICT as complementarities) in Germany and the Netherlands. The results showed, that ICT capital has positive impact on labor productivity, being more productive if complemented by innovational efforts (thus the complementarity between ICT and innovations was proved). Interestingly, they found direct and positive influence of product and process innovations on TFP in Germany, but not in the Netherlands. Moreover, it appeared, that continuous innovation effort brings better effects as for productivity, than innovating occasionally. Cainelli et al. (2006) found, that investments in ICT, treated as innovation activities, have positive impact on productivity and there is self-reinforcing mechanism between innovations and productivity in place in the service sector. It appeared, that ICT investments played crucial role in explaining the link between innovations and productivity, while the role R&D was much weaker. Edquist and Henrekson (2016) analyzed effects of R&D and ICT on TFP using Swedish industry-level data. They found that R&D influences TFP immediately, exhibiting short-run spillover effects between sectors, while impact of ICT on TFP becomes significant with delay of 7–8 years and requires complementary changes to make ICT investments productive. This result is in line with the "lag" and "ICT complementarities" hypothesis discussed in the productivity paradox literature. It also shows that companies should balance their involvement in different type of investments, to secure the base for productivity growth both in the short and long run.

[7]Their analysis covered service companies in the Madrid area. They emphasized, that impact of ICT (as a GPT) is much greater than just the pure effect of capital input related to ICT investment, because ICT facilitate complementary innovations.

However, Diaz-Chao et al. (2015) analyzing data from SMEs based in the province of Girona (Spain), argued that co-innovation sources of company performance (defined within a new firm paradigm), did not affect labor productivity, which may mean that the relationship between ICT, innovations and productivity are weaker in smaller firms than in the large ones. Relationship between factors defining "new firm paradigm" and innovations in Swiss and Greek companies was analyzed by Arvanitis et al. (2013). They found that ICT had positive impact on process innovations and R&D activity in Swiss companies, while in Greek firms ICT affected product and process innovations but not R&D activity. However, in both countries new organizational changes (e.g. team work, job rotation, and to some extent decision decentralization) positively affected innovations. Noticeable differences were reported in case of human capital—in the Swiss companies human capital enhanced product innovation and R&D activity, while in Greece it did not play any role as for innovations.

Studies based on the extended CDM framework also reported generally positive relationship between ICT, innovations and firm's productivity. The previously mentioned work of Polder et al. (2009), which focused on product, process and organizational innovations in the Netherland's companies, found that R&D affected product innovation in manufacturing firms, while ICT investments and its use had positive impact on all types of innovations mainly in the service sector. Their analysis also revealed that product and process innovations affect TFP only when they are performed together with organizational innovation. Nguyen Thi and Martin (2011) research study for Luxembourg manufacturing and services companies showed that not all investments in ICT translates into improvement of innovative performance, while better organizational arrangements, conditional to ICT, led to increase of labor productivity. Hall et al. (2012), who analyzed Italian companies, argued that while both R&D and ICT had positive impact on innovation and productivity, R&D was more important for innovation, while ICT investments were essential for productivity. The example of research study in emerging economy, based on augmented CDM model, is work of Aboal and Tacsir (2015). They used data from Uruguay companies to point out that productivity in the service sector is positively influenced by investments in ICT and other types of innovation activities, while in the manufacturing sector only ICT affects productivity. Moreover, they proved that ICT investments are subject to the economy of scale—which explains higher intensity of investment activities in large (especially foreign) companies.

3 ICT, Innovations and Productivity in Poland: General Overview

After severe economic slowdown in the first half of 1990s, which was a result of transition from centrally-planned to market economy, Poland has recovered and became one of the best performing countries in the group of post-communist

Table 1 Growth rates in Poland, CEE region and EU-15 countries (1995–2014)

Growth rate[a]				Contribution of production factors to GDP growth			
	PL	CEE	EU-15		PL	CEE	EU-15
GDP	4.10	2.86	1.92	Labor quality	0.17	0.26	0.26
Labor quality	0.28	0.44	0.41	Labor quantity	0.23	−0.24	0.41
Labor quantity	0.42	−0.34	0.66	ICT capital services	0.71	0.92	0.54
ICT capital	23.6	19.2	11.6	Non-ICT capital services	1.15	1.15	0.76
Non-ICT capital	3.16	3.13	2.24	TFP growth	1.94	0.86	0.00

Source: Own calculations based on the Total Economy Database. Average for 1995–2014
[a]Log change

economies, which joined European Union in and after 2004. Since 1995 Polish economy has noted significant GDP growth—noticeably higher that in the CEE and EU-15 countries (Table 1).[8] Labor quality grew slower in Poland than in CEE and EU-15 region. However, having this in mind, it should be stressed that contribution of labor quality to GDP growth was relatively high in Poland.

Employment growth and contribution of the labor quantity to economic growth in Poland was below EU-15 level, but positive, while average results for the CEE region were negative in both cases (Table 1).

CEE countries (including Poland) noted higher growth of capital in comparison with EU-15, especially in case of ICT-capital (rate of growth of ICT-capital in Poland was twice as high as in EU-15 countries). This leads to two important conclusions. The first one—CEE region focused on capital-deepening activities trying to rebuilt a solid base for economic growth. The other one—economic agents (in private and public sector) understood the importance of ICT-capital, and, as there has been a gap in ICT infrastructure between CEE and more developed countries, they tried to catch up by investing in the ICT infrastructure.

This commitment brought positive results in the form of contribution of both types of capital to GDP growth in the CEE region. But more importantly, it created spillover effects measured by TFP growth, which revealed mainly in Poland, and to a lesser extent, in other CEE countries. Contribution of TFP accounted for 47% of GDP growth in Poland and 30% in the CEE region, while for EU-15 it was non-existent. TFP captures innovativeness and ICT-driven spillover effects—unfortunately, this data does not allow for assessment which effect—related to ICT or innovations—had bigger impact on creating spillover effect in CEE region, including Poland. However, as the innovativeness of the Polish economy is relatively low, it may be argued, that ICT and ICT-complementary factors have played

[8]CEE region and EU-15 countries are used as a point of reference. CEE region does not include Estonia, Latvia, Lithuania and Croatia as for discussion on the sources of GDP growth due to a lack of data on ICT capital in the Conference Board 2015 database.

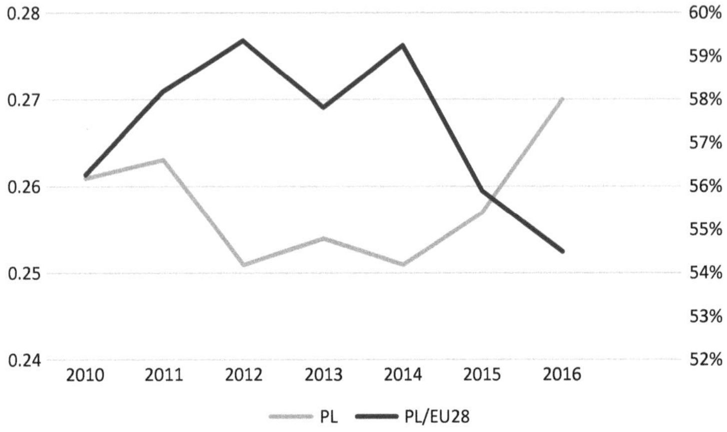

Fig. 1 Summary Innovation Index in Poland (2010–2016). Source: Based on (EIS 2017, p 90)

crucial role in enhancing economic performance in Poland since 1995, even if Poland has not acquired technological advantage in the area of ICT.[9]

The national innovation system in Poland is of the catching-up type (Weresa 2012), and innovations' performance does not reflect the success of the Polish economy. Poland is categorized as a moderate innovator in the EU region,[10] with relatively minor fluctuations (in absolute and relative terms) between 2010 and 2016[11] (Fig. 1). There were only three countries in the European Union in 2016 (Bulgaria, Croatia and Romania), which results were worse than the Polish score. Detailed analysis of Summary Innovation Index depicts rather miserable picture of Poland's performance. It was below EU-28 average in all dimensions of SII (EIS 2017).

Also OECD review on key trends in Science, Technology and Innovation (STI) policies reveals relatively poor innovation's performance in Poland (OECD 2014). Polish results were below median for OECD countries for 18 out of 22 indicators. One of main weaknesses of the Polish innovation system in 2007–2014 was insufficient innovative activity in the business sector—private spending on R&D accounted for less than 1/20 of OECD median. Development of electronic communication (measured by the number of mobile broadband users per inhabitant and internet network density per capita) and high internationalization of patenting activity (number of joint patent applications with inventors from abroad was far above the OECD median) were pointed out as the only strengths.

[9]Even though Revealed Technological Advantage (RTA) in ICT increased in 2009–2011, it was still below OECD median and reference RTA value.

[10]According to Summary Innovation Index (SII) methodology, which is based, since 2017, on 27 indicators grouped within 10 dimensions.

[11]Polish performance relative to EU average varied between 55% in 2016 and 59% in 2012 and 2014, which supports Weresa's argument, that no convergence process has been in motion.

This poor innovation potential on macro level resulted, to large extent, from the approach to innovation policy that was in place during the times of centrally-planned economy system in Poland. There was no support to private entrepreneurs to invest in innovation activities, intellectual protection rights were underdeveloped and not working sufficiently, and innovation effort was administered centrally, thus focusing on selected sectors of the economy (Klonowski 2011).

Relatively weak macro-base of innovation system in Poland translates into innovation activities at the firm level. In 2012–2014 innovative companies comprised 17.5% of all enterprises in the manufacturing industry and 11.4% in the service industry. Innovations were introduced mainly in large companies, operating in the basic pharmaceutical products, and in the insurance, reinsurance and pension funding divisions. The most popular were the process (introduced by 12.9% industrial firms and 8.4% service companies), and product innovations (11.7% and 6.8% respectively), followed by organizational (8.4% and 9.7% respectively) and marketing (7.6% and 7.9% respectively) innovations (CSO 2015).

Growth theory literature argues, that capital deepening and improvement in labor quality should enhance labor productivity—indeed, data shows significant improvement of GDP per person employed between 1995 and 2015 (Fig. 2). In this period labor productivity in UE-15 was much higher than in CEE countries, however this gap has been decreasing—in 1995 it reached 46,480 USD, while in 2015 it was 35,091 USD (as for Poland this gap amounted to 46,336 USD and 30,988 USD respectively).

It was a result of much faster increase of GDP per person employed in the CEE region (84% growth between 1995 and 2015—in Poland: 96%), than in EU-15 (19%). There is no doubt, that labor productivity growth in Poland and other CEE countries was driven to large extent by increasing stock of ICT-capital, more intensive use of ICT, as well as innovation activities undertaken by companies. Of course, we have to bear in mind that labor productivity gains in Polish companies are also linked to the process of massification of higher education, which translated into generally higher level of educational attainment (and better quality of human capital) of available workforce. This has direct effect on productivity, but

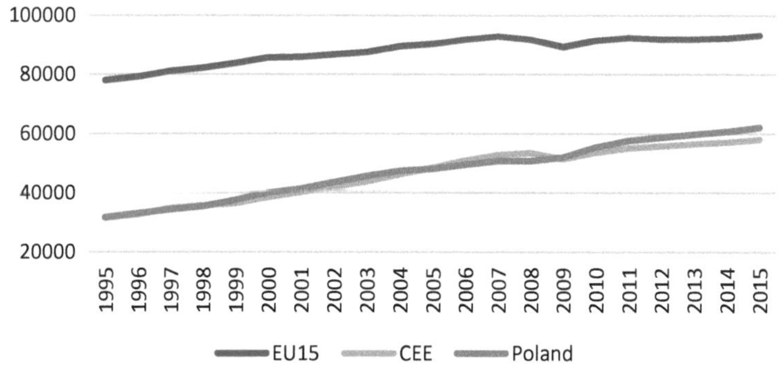

Fig. 2 Labor productivity per person employed (in 2014 US$—converted to 2014 price level with updated 2011 PPPs). Source: Total Economy Database

may also be perceived as a factor enhancing probability of carrying out R&D activities and innovativeness, as companies may be inclined to invest in R&D to reach long-term productivity gains, taking into account the persistence hypothesis (Huergo and Moreno 2011).

Interestingly, GDP per person employed in Poland and CEE region followed similar path for many years, but since 2009 Polish labor productivity has outrun the CEE level. Nevertheless, in 2015 it amounted to only 67% of EU-15 average.

4 Data and Methodology

Data for the analysis of relationship between innovations, ICT and labor productivity was collected within the PAPI survey of 1000 Polish companies. Interviews were conducted in the first half of 2015 in the framework of the research project "Impact of Information and Communication Technologies on productivity—macro and micro analysis". Respondents covered the representatives of management personnel. The survey was administered to a representative sample of Polish enterprises that use ICT. The sample was stratified by company size (small, medium and large entities), sector (manufacturing, services), and region (NUTS II) (Table 2).

The sample was dominated by private-owned companies which operated mainly in the service and manufacturing industries, and some of them covered both sectors[12] (Table 2). The share or agriculture-based businesses was extremely small, and thus irrelevant for further analysis in the sectoral dimension. Like in all developed and emerging economies, majority of firms were classified as Small and Medium-sized Enterprises (SMEs), with total employment below 250 persons. Within SMEs, micro enterprises constituted the highest number of companies. And even though the median employment reached 20 persons, which just confirms that many firms were rather small businesses, large number of enterprises operated on the international markets and only 13% declared local markets coverage. It means that not only large Polish companies enter international (global) markets, looking for new customers and trying to adapt to global competitive game.

Unsurprisingly, managers were (on average) better educated than the line workers, however there was a minor share of companies in which most of management did not possess even secondary education. Mean gross salary, which captured the labor productivity in the surveyed enterprises, accounted for 2856 PLN and was lower than average salary in the Polish economy in 2015 (3900 PLN). Most companies offered wages between 1750 PLN (minimum wage) and 3000 PLN. There were businesses that offered salaries below the minimum wage level, which would indicate a relatively large share of part-time employment in these enterprises (Table 2).

Griffith et al. (2006) argued that all companies make, to some extent, innovative effort, but below a certain threshold they are incapable of measuring this effort and

[12]As a result, total share of companies by type of business is higher than 100.

Table 2 Descriptive statistics—company characteristics

Size	Mean: 165; Median: 20; Micro: 41.0%; Small: 18.7%; Medium: 21.9%; Large: 18.4%
Ownership	Private: 94.9%; State-owned: 5.1%
Sector	Agriculture: 0.3%, Manufacturing industry: 26.6%; Service industry: 79.8%[a]
Market coverage	Local: 13%; Regional: 18.4%; National: 28.8%; International: 39.8%
Education attainment—management	Majority with university degree: 74.4% Majority with secondary education: 24.2% Majority below secondary education: 1.4%
Education attainment—workers	Majority with university degree: 25.9% Majority with secondary education: 58.7% Majority below secondary education: 15.4%
Average gross salary (in PLN)	Mean: 2856; Median: 2520; ≤1750[b]: 9.4%; (3000–4000): 22.8%; (1750–3000): 53.2%; ≥4000: 14.6%;

[a]As some companies operated both in manufacturing and service industries, total share is higher than 100
[b]Minimum salary in Poland in 2015

report on it. In other words, innovations' occurrence is a matter of definition used to describe the innovation activities. Moreover, it should also be taken into account, that companies and managers may perceive the innovation effort and its results differently (the same solution in one company may be perceived as innovation, while in the other enterprise it will be seen as a typical procedure—company size seems to be important determinant in this case). In our study the innovative effort (innovation capacity) was measured on the basis of objective (existence of R&D department in a company) and subjective assessment of innovation performance by the managers.

Less than ¼ of surveyed businesses had internal R&D department. However, almost 40% of managers reported, that their companies introduced solutions that might be perceived as innovative ones within previous 24 months (Table 3). To understand better the interactions between innovations, ICT and labor productivity we referred to the classification of innovations in the Oslo Manual (2005), however we introduced one modification. Oslo Manual (2005) defines process innovation as implementation of new or significantly improved production or delivery method, while organizational innovation encompasses implementation of new organizational method in the firm's business practices, workplace organization or external relations. Both types of innovations are aimed at reducing costs (production, transaction or delivery costs), and thus increasing productivity (although organizational innovations are supposed to have bigger influence on labor productivity, also process innovations play important role in this area). Because of these similarities, it was decided to combine these two types of innovations into one variable (labeled "process-organizational innovations"), and then to use this variable in the econometric analysis. Surveyed companies introduced mainly product, and process-organizational innovations. Marketing innovations were less popular (Table 3).

Table 3 Innovations, ICT use and organizational change capacity

Innovations: 38.6%	Product: 27.8% (72%); Process-organizational 25.1% (65%); Marketing: 16.1% (41.6%)[a]
Role of ICT for introducing innovations	Impossible without ICT: 22.6%; Difficult without ICT: 36.2%; ICT as enabling factor: 30.8%; ICT was not important: 8.7%; Hard to say: 1.7%
ICT use—UICT variable	0: 6.4%; 1/7: 34.0%; 2/7: 20.9%; 3/7: 14.4%; 4/7: 7.3%; 5/7: 6.0%; 6/7: 4.1%; 1: 6.9%
Employees' participation in changes within enterprise	Always: 8.3%; Very often: 21.2%; Often: 33.7%; Rarely: 21.2%; Very rarely: 9.9%; Never: 5.7%
Flexible working time arrangements	Always: 7.5%; Very often: 21.7%; Often: 29.9%; Rarely: 19.4%; Very rarely: 13.5%; Never: 8.0%
Employees' ICT/digital skills	High: 49.6%; Satisfactory: 48.5%; Low: 1.9%
Candidates' ICT/digital skills	All candidates: 28.1%; Most candidates: 55.9%; Some: 13.9%; Almost none: 2.1%
R&D Department: 23.0% Investing in ICT in last 24 months: 50.2%	
Motivation pay system—managers: 57.3% Motivation pay system—line workers: 62.3%	
Development of employees' teams: 40.4%	

[a]Companies might point out many types of innovations, that were implemented. Data in parentheses shows the percentage of enterprises introducing selected types of innovations in the group of businesses which implemented any innovation

These results differ from the one reported by the Polish Central Statistical Office, however it may be explained by the fact that sample in our survey covered only these companies which take advantage of Information and Communication Technologies. ICT utilization was assessed by two variables: access to computers and the type of utilized software.

Interviews were conducted only in these companies that used ICT in at least two out of nine following business processes: office management, accountancy, HR management, supply management, production management, CRM software, ERP software, CAD/CAM systems, and CNC systems. By this procedure enterprises which did not use ICT at all or used it in very limited way (that could not lead to organizational change, and thus productivity increase), were excluded from the analysis. The variable UICT, which is scaled between 0 and 1, depending on the number of business processes supported by ICT, was constructed for the purpose of econometric modelling.[13]

Majority of companies took advantage of ICT in three or four types of business processes. Only small percentage of firms used ICT in more than five types of processes. However, it should be stressed, that ICT investment activity was quite high—around half of companies invested in ICT within previous 24 months. This supports the conclusions stemming from analysis of sources of GDP growth

[13]The variable takes on value 0 if two business processes are supported by ICT, 1/7 if three processes are support, 2/7 in the case of supporting four business processes, etc.

presented in the previous section of the chapter. It also seems that ICT is perceived by managers as important (complementary) factor in the processes of implementation of innovations (Table 3).

Readiness to organizational change (a co-innovative source of productivity within a new firm paradigm) seems to be relatively high in Polish companies. Employees' teams are created relatively often. Moreover, in many enterprises employees participate actively in implementation of new solutions, and use flexible time arrangements. Interestingly, motivation pay systems were reported to be more common in the group of line workers than managers. ICT-related needs are also important in the human resources management processes—lot of emphasis is put on digital skills in the recruitment phase, which translates into attracting highly qualified human capital (thus employees' ICT skills are assessed as more than satisfactory) (Table 3).

5 Results and Discussion

Since the survey does not contain financial data information (this kind of data is sensitive and Polish managers are usually not willing to share such information within survey-type studies) it was not possible to use the CDM model to assess the impact of different types of innovations and ICT utilization on productivity in the Polish companies. Therefore, the parameters of the simultaneous discrete choice model was estimated, which put an emphasis on the role of co-innovative productivity sources as for productivity gains (Fig. 3).

In the first step, parameters of the multivariate probit model were estimated in order to identify determinants of companies' innovativeness. It was expected, that propensity of a firm to innovate shall depend on such factors like possession of own

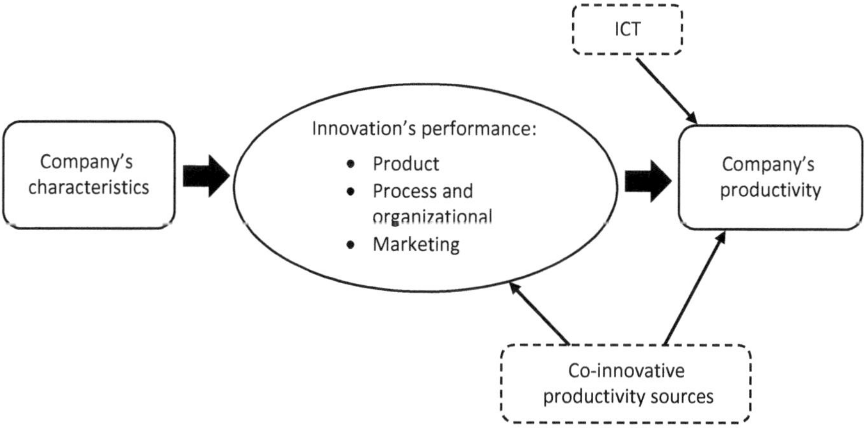

Fig. 3 Relations between innovations, ICT and productivity

Research and Development department (which is dependent on the size of a firm), quality of human capital, innovation-friendly environment within a company, and ICT investments. In the next step, parameters of classical linear regression model were estimated to explain productivity as a function of innovations, ICT use and introduction of complementary factors.

We distinguish three types of innovations (product, process-organizational, and marketing). The rationale for combining process innovations and organizational innovations was presented in the previous section. There is no doubt, they should positively influence company's productivity. Product innovations result from using new knowledge or technologies or implementation of new combination of these technologies and knowledge. So it also should have positive impact on labor productivity while new products may be produced faster and cheaper. It seems however, that the weakest impact on productivity is related to marketing innovations, as they focus more on customer needs and market positioning than on production processes.[14]

The appropriate tool for identifying innovations' drivers in such case is to use multivariate probit model. However, as different types of innovations are determining labor productivity, the endogeneity problem may arise. Because of this, we use the conditional expectations of unobservable latent variables in Eq. (2). As the firm-related factors of productivity, we distinguish scale of using ICT technology. We expect that this relation is stronger in large firms, possessing R&D department, which provide their workers with more flexibility in their working time management. As a result, parameters of the following model were estimated:

$$IN\_k_i^* = \mathbf{x}_i^{INk}\_\beta_k + \varepsilon_{ki}, \tag{1a}$$

$$IN\_k_i = I\{IN\_k_i^* > 0\}, \quad k = PROD, PROC, MARK \tag{1b}$$

$$PR_i = UICT_i\beta_1 + UICT_i\mathbf{x}_i^{COMP}\beta_2 + IN\_PROD_i^*\beta_3 + IN\_PROC_i^*\beta_4 + IN\_MARK_i^*\beta_5 + \varepsilon_{1i}, \tag{2}$$

where IN_PROD_i, IN_PROC_i and IN_MARK_i are dummy variables, which take on value 1 for firms, which introduced product innovation, process-organizational innovation or marketing innovation over last 2 years and 0 otherwise. $\mathbf{x}_i^{IN_PROD}$, $\mathbf{x}_i^{IN_PROC}$ and $\mathbf{x}_i^{IN_MARK}$ consists of variables, which are used as the determinants of the propensity to introduce product, process-organizational, and marketing innovations respectively.

Variable $UICT_i$ takes on values between 0 and 1 depending on the extent of using ICT technology in an enterprise. $\mathbf{x}_i^{COMP}$ consists of variables, which should have

[14]However, Kijek (2013) showed that product and marketing innovations are complementary, in the sense that product innovations induce marketing innovations. This relationship did not hold for process and marketing innovations, while technological innovations had only limited impact on introduction of marketing innovations in the Polish manufacturing firms.

Table 4 Definitions of explanatory variables

Variable	Definition
UEM	Dummy; 1, if majority of management of the company possess university degree
RD	Dummy; 1 for firms with their own Research and Development department
INF_CAP	Dummy; 1 for firms, which require ICT skills from all new employees
INVEST_ICT	Dummy; 1 for firms investing in ICT in last 24 months
SERVICES	Dummy; 1 for firms from the service industry
WORK_CH	Dummy; 1, if employees participate actively in changes (inside an enterprise)
MOT_WORK	Dummy; 1, if company introduced motivation pay system
Size	Logarithm of a number of workers in a firm
FWT	Dummy; 1 for firms, that take an advantage of flexible working time for their employees
WT	Dummy; 1 for firms, in which employees' teams are created
UEW	Dummy; 1 for firms, where dominating employees' education level is the university education
ORG	Synthetic index measuring the readiness to organization change of a given company

moderating impact on the relations between using ICT technology and productivity. We expect that $\beta_3 > 0$, $\beta_4 > 0$ and $\beta_1 + \sum_{j=1}^{K} \beta_{2j} > 0$.

Table 4 gives definitions of explanatory variables, while the results of the estimation of the parameters of the multivariate probit model explaining the propensity to introduce different types of innovations and results of the estimation of the productivity equation are presented in Table 5.

Probabilities of the implementation of the product, process-organizational, and marketing innovations are significantly larger for firms with internal R&D department in comparison to these, which do not have such department. The study also revealed positive interaction between ICT investments and innovations, especially in case of marketing and product innovations,[15] and to a lesser extent in case of process-organizational innovations. Nevertheless, it confirms general complementary relationship between ICT and innovations in Polish companies, which has been reported for other countries (Arvanitis et al. 2013; Gago and Rubalcaba 2007).

The positive relationship between employees' educational attainment and firm-level innovations has been revealed in many empirical studies (van Uden et al. 2014; Smith et al. 2011), in which human capital was treated as the main factor of innovation capacity. Our results reveal similar, however a bit more complex link. Managers' skills influenced only product innovations (if majority of managers had university diploma, probability of introducing product innovations was significantly larger than in firms which hired less educated management personnel),

[15]These results are in line with Spiezia's (2011) findings, arguing ICT is enabling factor especially of product and marketing innovations.

Table 5 Estimation results

Explanatory variable	Propensity to introduce product innovations—estimate	Propensity to introduce process-organizational innovations—estimate	Propensity to introduce marketing innovations—estimate	Productivity equation—estimate
Intercept	−2.37***	−2.10***	−2.23***	7.86***
RD	1.06***	0.59***	0.66***	–
UEM	0.42***	–	–	–
INF_CAP	0.29***	–	0.29***	–
INVEST_ICT	0.23**	0.17*	0.29***	–
WORK_CH	0.35**	0.24*	–	–
MOT_WORK	0.29***	0.17*	–	–
Size	0.17***	–	0.19***	0.04***
SERVICES	–	–	−0.38*	−0.13***
$\hat{INPROD}_i^*$	–	–	–	0.10***
$\hat{INPROC}_i^*$	–	–	–	0.03*
UICT	–	–	–	−0.44***
FWT*UICT	–	–	–	0.50***
WT*UICT	–	–	–	0.13**
UEW*UICT	–	–	–	0.13**
ORG*UICT	–	–	–	0.24***

*, **, *** denote significance respectively at the 0.1, 0.05, 0.01 level of significance

while digital competences of candidates were significant for product and marketing innovations (if all recruited employees possessed ICT skills, probability of introducing these types of innovations was larger by 0.10 than in case firms with less ICT-skilled new employees).

Higher probability of introducing product, as well as process-organizational innovations by, respectively, 0.08 and 0.06 were reported in companies which introduced motivation pay systems (Table 5). Enterprises, which encouraged employees to active participation in change processes, were characterized by higher probability of introducing product innovations, and process and organizational innovations by, respectively, 0.11 and 0.08. Probability of introducing marketing innovation was lower by 0.11 for firms operating in service industry sector, comparing to firms from the manufacturing industry branches.

The results showed the probability of introducing product and marketing innovations increases with the size of an enterprise (Table 5). Discussion on the relationship between firm size and innovation activities was initiated by Schumpeter, who argued that because of some monopoly power large companies may have a positive effect on innovations. However, this theoretical idea has not found empirical confirmation (Symeonidis 1996). Small and large firms may have different innovation strategies—large companies put more attention to process R&D and market expansion, while small ones focus on product R&D (Vaona and Pianta 2006; Plehn-Dujowich 2009). Despite this, it is still argued that large firms are often better than the small ones in introducing innovations, which seems to be the case in our sample of Polish companies.

In the following step conditional expectations of unobservable latent variables were calculated. These conditional expectations were introduced to the productivity equation. In our approach labor productivity was proxied by (log of) average gross wage of full time employees in each company, which is a solution applied in empirical studies by Lallemand et al. (2009), Faggio et al. (2010), and Mahy et al. (2011).

In order to eliminate the problem of heteroscedasticity, which arises when cross-section data is analyzed (especially in the case of wages), White heteroscedasticity—corrected estimator was used. In general, results revealed positive impact of innovations and ICT on labor productivity, with important role of co-innovative sources of productivity. Firstly, product innovations as well as process-organizational innovations proved to be productivity enhancing factors ($\beta_3 > 0$, $\beta_4 > 0$), while marketing innovations appeared to be statistically insignificant, which is in line with our expectations and results of other research studies. Secondly, ICT itself did not bring productivity gains but rather loses ($\beta_1 < 0$). However (and thirdly), combining ICT with other factors—labeled as ICT complementarities, or more broadly as co-innovative sources of productivity—led to productivity gains, and thus confirmed the initial assumption, that $\beta_1 + \sum_{j=1}^{K} \beta_{2j} > 0$. Except readiness to organizational change, other complementarities included high quality of employees' human capital and flexible work organization described by introduction of flexible working time arrangements and creation of employees' teams. Our study also revealed that labor productivity is influenced by type of business—companies in the service industry were characterized by lower productivity level, comparing to manufacturing industry firms (Table 5).

6 Conclusions

The chapter examined a joint link between ICT, innovations, and productivity in one of the best performing emerging economies in the CEE region. This firm-level analysis, based on unique data collected within the survey of 1000 Polish companies in the first half of 2015, combined two empirical research streams—CDM model and the new firm paradigm. We developed and estimated a simultaneous discrete choice model—we firstly identified determinants of firms' innovativeness, and then analyzed which types of innovations and co-innovative productivity sources linked to ICT usage are crucial for productivity gains.

The study revealed that there is positive link between all analyzed types of innovations in Polish companies, ICT investments, and R&D activities proxied by possession of internal R&D department. Thus, we proved innovations and ICT are complementary factors. The relationship between innovations and human capital as well as firms' organizational readiness appeared to be more complex. Managers' skills are important to introduce product innovations, while ICT skills of candidates significantly improve propensity to implement product and marketing innovations. At the same time higher participation of employees in decision-making processes and motivation pay systems enhance product and process-organizational innovations.

We showed that these two types of innovations influence positively labor productivity in the Polish firms. It appeared that the impact of marketing innovations on productivity is insignificant, but in line with expectations, as they primarily influence customer needs and market positioning than production processes. The study also confirmed crucial role of organizational change to reap benefits of ICT use in companies—our estimates pointed out that ICT alone do not translate into higher productivity, however if complemented by other mediating factors (co-innovative productivity sources) Information and Communication Technologies bring labor productivity gains in the Polish enterprises.

This conclusion is especially important from the point of view of policy-making processes in Poland (as well as other CEE countries). If interactions between different types of innovations, ICT and productivity are of similar nature in developed and developing countries, emerging economies should take advantage of available programs and solutions developed by high-income economies. We argue, that to promote innovations and labor productivity growth in Poland, support schemes should link investments in R&D and ICT with investments in intangible factors—co-innovative productivity sources. Unfortunately, such approach has not been widely used in operational programs that have been implemented in Poland within EU cohesion policy scheme.

Acknowledgments This chapter was prepared within the framework of research project "Impact of Information and Communication Technologies on productivity—macro and micro analysis" at the Department of Economic Policy, University of Lodz, financed by the National Science Centre (contract number DEC-2013/11/B/HS4/00661).

References

Aboal D, Tacsir E (2015) Innovation and productivity in services and manufacturing. The role of ICT investments. IDB working paper series, 658

Arvanitis S, Loukis EN, Diamantopoulou V (2013) Are ICT, workplace organization and human capital relevant for innovation? A comparative study based on Swiss and Greek micro data. KOF working papers, 333

Bertschek I, Engelsatter B, Szczygielski K (2010) The role of technology for service sector performance. SERVICEGAP review paper, 3

Black SE, Lynch LM (2004) What's driving the new economy? The benefits of workplace innovation. Econ J 114(493):97–116

Brynjolfsson E (2005) Seven pillars of productivity. Optimize, May 2005

Brynjolfsson E, Hitt LM (2003) Computing productivity: firm-level evidence. Rev Econ Stat 85 (4):793–808

Brynjolfsson E, Saunders A (2010) Wired for innovation. How technology is reshaping the economy. MIT Press, Cambridge

Cainelli G, Evangelista R, Savona M (2006) Innovation and economic performance in services: a firm-level analysis. Camb J Econ 30:435–458
Cardona M, Kretschmer T, Strobel T (2013) ICT and productivity: conclusions from the empirical literature. Inf Econ Policy 25:109–125
Crepon B, Duguet E, Mairesse J (1998) Research, innovation and productivity: an econometric analysis at the firm level. Econ Innov New Technol 7(2):115–158
CSO (2015) Działalność innowacyjna przedsiębiorstw w latach 2012–2014. Główny Urząd Statystyczny, Urząd Statystyczny w Szczecinie, Warszawa
Curley M, Salmelin B (2013) Open innovation 2.0: a new paradigm. http://ec.europa.eu/information_society/newsroom/cf/dae/document.cfm?doc_id=2182
David P (2002) Understanding digital technology's evolution and the path of measured productivity growth: present and future in the mirror of the past. In: Brynjolfsson E, Kahin B (eds) Understanding the digital economy. Data, tools, and research. MIT Press, Cambridge, pp 49–95
Diaz-Chao A, Sainz-Gonzalez J, Torrent-Sellens J (2015) ICT, innovation, and firm productivity: new evidence from small local firms. J Bus Res 68:1439–1444
Dimelis SP, Papaioannou SK (2010) FDI and ICT effects on productivity growth: a comparative analysis of developing and developed countries. Eur J Dev Res 22(1):79–96
Edquist H, Henrekson M (2016) Do R&D and ICT affect total factor productivity growth differently? IFN working paper, 1108
EIS (2017) European innovation scoreboard 2017. European Union
Faggio G, Salvanes KG, Van Reenen J (2010) The evolution of inequality in productivity and wages: panel data evidence. Ind Corp Chang 19(6):1919–1951
Freeman R, Lazear E (1995) An economic analysis of work councils. In: Rogers J, Streeck W (eds) Work councils. University of Chicago Press, Chicago, pp 27–52
Gago D, Rubalcaba L (2007) Innovation and ICT in service firms: towards a multidimensional approach for impact assessment. J Evol Econ 17(1):25–44
Goya E, Vayá E, Suriñach J (2013) Do Spillovers matters? CDM model estimates for Spain using panel data. Search working paper, 4/28
Gretton P, Gali J, Parham D (2004) The effects of ICTs and complementary innovations on Australian productivity growth. In: The economic impact of ICT: measurement, evidence and implications. OECD Publishing, Paris, pp 105–130
Griffith R, Huergo E, Mairesse J, Peter B (2006) Innovation and productivity across four European countries. Oxf Rev Econ Policy 22(4):483–498
Hall BH, Mairesse J, Mohnen P (2010) Measuring the returns to R&D. In: Hall BH, Rosenberg N (eds) Handbook of economics of innovation. North-Holland, Amsterdam, pp 1033–1082
Hall BH, Lotti F, Mairesse J (2012) Evidence on the impact of R&D and ICT investments on innovation and productivity in Italian firms. Econ Innov New Technol 22(3):300–328
Hempell T, van Leeuwen G, van der Wiel H (2006) ICT, innovation and business performance in services: evidence for Germany and Netherlands. ZEW discussion paper, 04-06
Huergo E, Moreno L (2011) Does history matter for the relationship between r&d, innovation and productivity? Ind Corp Chang 20(5):1335–1368
Jorgenson DW, Stiroh KJ (2000) Raising the speed limit: U.S. economic growth in the information age. Brook Pap Econ Act 231(1):125–236
Jorgenson DW, Vu K (2010) Potential growth of the world economy. J Policy Model 32 (5):615–631
Jung J, Mercenier J (2014) Routinization-biased technical change and globalization: understanding labor market polarization. Econ Inq 52(4):1446–1465
Kijek T (2013) An empirical analysis of the relationship between technological and marketing innovations: a case of polish manufacturing firms. Oeconomia 12(2):15–25
Kleis L, Chwelos P, Ramirez R, Cockburn I (2012) Information technology and intangible output: the impact of IT investment on innovation productivity. Inf Syst Res 23(1):42–59

Klonowski D (2011) Private equity in Poland. Winning leadership in emerging markets. Macmillan, New York
Koellinger P (2005) Why IT matters – an empirical study of e-business usage, innovation and firm performance. German Institute for Economic Research discussion paper, 495
Lallemand T, Plasman R, Rycx F (2009) Wage structure and firm productivity in Belgium. In: Lazear EP, Shaw KL (eds) The structure of wages: an international comparison. University of Chicago Press, Chicago, pp 179–215
Mahy B, Rycx F, Volral M (2011) Wage dispersion and firm productivity in different working environments. Br J Ind Relat 49(3):460–485
Milgrom P, Roberts J (1990) The economics of modern manufacturing: technology, strategy and organization. Am Econ Rev 80(3):511–528
Mohnen P, Hall BH (2013) Innovation and productivity: an update. Eur Bus Rev 3(1):47–65
Nguyen Thi TU, Martin L (2011) The relationship between innovation and productivity conditional to R&D and ICT use. An empirical analysis for firms in Luxembourg. Development and Policies Research Center (DEPOCEN), 11
OECD (2014) OECD science, technology and industry outlook 2014. OECD, Paris. https://doi.org/10.1787/sti_outlook-2014-en
Oliner SD, Sichel DE (2000) The resurgence of growth in the late 1990s: is information technology the story? J Econ Perspect 14(4):3–22
Oslo Manual (2005) Guidelines for collecting and interpreting innovation data, 3rd edn. https://doi.org/10.1787/9789264013100-en
Pilat D (2006) The ICT productivity paradox: insights from micro data. OECD Econ Stud 38
Plehn-Dujowich JM (2009) Firm size and types of innovation. Econ Innov New Technol 18 (3):205–223
Polder M, van Leeuwen G, Monhen P, Raymond W (2009) Productivity effects of innovation modes. Statistics Netherlands discussion paper, 09033
Schumpeter J (1994) Capitalism, socialism and democracy. Routledge, London
Smith A, Courvisanos J, Tuck J, McEachern S (2011) Building innovation capacity: the role of human capital formation in enterprises—a review of the literature, NCVER occasional paper
Solow R (1987) We'd better watch out. New York Times Book Review, July 12
Spiezia V (2011) Are ICT users more innovative?: an analysis of ICT-enabled innovation in OECD firms. OECD J Econ Stud 2011(1):1–21
Stiroh KJ (2002) Information technology and the U.S. productivity revival: what do the industry data say? Am Econ Rev 92(5):1559–1576
Stojcic N, Hashi I (2014) Firm productivity and type of innovation: evidence from the community innovation survey 6. Croat Econ Surv 16(2):121–146
Symeonidis G (1996) Innovation, firm size and market structure: Schumpeterian hypotheses and some new themes. OECD Economics Department working paper, 161
Szczygielski K, Grabowski W (2014) Innovation strategies and productivity in the Polish services sector. Post-Communist Econ 26(1):17–38
Van Ark B, Piatkowski M (2004) Productivity, innovation and ICT in Old and New Europe. IEEP 1(2–3):215–246
Van Uden A, Knoben J Vermeulen P (2014) Human capital and innovation in developing countries: a firm-level study. Institute for Management Research working paper, Radbound University Nijmegen
Vaona A, Pianta M (2006) Firm size and innovation in European manufacturing. Kiel working paper, 1284
Weresa MA (2012) Systemy innowacyjne we współczesnej gospodarce światowej. Wydawnictwo Naukowe PWN, Warsaw
Yang S, Brynjolfsson E (2001) Intangible assets and growth accounting: evidence from computer investments. MIT Center for Digital Business, p 136

Lukasz Arendt holds PhD in Economics from the University of Lodz. He is a Director of the Institute of Labour and Social Studies in Warsaw, and Assistant Professor in the Department of Economic Policy at the University of Lodz.

Lukasz Arendt's professional interests are focused on ICT-driven changes in economy at macro and micro level, labour market and lifelong learning. His research activities concentrates on the issues of digital divide, ICT impact on productivity in transition countries—especially in the area of ICT complementarities, Skill Biased Technical Change and polarisation hypotheses, and skills mismatch in modern labour markets.

He has participated as an expert in many national and international research projects. He is a co-author of a system of diagnosis skills needs at local and regional labour markets developed for the Ministry of Labour and Social Policy. Recently Lukasz Arendt has been involved in the project "Establishing the integrated forecasting and information system providing employment forecasts" that provides the employment forecast in Poland till 2022.

Wojciech Grabowski is an assistant professor in the Department of Econometric Models and Forecasts at the University of Lodz. He specializes in limited dependent and qualitative variables models and applies these methods in economics of innovation and entrepreneurship as well as in modelling macroeconomic phenomena (currency crisis, foreign exchange interventions). He published several papers in high-quality international and Polish journals, co-authored a few books and participated in many prestigious economic conferences. He has been working as co-investigator in several research grants. Currently, Dr. Grabowski is involved, as a project manager, in the grant financed by the Polish National Science Centre.

Equipment Lifecycle Management Framework

Pedro Alexandre Ferreira Fernandes
and Carlos Alberto Galamba Palma Pinto

Abstract With the recent evolution of Information and Communication Technologies, the widespread usage of applications to assist in multiple tasks has increased exponentially in the industry. This was made possible by switching from the older paper based management to dematerialized management. For most companies this is achieved using simple databases and applications that manage daily work. However, for some companies this is not enough. Those who deal with a large number of warehouses, most of them with a large storage space, have a need for more complex and evolved management applications, specifically in the area of logistic operations. To address this need, applications were developed to manage equipment and parts in a warehouse, also supported on mobile devices, helping the companies to plan and manage the operations more efficiently and achieving significant cost benefits. These management applications are essential in modern company's business, but for the most part are still limited to managing operations in the warehouse, only controlling the incoming, the stock and the outgoing of the equipment.

For other companies that have the requirements for further interventions on equipment outside the warehouse, such as managing procurement, maintenance or occurrences, it is vital to be able to follow the equipment lifecycle to anticipate and react efficiently to any situation that might occur. To harness this benefit, the companies need an application that can compile all the information needed from the suppliers and can directly collect data from the equipment while its being used, including, for example, life expectancy of the equipment, the effects of working in different environments and the number of hours of usage, among others. This information, when merged in a defined framework, helps to specify and forecast the equipment needs, allowing for example to plan equipment replacement or the need of preventive maintenance.

P.A.F. Fernandes (✉)
Nokia Solutions and Networks Portugal, S.A., GS DE Tools, Networks, Edifício Horizonte, Estrada do Casal do Canas, Alfragide, 2720-092 Amadora, Portugal
e-mail: paffernandes@gmail.com

C.A.G.P. Pinto
Universidade Atlântica, Barcarena, Portugal
e-mail: alberto.palma.pinto@gmail.com

A. Dias et al. (eds.), *Modeling Innovation Sustainability and Technologies*,
Springer Proceedings in Business and Economics,
https://doi.org/10.1007/978-3-319-67101-7_18

In this chapter we propose a definition of an Equipment Lifecycle Management Framework, thus enabling the design and creation of an application that will be able to deliver all the benefits of the Operational Dematerialized Management.

Keywords Equipment lifecycle • Management framework • Dematerialized management • Logistics

1 Introduction

Equipment management and repair has a significant economic impact in most companies. A study by Deloitte (2006) covering 120 industrial companies in North America, Asia Pacific and Europe, found that the spare part sales and the after-sales service account for 40% of sales of these companies.

The major equipment suppliers (OEMs—Original Equipment Manufacturers) are seeking to develop the after-sales potential of services by giving them increasingly relevance in their business model (Oliva and Kallenberg 2003). Major operators are nonetheless still doing maintenance activities, including the part and equipment needs planning, with the objective of reduce the cost of equipment ownership inherent in outsourcing these services.

To achieve this objective a Dematerialized Management software is often used by these companies to help manage all the logistic and operational information, processes and resources related with the equipment, normally inside the universe of a warehouse. But to further increase the saving in operations it is needed to go beyond the warehouse and follow the equipment during their lifecycle, thus using an Operational Dematerialized Management system to manage the information and processes.

2 Operational Dematerialized Management

The Operational Dematerialized Management is the usage of information technology to replace manual processes of collecting and providing operational support information to everyday management and decision taking. The information managed in operations management is extensive, contextual (depending on infrastructure, services and resources used) and situational (many exceptions exists in terms of incidents, loss of resources, changes to services), making this exchange of information expensive, even more when based on paper. The Operational Dematerialized Management systems are changing this paper based reality, which is still present in the overwhelming majority of companies, through the usage of technological mechanisms, including mobile platforms of communication and information (e.g. Smartphone), RF-ID tags, Cloud systems, GPS tracking, to name a few.

However, the systematic usage of operational dematerialized processes in companies is still far from being a common reality because most companies don't have a coherent strategy of operational dematerialized management. Their operational management applications are usually disconnected from the rest of the enterprise applications and not aligned with the operational processes implemented in the company.

3 The Operational Management Concept

In the logistics sector, transportation companies are one of many that can collect great benefits of the Operational Dematerialized Management concept, due to the permanent geographical dispersion of resources used and the high human component involved along with the constant interaction necessary with the respective management processes.

According to Cheng et al. (2003) the maintenance strategy in a transportation system must focus on security. In its empirical research, the authors conclude that preventive maintenance (based on the expectation of life) should be a top emphasis on corrective maintenance (fault-based. In this perspective, the interval of time between preventive maintenance actions depends on the life span of the equipment and the cost of the maintenance activity. However, preventive maintenance action cannot prevent faults to occur (random), making the total cost of maintenance dependant on the amount/ratio of preventive and corrective maintenance activities. Also the stock of parts depends heavily on the maintenance strategy. So, the selection of an appropriate strategy becomes complicated because the operator must take into account hardly measurable variables (safety, quality) along with the readily measurable ones (cost of maintenance, cost of inventory, cost of missing parts). Cassady and Maillart (1998) studied the principles for minimizing the maintenance cost by seeking the best replacement cycle, taking into account the preventive maintenance costs and corrective maintenance costs. Sarker and Haque (2000) suggest that the systems failure rate increases with its amortization and propose that the definition of a preventative maintenance period can minimize the cost when corrective maintenance costs outweigh the preventive maintenance costs.

Cheng et al. (2003) studied the importance of three factors and respective sub-factors, proposed as influencing the selection of the maintenance strategy such as: quality and efficiency (with the sub-factors high quality maintenance, increase the maintenance of technical efficiency, maintenance repair parts ready for use and reduction of vehicle failure rate); cost and reliability (with the sub-factors rolling stock maintenance in good condition, reduced maintenance costs and immobilization) and security (with the sub-factors ensure the safety of passengers and workers and ensure the safety of vehicles).

In summary, the authors identified the main factors that influence the cost of maintenance, most of them are information obtained from the equipment lifecycle like life expectancy, reliability, preventive maintenance effects and maintenance costs. Based on this factors and information, is very important to define the best maintenance strategy, for which an Equipment Lifecycle Framework is necessary, allowing to easily define the important information and the equipment lifecycle stages that influence the maintenance strategy, identifying the best action plan that lead to have the most efficiency, reliability, quality and security which translates into greater savings.

4 Equipment Lifecycle Management Framework

The concepts used to define the Equipment Lifecycle Management Framework are the warehouse management process, maintenance strategies and equipment categories. These concepts are used in most companies that have warehouse and repair systems in place to enable their business and are explained in the following paragraphs.

4.1 Warehouse Management

Companies with strong logistic needs usually have multiple warehouses that store products and equipment, were the management is done in a closed loop only, i.e. considering the products since the entry in the warehouse (e.g., new products from the factories or damaged products returned) until the warehouse exit (e.g., products sold to customers or sent to recycling). But for the equipment that is used internally and for supporting the company business, a different approach is normally used where operational and/or maintenance warehouses are used to manage the equipment and parts. These are usually next to maintenance workshops that may have different maintenance and repair capabilities, depending, for example, on the specialized equipment available in each one. This mean that the part needs of each workshop varies from one to another, compelling the use of a main centralized warehouse to manage all the parts and sorting the part delivery by each workshop needs (Fig. 1).

This maintenance warehouse network should address all the need of parts of the workshops in an efficient and timely manner, delivering the parts to the right warehouses at the right time. For this to be possible, the need of specific parts have to be foreseen and anticipated because the part suppliers can take too long to deliver if the part is requested only when needed. To avoid late deliveries the central warehouse stocks a calculated quantities of each part to respond to the maintenance needs, but this brings additional costs on the operation due to the initial investment on parts, the cost of storage space occupied by parts instead of other products, higher warehouse management effort and the possibility of having parts that are never needed and consequently never used. Using an Equipment Lifecycle Management Framework and a correct maintenance strategy enables the development of

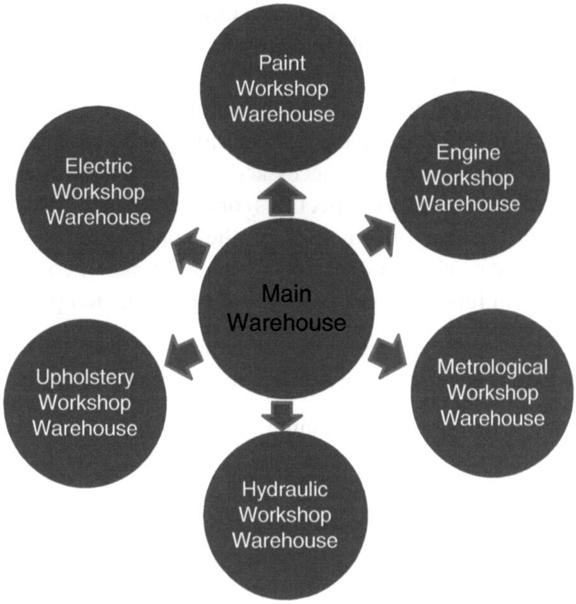

Fig. 1 Simple example of a maintenance warehouse network

a maintenance management software application that will help companies to avoid these additional costs and to achieve other improvements in the end to end maintenance process.

4.2 Maintenance Strategies

A study on maintenance strategies was done by Arts (2013) with the objective of having a deep understanding of the equipment lifecycle. Arts (2013) typifies the provision of maintenance services through a set of maintenance strategies, with different implications in terms of equipment management. These strategies are mainly:

- Rehabilitation or replacement of equipment for newer more advanced versions with the objective of improving the system performance. The main purpose of this strategy is the improvement of efficiency by upgrading obsolete (or near obsolete) equipment.
- Corrective maintenance or replacement of damaged parts. This strategy is appropriate for parts that do not wear out (e.g. electronic components) and can be subdivided into two main scopes:
  - Maintenance based on usage—the total usage of the equipment is measured and maintenance is done when a given level is reached. The usage can be measured in various ways depending on the nature of the equipment, but time

and mileage are the most common measures. As the usage of the equipment is usually scheduled, the time to maintenance is also predictable and is possible to perform the replacement of several parts of the same type in the same equipment maintenance operation.
 - Maintenance based on condition—the equipment is inspected and the maintenance done if necessary. The equipment/part condition can be measured periodically (inspections) or continuously via sensors (condition monitoring). The way of measuring the state depends on the nature of the device (e.g. visual inspection for metallic parts, contamination by ferrous particles in lubricating moving parts, lubricants temperature in engines).

- Preventive maintenance or equipment maintenance before malfunction. This strategy is most appropriate to parts that wear out (typically mechanical).

For maintenance planning, the need for inspection can be defined periodically (during inspection operations) or continuously if there is automated continuous monitoring system and the type of maintenance depends on the parts involved (corrective to the electronic and preventive for heavy machinery, for example). There is always a reasonable uncertainty in maintenance operations, particularly in terms of time (when performing) and content (what to keep or replace), but can be reduced with the proper monitoring and planning.

In the case of maintenance based on usage, interventions can be scheduled in accordance of the equipment condition through regular inspections, allowing an effective use of resources, but regarding corrective maintenance, even with continuous monitoring of the equipment condition, there will be unpredictability in terms of time, leading to waste of time and resources.

With all the factors involved, it becomes difficult to combine these different forms of maintenance in order to minimize costs, particularly in the case that large-scale equipment where various technologies coexist, requiring the usage of multiple types of maintenance strategies.

4.3 Equipment Categories

Each maintenance strategy has to be defined considering the equipment category to assure the best impact in savings and efficiency. The equipment is composed of various systems and subsystems, parts and components, each often with multiple technologies combined, so defining clear categories to manage the warehouse is of utmost importance.

The systems, subsystems, equipment, parts and components are defined in the following categories:

- Rotable/park parts—correspond to a sufficiently large subsystem, thus having a specific maintenance strategy. Each part (e.g. motors, bogies.) has an individual follow-up of its usage and is maintained by a set of specialized human and

maintenance resources. Maintenance based on usage is the most appropriate maintenance strategy.
- Serviceable parts—are parts that can be repaired after replacement and can be re-used (e.g. Compressor, pump) but are not followed individually in terms of usage or have specialized resources in its repair (a single shop can repair a wide range). Corrective maintenance and based on condition is the most appropriate maintenance strategy.
- Consumables—parts that are usually sent for recycling after its replacement (e.g. springs, lamps). Corrective maintenance and based on condition is the most appropriate maintenance strategy.

The inherent characteristics of each equipment category is what most influences the definition of the maintenance strategy, taking in consideration the life expectancy, the preventive maintenance intervals, the work environment, expected workloads and usage time.

In summary, companies need to have this multiple factors and information combined, analysed and inferred, enabling the definition of the best maintenance strategy and planning possible with data available. To support this need, the Equipment Lifecycle Management Framework was created.

4.4 *Equipment Lifecycle Management Framework*

The objective of the framework is to create a knowledge base towards creating a customized algorithm (each company need a different approach) that can be implemented on a software application in order to combine, analyse and infer all the factors and equipment data, delivering the best maintenance strategy and plan for each equipment.

The framework follows the equipment through its complete lifecycle, since acquisition until recycling, as Fig. 2 shows. In summary, the framework steps are the following:

- Acquisition—When a need is detected, the purchase department starts the procurement process to acquire the equipment from the best possible supplier. When received the equipment is received in the company warehouse.
- Warehouse Entry—When the equipment is received all its information must be inserted in the warehouse management system. At this stage, the following information of the equipment must be registered:
  - Life Expectancy
  - Hours/Km recommended between scheduled maintenance
  - Environmental effect rates, defined from 1 to 5 (1—Very Bad; 2—Bad; 3—Average; 4—Good; 5—Very Good)
  - Workload effect rates defined from 1 to 5 (1—Very Low; 2—Low; 3—Average; 4—High; 5—Very High)

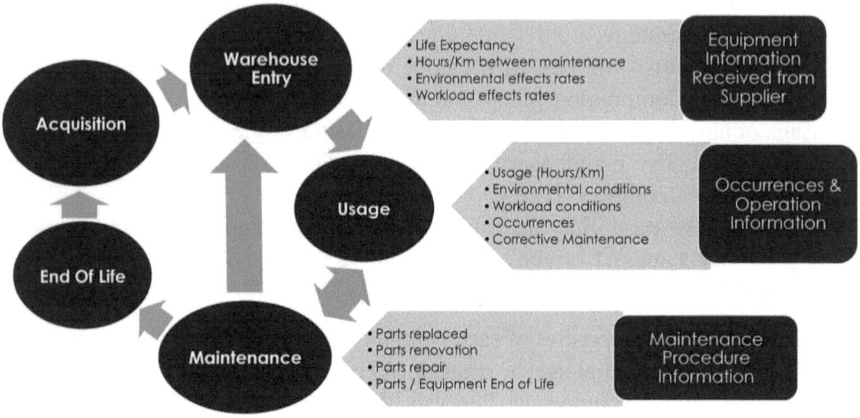

Fig. 2 Equipment Lifecycle Management Framework

- Usage—The equipment is in use and the usage information is collected and registered in the warehouse management system. The main information collected is:
  - Hours/Km of use
  - Environmental conditions
  - Workload conditions
  - Occurrences
  - Malfunctions (corrective maintenance)
- Maintenance—Preventive and corrective maintenance is performed on the equipments and registered on the warehouse management system. The information collected should be:
  - Parts replaced
  - Parts renovated
  - Parts repaired
  - Equipment/parts that have met the End of Life/sent to recycling
- End of Life—When the end of life of a part is reached, the part is sent to recycling. If the part is still needed, a new acquisition may be requested.

In the framework, the Equipment Lifecycle starts by its acquisition. The supplier delivers the equipment together with the information needed to be registered in the Warehouse Management System. This allows the application to define the maintenance strategy and an initial maintenance planning. To the life expectancy and the time between preventive maintenance is added or reduced the index associated to the environmental and workload effects, for example, if the rates are both Average, no variation factor is applied, but if the effects are Bad, a negative variation factor would be applied making the life expectancy shorter and reducing the time between maintenances. During the usage of the equipment also the information registered is used to update the life expectancy and maintenance schedule. If the conditions get better, for example, a positive variation factor is applied and the life expectancy increased. When the usage reached the define value (hours or Km), the maintenance request is triggered and the equipment sent to the workshop. When the parts are replaced, the life expectancy and time between maintenance is reset always considering the most recent environmental and workload conditions. This also happens with the renovated and repaired parts, but is after applied a "used part" negative factor because in fact the part is not brand new. In the end of the maintenance operation the equipment has a physical or virtual warehouse entry, i.e., if the equipment was substituted by other one, it is stored until needed; if not he is registered in the system as back in the warehouse and immediately after registered as being in use, not physically entering the warehouse. The disposed parts are sent to recycling and if the warehouse has few of these parts, a new acquisition is requested, restarting the framework cycle.

4.5 Future Work

To allow seamless integration of the framework in multiple enterprise environments, the creation of standard algorithms that covers the most enterprise realities possible is necessary. The availability of these standards for the most usual situations allows efficient and quick implementations but for more specific cases, the framework will still need to be adjusted to the factors affecting each company reality. The algorithms need to be implemented in a software application to allow testing in enterprise environment. It can then be analysed while in utilization to understand its strengths and shortcomings, enabling the understanding of its real value. Also enables the identification of areas and issues where it can be adjusted and further improved.

References

Deloitte (2006) The service revolution in global manufacturing industries, s.l. Deloitte Research

Pedro Fernandes holds a Bachelor's Degree on Information Technology from Atlântica University. He is a Certified SCRUM Product Owner at Nokia Networks and has 17 years of experience in telecommunications industry.His professional interests are focused on software development applied to operational improvements. He is involved in the development of software tools for the mobile telecommunications industry, specialized on mobile network planning and optimization, where he uses his experience in driving operational improvements in the day to day operations and delivering quality services to telecommunications operators.He developed the "Equipment Lifecycle Management Framework" paper as a final dissertation for his Bachelor's Degree on Information Technology, in collaboration with Carlos A. Galamba Palma Pinto. The objective was to create a framework that enabling the design and creation of tools and applications to drive operation efficiency improvements on the operational equipment lifecycle. Recently at Nokia, Pedro Fernandes has been involved in the Automation, Virtualization and Analytics (AVA) project, a cloud-based cognitive services platform that combines big data storage, intelligent analytics and extreme automation to enable operators to move from reactive network operations to a cognitive approach in which degradations are predicted and changes are rapidly implemented through automation.

Carlos Pinto Doutorate in Management—Specialized in Technology and Operation Management by Instituto Universitário de Lisboa (ISCTE); Master Degree in Enterprise Management—Specialized in Information Management by Universidade Católica Portuguesa (UCP); Degree in Computer and Electrotecnic Engineering by Instituto Superior Técnico (IST); Degree in Telecomunications and Electronic Engineering by Instituto Militar dos Pupilos do Exército; was previously a Professor at Instituto Superior de Gestão Bancária and Service Manager at Philips Portuguesa. Is actually a professor at Universidade Atlântica on the Information Technology and Systems and the Systems Managing and Computing degrees and also is a Technical Manager at CP Comboios de Portugal where he works in the Inovation and Auditing areas.

The IT Audits in the Spanish Business Sector: Longitudinal Analysis (2001–2011)

Alfonso Infante-Moro, Juan-Carlos Infante-Moro, Francisco-José Martínez-López, Mercedes García-Ordaz, and Albertina Dias

Abstract In a society in which Information Systems and Information are considered like important assets for most companies, the realization of IT audits to measure its efficiency and prevent IT support breakdowns, is required. To this aim, this paper analyzes the evolution of the IT audit practice in Spanish companies, reporting the percentages of realization and the amount of professionals that execute such audit, and how it correlates with the importance that these companies give to Internet and the use of e-commerce. Our longitudinal study, encompassing the period 2001–2011, has shown substantial increase in the awareness of Spanish companies towards the realization of these audits: almost doubling the percentage of companies that performed them in this decade. We have also found that the realization of IT audits usually involves several types of professionals of the same company. Finally, we were able to analyze the association between the adoption of Internet and the use e-commerce in surveyed companies, and the realization of these IT audits.

Keywords IT audit • Spanish business sector • Spanish companies • IT security • Information systems • ICT

1 Introduction

In a society where Information Systems and Information Technology play an increasingly important role in business, it is necessary to conduct IT audits to measure efficiency and prevent computer problems in the company, but not all companies perform them.

A. Infante-Moro (✉) • J.-C. Infante-Moro • Francisco-José Martínez-López • M. García-Ordaz
GITICE Research Group, University of Huelva, Huelva, Spain
e-mail: alfonso.infante@uhu.es; juancarlos.infante@uhu.es; francis@uhu.es; ordaz@uhu.es

A. Dias
GITICE Research Group, New Atlantic University, Barcarena, Portugal
e-mail: tina.melo.dias@gmail.com

A. Dias et al. (eds.), *Modeling Innovation Sustainability and Technologies*,
Springer Proceedings in Business and Economics,
https://doi.org/10.1007/978-3-319-67101-7_19

In this paper we study the evolution of the IT audit's implementation within the Spanish business sector, both in its percentages and the professionals that perform it, and its link with the importance these companies give to the Internet and the use of e-commerce.

For this, we performed a longitudinal study from research group GITICE, which began 20 years ago and is still working today. Specifically, we performed macrosurveys on Information Systems in Spanish companies with 211 variables every 10 years: 1991–1992, 2001–2002 and 2011–2012. And for this paper we have selected six variables corresponding to 2001–2002 and 2011–2012.

In the next point, we frame the IT audit through a state of the art viewpoint. And we continue with the methodology and analysis of results in three points: 'IT audit', 'Relationship between Importance of Internet and IT Audit' and 'Relationship between e-commerce and IT Audit', concluding with the main features in the evolution of such IT audits' implementation in Spanish companies.

2 State of the Art

The technological development in companies is increasing and it could create problems within them, as this technology is present in many of their activities (Infante Moro et al. 2014a, b; Luna Huertas 2013). And these problems, created by computer failures, are caused mainly by:

- Incorrect use of computers and computerized information, and poor storage of such information;
- Lack of security measures for the software's upkeep;
- Lack of a contingency policy for computer problems;
- Low speed and lack of optimization of computer networks;
- Security problems in e-commerce.

Therefore, companies act preventively through IT audits to determine if information systems safeguard business assets. In short, IT audits are preventive protocols, carried out in order to:

- Promote the efficiency of technological resources;
- Establish a preventive maintenance policy of Information Systems for the use of equipment;
- Establish a policy of using Information Systems for their employees;
- Analyze the efficiency of computer networks;
- Implement a policy of online security;
- Create an action plan in case of computer problems.

These are all carried out by professionals, either IT internal auditors who belong to their own company, IT auditors relating to the audit of accounts and IT external auditors.

All this information is the work of the authors who are mentioned in this paragraph, in which we focus more on this study, in the degree of implementation of computer audits and in the typology and origin of the professionals who perform it, something that has been studied for more than 45 years, as the first association of professionals on this topic was created in 1966, to which we belonged for some time and have dedicated some of our research (Martínez López 1991). In this area, in our country, we can highlight the works of Poveda (1995), Piattini (Piattini and Del Peso 2001; Piattini Velthuis 2008) or Calvo-Manzano (Calvo-Manzano et al. 2002; Muñoz et al. 2012). And, in academia, stand Ferreira Rodrigues et al. (2013), Martins (2013), Ibarra (2014), Térmens (2014) and Faura et al. (2015).

3 Methodology

In the field of social sciences, it is necessary to have tools that provide primary and specifical information that will be studied, to allow us to work with empirical data. For this, we turn to the process of survey, as it represents one of the most important tools for advancing in the field of social sciences.

On this basis, and in order to fulfill the proposed goal stated above, we selected a part of the survey performed in the Spanish business sector (which we named in the introduction), in late 2001 and early 2002 and late 2011 and early 2012, that consists of six variables: importance of Information Systems and Information Technology in companies, level of importance of the Internet in companies, use of e-commerce to buy and/or sell, performing IT audits and the people responsible for such conduct in companies.

In the survey that we call SPANISH COMPANIES, we use sampling theory to get statistically significant data on the study population, which consists of Spanish companies located in the top 25,000 by turnover included in the database 'España 25.000' (in surveys performed in all the years, we have acquired the data from the same institution, the journal 'Fomento de la Producción').

In total, 395 companies participated in 2001–2002 and we did 411 surveys in 2011–2012, whose answers were validated in 2013. There was a lack of significant response due to the minimum number of variables, so qualitative variables were used which were then analyzed through non-parametric statistical model: X^2 association test (two qualitative variables).

The overview of the research can be seen in Table 1:

Table 1 Technical details of the research

	Spanish companies	
Survey	2001–2002	2011–2012
Hypothetical universe	Spanish companies	Spanish companies
Target population	Spanish companies, whose turnover exceeds 1 million euros annually	Spanish companies, whose turnover exceeds 2 million euros annually
Population mark	Database "España 25.000", *Fomento de la producción*	Database "España 25.000", *Fomento de la producción*
Sampling error	±0.05	±0.049
Confidence level	95.50%	95.50%
Hypothesis parameters	Z = 2 and P = Q = 0.5	Z = 2 and P = Q = 0.5
Sample size	395 companies	411 companies
Sampling procedure	Random sampling without replacement (random numbers)	Random sampling without replacement (random numbers)
Method survey	Questionnaire by email, postal mail, telephone contact, fax and in person	Questionnaire by email, postal mail, telephone contact, fax and in person

4 Discussion and Results

This study is divided into three main points: 'IT audit', 'Relationship between Importance of Internet and IT Audit' and 'Relationship between e-commerce and IT Audit'

4.1 IT audit

The high importance of Information Systems and Information in Spanish companies is observed even in 2001–2002. It remained relatively constant, as 'essential' or 'a lot' in the last decade in the Spanish business sector (see Table 2).

So these companies have had to perform IT audits to measure the efficiency of these Information Systems and prevent computer problems, as more and more functions are used within these companies (Infante Moro et al. 2014a, b).

In 2001–2002, only approximately 30% of the sector performed IT audits, which has changed in 2011–2012 with an almost double increase (exactly 26.82%) (see Table 3). Companies are increasingly aware of the importance of performing this task.

These IT audits are performed by IT internal auditors who belong to their own company, IT auditors relating to the audit of accounts and IT external auditors. In 2001–2002, concerning the performance of these audits, the Spanish companies showed a preference for IT internal auditors who belong to their own Companies and IT external auditors (see Table 4).

Table 2 Importance of information systems and information technology in the Spanish business sector

Level of importance	Spanish companies	
	2001–2002 (%)	2011–2012 (%)
Essential	67.59	69.83
A lot	24.30	19.46
Normal	6.58	9.00
Low	1.27	0.73
Nothing	–	–
No answer	0.25	0.97

Table 3 Companies that perform IT audits

Perform IT audits	Spanish companies	
	2001–2002 (%)	2011–2012 (%)
Yes	29.87	56.69
No	70.13	42.09
No answer	–	1.22

Table 4 Professionals who perform IT audits

Level of importance	Spanish companies	
	2001–2002 (%)	2011–2012 (%)
1	43.97	22.84
2	16.38	18.53
3	39.66	32.76
1–3	–	5.60
1 and 2	–	7.33
1 and 3	–	8.19
2 and 3	–	4.74

1—IT internal auditors who belong to their own company
2—IT auditors relating to the audit of accounts
3—IT external auditors

This trend did not change during this decade, since a preference for these professionals still prevails, but the performance of various IT audits by different types of professionals in the same company appears. Although very few companies (only 25.86%) choose these varied options (see Table 4).

4.2 *Relationship Between Importance of Internet and IT Audit*

The importance of the Internet in companies has also increased in the last decade, and proof of this is the growth of companies that consider the Internet essential in their operations, by almost 30% (see Table 5).

Table 5 The importance of the Internet in Spanish companies

Level of importance	Spanish companies 2001–2002 (%)	2011–2012 (%)
Essential	12.91	42.58
A lot	24.05	32.60
Normal	42.78	22.63
Low	18.48	1.70
Nothing	1.27	–
No answer	0.51	0.49

Table 6 Relationship between importance of internet and performance of IT audits in Spanish companies

Level of importance given to the internet	Companies that perform IT audits, Spanish companies, 2001–2002: Yes	No	2011–2012: Yes	No
Essential	49.0%	51.0%	59.2%	40.8%
A lot	31.6%	68.4%	56.0%	44.0%
Normal	28.4%	71.6%	54.9%	45.1%
Low	17.8%	82.2%	71.4%	28.6%
Nothing	40.0%	60.0%	–	–
Degree of freedom	4		3	
Confidence level	5%		5%	
Critical X^2	9.49		7.81	
Calculated X^2	14.501		1.129	
Correlation	Yes		No	
Margin of error	0.006		0.77	

But this level of importance that companies give to the Internet currently has no connection with the performance of IT audits, although this relationship existed in 2001–2002, where it was observed that the higher the level of importance given to the Internet, the lower the number of companies that do not perform IT audits (see Table 6).

The performance of IT audits has become more widespread today, and even companies that give low importance to the Internet, mostly carried out such audits (see Table 6).

And we can also confirm there is no relationship between the importance that these companies give to the Internet and the election (by the company) of the professionals who perform these audits (see Table 7).

Also noteworthy in 2001–2002 are the choice of 'IT internal auditors who belong to their own company' in companies where the importance of the Internet is considered essential or normal, and the choice of 'IT external auditors' in companies where the importance of the Internet is considered a lot.

Whereas in 2011–2012, companies where the Internet is considered essential are opting for 'IT external auditors' and 'IT internal auditors who belong to their own company'. In addition, to start, the option of these IT audits will be performed by

Table 7 Relationship between importance of internet and professionals who perform IT audits in Spanish companies

Level of importance of the internet	Professional who performs IT audits in company									
	Spanish companies									
	2001–2002			2011–2012						
	1	2	3	1	2	3	1– 3	1 & 2	1 & 3	2 & 3
Essential	50.0%	20.8%	29.2%	27.2%	18.4%	33.0%	4.9%	5.8%	6.8%	3.9%
A lot	40.0%	13.3%	46.7%	23.0%	13.5%	27.0%	5.4%	8.1%	16.2%	6.8%
Normal	45.8%	16.7%	37.5%	16.0%	26.0%	38.0%	8.0%	10.0%	–	2.0%
Low	41.7%	16.7%	41.7%	–	20.0%	60.0%	–	–	–	20.0%
Nothing	–	–	100.0%	–	–	–	–	–	–	–
Degree of freedom	8			18						
Confidence level	5%			5%						
Critical X^2	15.51			28.87						
Calculated X^2	4.972			24.575						
Correlation	No			No						
Margin of error	0.761			0.137						

1—IT internal auditors who belong to their own company
2—IT auditors relating to the audit of accounts
3—IT external auditors

different types of professionals in the same company, but through no relation to the importance that companies give to the Internet (see Table 7).

4.3 Relationship Between e-Commerce and IT Audit

Regarding e-commerce, an increase is also observed in this activity. In its role of both purchasing and selling, it has grown in the last decade (see Table 8).

4.3.1 Companies That Use e-Commerce to Buy

This growth in the use of e-commerce to buy can be studied as a possible relationship between this activity and the performance of IT audits by the company due to the use of Information Systems in economic transactions and the importance of these to work correctly. A relationship does exist: most companies do not use e-commerce to buy, and do not perform these audits. A situation that is disappearing over time, but is emphasizing its use with such embodiment (see Table 9).

But of course, there is some relationship between the use of electronic commerce to buy and the choice (by the company) of professionals who perform such audits (see Table 10).

Table 8 Use of e-commerce to buy and sell in Spanish companies

e-Commerce	Spanish companies 2001–2002 (%)	Spanish companies 2011–2012 (%)
To buy	23.54	60.44
To sell	21.01	43.46

Table 9 Relationship between e-commerce to buy and IT audits in Spanish companies

	IT audits			
	Spanish companies			
	2001–2002		2011–2012	
Use of e-commerce to buy	Yes	No	Yes	No
Yes	40.9%	59.1%	65.7%	34.3%
No	26.5%	73.5%	44.0%	56.0%
Degree of freedom	1		1	
Confidence level	5%		5%	
Critical X^2	3.84		3.84	
Calculated X^2	7.009		18.526	
Correlation	Yes		Yes	
Margin of error	0.008		0	

Table 10 Relationship between e-commerce to buy and professionals who perform IT audits in Spanish companies

	Professionals who perform IT audits in the company									
	Spanish companies									
	2001–2002			2011–2012						
Use of e-commerce to buy	1	2	3	1	2	3	1–3	1 & 2	1 & 3	2 & 3
Yes	52.6%	13.2%	34.2%	22.4%	19.3%	32.9%	5.6%	6.8%	10.6%	2.5%
No	39.7%	17.9%	42.3%	23.2%	17.4%	31.9%	5.8%	8.7%	2.9%	10.1%
Degree of freedom	2			6						
Confidence level	5%			5%						
Critical X^2	5.99			12.59						
Calculated X^2	1.746			9.708						
Correlation	No			No						
Margin of error	0.418			0.137						

1—IT internal auditors who belong to their own company
2—IT auditors relating to the audit of accounts
3—IT external auditors

4.3.2 Companies That Use e-Commerce to Sell

Regarding the use of e-commerce to sell, this relationship exists only in 2011–2012, where there are significant differences in the performance of IT audits between those using e-commerce to sell and those who do not (see Table 11).

And with all of this, we found no relationship between the use of e-commerce to sell and the election (by the company) of the professionals who perform such audits (see Table 12).

Table 11 Relationship between e-commerce to sell and IT audits in Spanish companies

	IT audits			
	Spanish companies			
	2001–2002		2011–2012	
Use of e-commerce to sell	Yes	No	Yes	No
Yes	37.3%	62.7%	67.4%	32.6%
No	27.9%	72.1%	50.0%	50.0%
Degree of freedom	1		1	
Confidence level	5%		5%	
Critical X^2	3.84		3.84	
Calculated X^2	2.803		12.118	
Correlation	No		Yes	
Margin of error	0.094		0	

Table 12 Relationship between use of e-commerce to sell and professionals who perform IT audits in Spanish companies

Use of e-commerce to sell	Professionals who perform IT audits in company									
	Spanish companies									
	2001–2002			2011–2012						
	1	2	3	1	2	3	1–3	1 & 2	1 & 3	2 & 3
Yes	45.2%	12.9%	41.9%	23.3%	18.1%	31.0%	3.4%	10.3%	8.6%	5.2%
No	43.5%	17.6%	38.8%	21.6%	18.9%	35.1%	7.2%	4.5%	8.1%	4.5%
Degree of freedom	2			6						
Confidence level	5%			5%						
Critical X^2	5.99			12.59						
Calculated X^2	0.381			4.548						
Margin of error	No			No						
Margin of error	0.826			0.603						

1—IT internal auditors who belong to their own company
2—IT auditors relating to the audit of accounts
3—IT external auditors

5 Conclusions

This study describes the evolution of the implementation of IT audits in the Spanish business sector, both in the percentages that it is conducted and the professionals that perform it, and its link with the importance that these companies give to the Internet and the use of e-commerce.

In a society in which Information Systems and Information Technology are considered essential for their companies, the performance of IT audits is required to measure their efficiency and prevent IT problems. We found that in the Spanish business sector the use of IT audits has experienced a boom in this decade, with the percentage of companies that performed it in 2001–2002 almost doubling, demonstrating these companies' increased awareness of their efficiency and the important role these tools play. Ratified through the choice of professionals who perform this task, throughout the decade the choice of IT internal auditors of their own companies and IT external auditors has prevailed, but also increasingly the companies that choose to carry out such audits by different types of professionals within the same company (25.86% of these companies chose this option in 2011–2012).

But we cannot link this increase in the percentage of Spanish companies that perform IT audits with the importance of the Internet within them today, because this awareness of IT audits makes them almost indifferent to it, which was significant in 2001–2002, where a greater performance of these audits would have been observed if these companies gave more importance to the Internet.

Instead, there is a relationship between the use of electronic commerce to buy and sell by the companies, and the performance of IT audits, except in 2001–2002, when there was no relationship between e-commerce to sell and the performance of these audits on them.

And, datawise, we must also stress there is no link between the importance of the Internet in companies and the use of e-commerce to buy and sell, and the choice of professionals who perform IT audits.

This research is limited to companies whose turnover exceeded an annual high amount (1 million per year in 2001–2002 and 2 million per year in 2011–2012), and this amount varies with the increase of these companies' turnover in this decade: companies with sizeable information systems which play a key role in their operations give us these high percentages of performing IT audits. This limitation may be removed in future studies, which may be extended to all companies (small, medium and big without depending on their level of turnover), and specific studies on sectors within Spanish business, for later comparison to Spanish companies in general, to better understand the situation of those IT audits in their markets.

References

Calvo-Manzano J, Carrillo J, Cuevas G, San Feliu T, Tovar E (2002) Introducción a la Auditoría Informática. Facultad de Informática de Madrid

Faura V, Momplet Badía R, López García JA (2015) Seguridad de la información más allá de la seguridad informática. Auditoría interna: publicación periódica del Instituto de Auditores Internos de España 31(108):20–22

Ferreira Rodrigues R, Fernandes Martins V, Souza Carmo CR (2013) Auditoria de sistemas de informática nas empresas modernas. Revista Científica Linkania Master 1(6):42–59

Fomento de la Producción. España 25.000 (2011 y 2001) http://www.fomentodigital.com/busqueda/empresa/espana25000.jsp. Accessed Oct 2011 and 2001

Ibarra JD (2014) Auditoría tecnológica con fines de seguridad Informática. Universidad Distrital Francisco José de Caldas

Infante Moro A, Martínez López FJ, García Ordaz M, Infante Moro JC (2014a) El teletrabajo en las grandes empresas del sector hotelero: El caso español. En: Turitec 2014: X congreso nacional turismo y tecnologías de la información y las comunicaciones. Universidad de Málaga (UMA), Escuela Universitaria de Turismo, pp 309–321. doi:10.13140/2.1.3571.2009

Infante Moro Á, Martínez López FJ, García Ordaz M, Infante Moro JC (2014b) Information exchange in big companies of the Spanish hotel sector: an inside and inter-organizational comparative analysis. En: TMS Algarve 2014: Management Studies International Conference. University of the Algarve, School of Management, Hospitality and Tourism, pp 209–220. doi:10.13140/2.1.2863.4400

Luna Huertas P (2013) La Percepción de la Importancia de los Sistemas de Información en la Empresa Española: un Análisis Longitudinal (1991–2012). Finanzas y Sistemas de Información, Andavira. ISBN 978-84-8408-702-1

Martínez López FJ (1991) La conceptualización de la auditoria informática y su implementación en el marco organizativo. Revista de Economía y Empresa 11(29–30):123–142

Martins I (2013) Auditoria dos sistemas de informação das instituições financeiras. Instituto Politécnico de Lisboa

Muñoz M, Mejia J, Calvo-Manzano Villalon JA, San Feliu Gilabert T (2012) Evaluación de los procesos software de una organización enfocando en la disminución de la resistencia al cambio. Comput Sci Eng 2(4):66–73

Piattini M, Del Peso E (2001) Auditoría Informática. Un enfoque práctico

Piattini Velthuis M (2008) Auditoría de tecnologías y sistemas de información. Ra-Ma

Poveda Maestre JP (1995) Valoración de los objetivos de control por los auditores informáticos españoles. Partida doble 56:50–57

Poveda Maestre JP, Alventosa Bernabéu LM (1994) Auditoría informática: controles de aplicaciones. Técnica Contable 46(548–549):585–592

Térmens M (2014) Las auditorías, una metodología para planificar la preservación digital. Experiencias en España. Ciência da Informação 41(1):140–142

Alfonso Infante-Moro, Ph.D. is Professor Ph.D. in Economics of Department of Financial Economics, Accounting and Operations Management at University of Huelva and member of Research Group "GITCE". Director of Virtual Education Unit at University of Huelva 2003–2013. Editor of "Campus Virtuales" and "International Journal of Information Systems and Software Engineering for Big Companies (IJISEBC)", scientific journals. He has published numerous research papers on various technological aspects of companies and he has participated in several research projects funded by European Union, related to ICT in companies and many others in national and regional convocations.

Juan-Carlos Infante-Moro, Ph.D. is Professor Ph.D. in Tourism Management of Department of Financial Economics, Accounting and Operations Management at University of Huelva. He is degree in Business Administration and is Master in Business Administration at University of Huelva. Specialized in Information Systems and Technologies. He has published numerous research papers on various technological aspects of companies.

Francisco-José Martínez-López, Ph.D. is Professor of Financial Economics and Accounting at University of Huelva. He is teaching for over 25 years. PhD in Economics and Business Administration (with honors). Member of the research group GITICE since 1993. He has been Rector of University of Huelva from 2005 to 2013. He has published numerous research papers on various technological aspects of companies. President of International congresses of ICT and companies. He has given lectures and masters and doctoral courses in various institutions European, American, African and Asian. Author of over 300 scientific papers.

Mercedes García-Ordaz, Ph.D. is Professor of Department of Financial Economics, Accounting and Operations Management at University of Huelva. She has published extensively on Information Systems and Tourism. Expert on merger of entities. She has won major awards like 'Arco Iris' Award on Research. She has published numerous books, book chapters and more than 50 scientific papers in national and international journals. She is director of Research Group GITICE, which is composed of professors from the Universities of Seville, Huelva, Pablo de Olavide and Granada. She has given numerous doctoral courses and has been research in more than 20 research projects, national and international.

Albertina Dias holds a Ph.D. in Economics, a Master in Statistics and Information Management and a Degree in Economics from the Nova University of Lisbon. She is a Chairman and CEO of the MIST association in Lisbon, and among several academic positions she is a Professor in the Department of Business at the New Atlantic University. Albertina has been working actively in the knowledge triangle having run her own company on innovative ICT between 2005 and 2013 (SbH, Ltd., developing multimedia interactive books, based on augmented reality technology and computer aided eBook interaction systems). In the past she has assumed several levels of responsibilities in companies such as Paulo Branco Holding (CEO, film industries), Oracle (Financial Analyst), Allianz (Financial Analysis), and Cap Gemini (Project Manager). Under the Program Equal e-Change (EU funded) she participated since 2005 for the presentation, demonstration and validation of innovative products in the areas of flexible security on employment and audiovisual and multimedia tools. She was responsible for the audiobook library of the Economists' College in PT and has produced several audiobooks under the economic content as for instance the "Portuguese Governmental Budget-2006", "2nd Annual Conference of Economists-2007", "SCUTS-An Economic Analysis, 2005", among others. Since 2008 she is an Independent Expert for the European Commission, acting as a Reviewer/Evaluator for the implementation of indirect actions under specific programmes. Since 2005, she is a fellow researcher of GITICE/ Universidad de Huelva in Spain (scientific domain of Economics and research line of ICT). In this institution she is developing research on Innovative Tools towards Regional Economic Development.

Part V
Environmental Sustainability Innovation Management: Case Studies in the Southern European Countries

The European Union (EU) action plan for the circular economy envisions a sustainable, low carbon, resource efficient and competitive economy where the value of products, materials and resources is maintained in the economy for as long as possible, and the generation of waste is minimized. The main idea of this action plan consists in preventing the scarcity of resources, by helping to create innovative businesses and technologies that may allow higher efficiencies both in producing and consuming patterns.

This EU initiative involves the need to create regulatory framework that can stimulate investments as well as the proper conditions for industrial innovation, and to remove obstacles from European legislation or its inadequate enforcement. Regarding the legislation on waste, adopted together with the action plan for the circular economy, the aims are to reduce landfills and promote conversion to reuse and recycle waste. In a similar way water may be seen as a natural raw material, being its wastes, the wastewater, important in terms of reuse after a severe treatment. There is a need for a comprehensive interaction studies among water, waste, soil, air and living ecosystems promoting more legislative proposals on related issues like waste, fertilizers and water reuse.

The urban development, both in terms of growing demographic density and of industrial concentration, along with economic growth, have generated vast amounts of domestic and industrial wastewater and solid wastes, as well as specific environmental pollution issues. In most developed countries the former practices of urban waste and water cycle management were environmentally unsustainable. Towards the end of the past century, the municipal rubbish dumps were closed and lots of untreated or partially treated wastewater just flowed into the streamlines and soil. These enclosure processes were not sustainable due to the uncontrolled disposal of waste in landfills without proper separation or previous treatment or ineffective water pollution control. The fact is that, in many cases, the process of sealing consisted in capping these waste landfills with soil from the surroundings. These procedures led to hazardous environmental consequences like high levels of pollutants that affected the atmosphere, soil and water resources. In terms of water there were implemented, in some cases, advanced solutions for sewage treatment, but mostly common procedures continued endangering the environment, at least is some particular ways.

As natural resources, including the primordial water, continue to become scarce these threats to the environment and consequently the human health are ever-present. This part of the book presents several real cases where trade-offs between economic development and environmental protection demand for concrete solutions in the short run. The purpose of this part is to discuss the future developments and limitations of adopting environmentally and economically sustainable solutions, considering real cases presented by experienced engineers and scientists.

Furthermore, these selected chapters focus on another critical trade-off between the environmental and energetic sustainability and the need for large amounts of capital needed for such investments.

Other discussions also are made when analyzing the spatial, demographic and socioeconomic characteristics of concrete regions and its common fiscal and regulatory framework for all countries related to the overall environment sustainability.

An important point is the soil pollution and rehabilitation/remediation. This part of the book focus on a geological and geotechnical point of view in more than one chapter stating the importance of the subject.

The Use of Geological Background Reference Values for Soil Evaluation and Remediation: The Trajouce Ecopark Case-Study

Graça Brito, Carlos Costa, Daniel Vendas, and Susana Dias

Abstract 2015 was proclaimed as the 'International Year of Soils' (IYS) at the 68th UN General Assembly. While soil is recognized as an essential, finite and non-renewable natural resource, it is increasingly degraded or irreversibly lost due to poor management and urban and infrastructure expansion, both in the EU and at a global level.

Soil contamination in particular has not only negative impacts on human health and ecosystems but also on economy, by the reduction in environmental services as a result of loss of natural soil capital and costs of land rehabilitation. The number of contaminated sites in the EU may exceed half a million. However, only a few EU Member States have specific legislation on soil protection. While the European Commission decided in May 2014 to withdraw the proposal for a Soil Framework Directive, the 7th Environment Action Programme, which entered into force on 17 January 2014, recognizes that soil degradation is a serious challenge and provides that by 2020 land is managed sustainably, soil is adequately protected and the remediation of contaminated sites is in good progress in the EU.

In Portugal draft legislation on the prevention of contamination and remediation of soils that seeks to eliminate major gap in the national legislative framework and comply with international commitments, was submitted for public consultation and is now being discussed. Among several innovations, for the definition of reference values for evaluation and remediation of soil, natural background values can be used instead of those adopted from the Ontario Standards [MOE (Soil, Ground Water and Sediment Standards for Use Under Part XV.1 of the Environmental

G. Brito (✉)
Earth Resources Department and GeoBioTec, Faculty of Science and Technology, NOVA University of Lisbon, Caparica, Portugal
e-mail: mgb@fct.unl.pt

C. Costa • D. Vendas
eGiamb, Consultoria Geoambiental, Lda, Caparica, Portugal
e-mail: carlos.costa@egiamb.pt; daniel.vendas@egiamb.pt

S. Dias
TRATOLIXO – Tratamento de Resíduos Sólidos, EIM, Trajouce, Portugal
e-mail: susana.dias@tratolixo.pt

A. Dias et al. (eds.), *Modeling Innovation Sustainability and Technologies*, Springer Proceedings in Business and Economics,
https://doi.org/10.1007/978-3-319-67101-7_20

Protection Act., April 15, 2011. Ministry of the Environment, 2011)], or other international standards, for the most common contaminants, if duly justified and accepted by the Environmental Portuguese Agency.

This paper reports a case study in which the conducted site investigation, sampling and interpretation of results confirmed the existence of a neat relationship between the presence of vanadium and geological background of the region where the Trajouce Ecopark (TRATOLIXO) operates, which allowed to classify the soil resulting from the treatment of several areas affected by the inadequate disposal of urban solid waste as suitable for environmental rehabilitation of the site [EGIAMB (Estudo das Concentrações Naturais de Vanádio na Envolvência do Ecoparque de Trajouce. Final Report, 2014)].

Keywords Soil quality • Reference values • Background geochemistry • Contamination • Remediation

1 Introduction

Geochemical background concentrations for toxic elements in the environment play a key role in environmental studies for establishing quality criteria for soils, sediments and waters and make political decisions in the evaluation of contaminated sites.

To prevent adverse effects of contaminated media in human health and ecosystem, several regulatory norms are based on toxicological studies (risk assessment) (USEPA 1989) and do not refer to geochemical data. Consequently, in cases where anomalous concentrations are due to geogenic sources, screening values for quality standards do not meet geochemical natural background values and remedial actions could be useless and very expensive (Galuzka and Migaszewski 2011; Cram-Heydrich et al. 2010).

In Portugal, it is suggested by the Environmental Portuguese Agency the adoption of the Ontario Standards (MOE 2011) as quality criteria for the evaluation of contaminated soils (sediments and waters) and remediation, however geochemical background can be used in cases where anomalous concentrations are due to natural sources instead of anthropogenic influence (APA 2015).

In this paper it is presented a case study of a waste management landfill site in Portugal where high concentration of vanadium in the former soils seem to have anthropogenic origin related to waste management and containment. Previous site investigation studies identified vanadium concentration above reference standards for the whole collected soil samples, however, as the site is located in a volcanogenic geological area, a geochemical background investigation was performed in order to evaluate the original source of the anomalous values and define protection actions for human health safety.

2 Methodology

For evaluating geochemical background it is common the adoption of one of the three approaches: (i) the direct approach; (ii) the indirect approach and; (iii) the integrated approach (Galuzka and Migaszewski 2011).

In this study it was used the direct approach that consisted in the adoption of the median values (or a range of values) obtained by direct data sampled on “pristine areas”, without anthropogenic influence. This method has the advantage of representing the real regional data and values are not subject to any processing.

The main difficulty pointed to the direct method is the sample selection criteria, the high costs associated to the field work and laboratory analysis and the need of expert knowledge (Galuzka and Migaszewski 2011; Cram-Heydrich et al. 2010).

The methodology used for evaluating the geochemical background trough the direct method enclosed four steps:

- Step 1: Characterization of geological environment
- Step 2: Definition of a representative soil sampling plan
- Step 3: Laboratory analysis
- Step 4: Statistical analysis and comparison to reference standards

3 Case Study

The Trajouce Ecopark waste landfill site is located in Cascais municipality, in an area of about 41 ha, and encloses two main waste infrastructures (Fig. 1):

(i) On the east side, an uncontrolled waste landfill, closed since 1997, with roughly 11 ha, marked as “Uncontrolled MSW Landfill” in Fig. 1.
(ii) On the southwest side, a controlled waste landfill (labelled as “Controlled MSW Landfill” in Fig. 1), which covers an area of about 8 ha;

To evaluate the natural concentrations of vanadium in the Trajouce Ecopark area it was analysed the Portugal Geological Map (Ramalho et al. 2001) and other scientific publications related to background studies in similar environments (Galuzka and Migaszewski 2011; Cram-Heydrich et al. 2010).

The investigation plan was designed with the support of these elements in order to ensure adequate samples geographical distribution in the various geological formations, particularly the “Lisbon Volcanic Complex” (CVL) (Ramalho et al. 2001).

Fieldwork was carried out in order to collect 30 soil samples representative of the surrounding Trajouce Ecopark. The collected samples were analysed for a set of 14 metals in a reference laboratory.

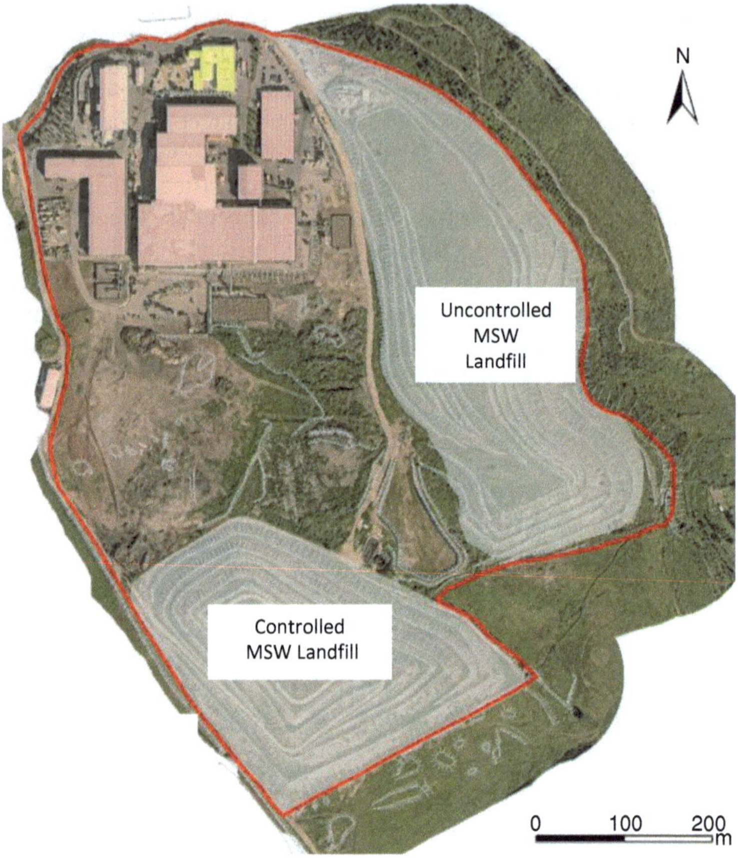

Fig. 1 Trajouce Ecopark (EGIAMB 2014)

Interpretation of the results was carried out based on statistical analysis of the chemical elements, correlation between geological subtract and comparison with Ontario standards (MOE 2011).

It was possible to conclude that the origin of vanadium in the Trajouce Ecopark soils is mainly related with geological background, namely with the volcanic materials of the "Lisbon Volcanic Complex" that occurs in the area (Ramalho et al. 2001; Pérez 1985; Silva et al. 1998).

Step 1: Characterization of Geological Environment

The Trajouce Ecopark is located (Fig. 2) on the sedimentary units of the Upper Cretaceous (Albian to Cenomanian), Bica (C2Bi) and Caneças (C2Cn) Formations, constituted by clayey limestones, partially covered by the volcanic materials of the "Lisbon Volcanic Complex" (CVL) (from Senonian), and structurally arranged in small syncline with axis oriented approximately, WSW-ENE, in alignment with the Alcabideche Syncline (Ramalho et al. 2001).

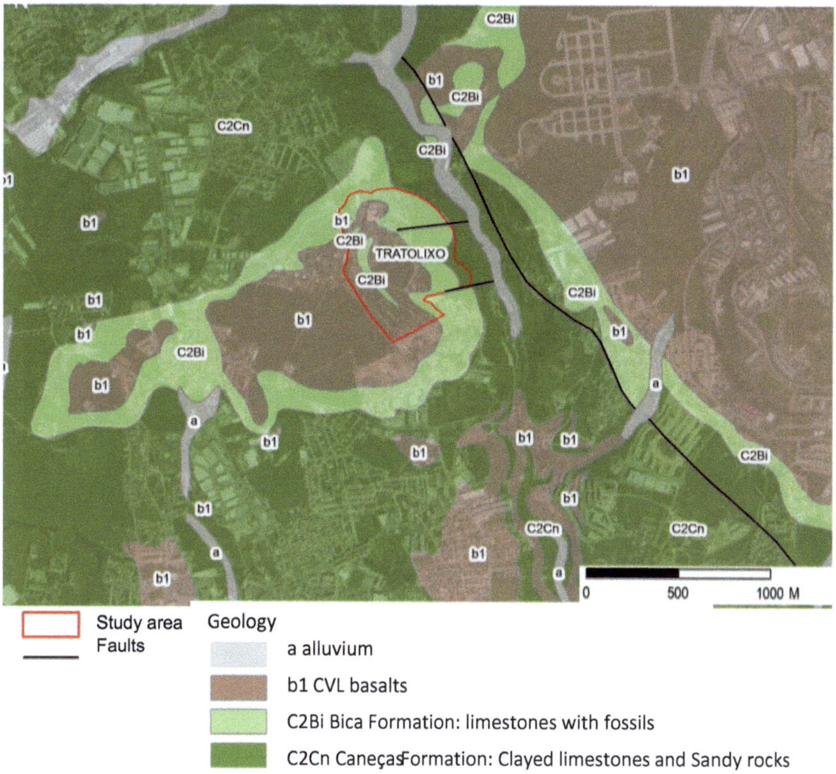

Fig. 2 Geological regional setting (EGIAMB 2014). Source: Portugal Geological Map, scale 1:50,000, sheet 34-C (Cascais) (Ramalho et al. 2001)

The present study allowed the structuring of information raised from three different sources: "Lisbon Volcanic Complex" Geochemistry, metal assays from Trajouce soil samples and metals in soils collected in the surroundings of Trajouce Ecopark.

This structure allows the understanding of the metal origin and how they incorporate soils, particularly in the surroundings of the Trajouce Ecopark.

3.1 *Lisbon Volcanic Complex*

The "Lisbon Volcanic Complex" (CVL) dates from the Upper Cretaceous, Senonian (about 70 Ma), and stands out as one of the major episodes of the Meso-Cenozoic igneous activity that accompanied the opening of the North Atlantic. CVL occupies a large area stretching from Lisbon, Cascais, Sintra, Mafra and Loures, with fair representation in the region between Lisbon, Sintra and Cascais (Ramalho et al. 2001).

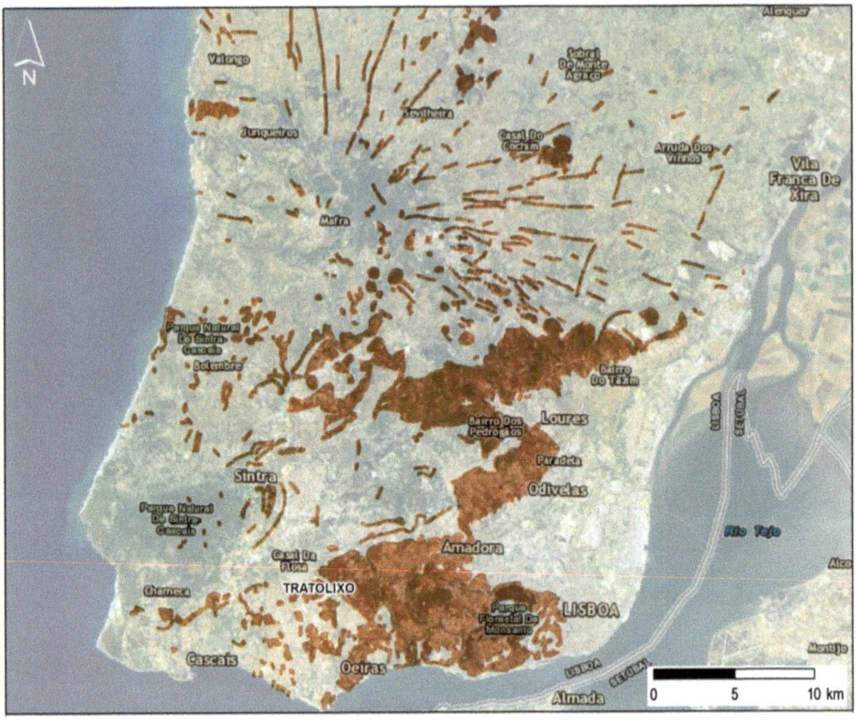

Fig. 3 Distribution of the Lisbon Volcanic Complex (CVL) (EGIAMB 2014). Adapted from AML Geological Map, scale 1:100,000, North and South sheets

It comprises various types of structures such as chimneys, dykes, lava flows and thresholds (among others) and is composed of various types of rocks from basalts and gabbro's the rhyolites, through the intermediate types such as trachyte's or trachybasalts, among others. Despite the variety of media types, the most abundant rock is basalt, giving to the CVL a fundamentally mafic or alkaline nature, which explains the essentially effusive character, subaerial, volcanism assigned to it (Ramalho et al. 2001). The basalts represent over 70% of outcropping material in the CVL. This study focused mostly on basalts due to the importance in the context of the CVL and in particular in the study area. The geographic occurrence of CVL (in the figure labelled as "Complexo Vulcânico de Lisboa") is presented in Fig. 3.

From the geochemical point of view, the CVL basalts, according to Perez (Pérez 1985), may be described as alkali-olivinic. They are rich in titanium ($TiO2 > 3\%$), (4–7%) and magnesium content between 6 and 11% depending on the degree of evolution.

The CVL geochemical analysis allows concluding about the presence in basalts of different metals with relatively high levels, such as vanadium, titanium, chromium or nickel.

These metals are mainly associated with the first minerals formed (according to the *Bowen Series*) such as the chromiferous spinels, olivines or pyroxenes.

According to *Goldish Series*, which represents the alterability of the igneous origin minerals, these minerals are the first to be altered by weathering processes (essentially chemical). Thus, it is consistent that residual soil basalt mass containing high quantities of metals have origin in the most alterable minerals.

Several other studies assessed the concentration of metals, particularly heavy metals in soils in the Lisbon region which included the vanadium analysis (Silva et al. 1998; Costa 2010).

Silva et al. (1998) refer that high levels of metals such as vanadium or chromium in soils appear to be related to the presence of a source of these same metals, in the case of the Lisbon region, the CVL basalts. This same hypothesis is again addressed and confirmed in COSTA studies (Ramalho et al. 2001) when analyzing a set of 26 soil samples collected in the municipality of Lisbon.

The analysis presented in COSTA (Costa 2010) and reinterpreted in the present study suggests that there is indeed a clear link between the levels of heavy metals in soil and the geological substrate, particularly in the case of CVL, and is expected a high presence of metals in the soils assigned to the CVL.

Finally, and based on these authors, an unusually high presence of heavy metals in soil of the Lisbon region is expected, particularly metals of the group most associated with basalts (Cr, Ni, Ti, V) in areas where basalts contribute to the soil formation or influence water quality by drainage and leaching of geochemical elements.

Step 2: Definition of a Representative Soil Sampling Plan

The research plan was intended to assess whether the concentration of heavy metals in soils of the Ecopark is anthropogenic or of natural origin, since the analytical determinations carried out on soil samples collected within the Ecopark, under the development of the rehabilitation plan, revealed the presence of vanadium in concentrations above the reference values (Ontario Guidelines), used at the remediation of the site, as recommended by Portuguese Environmental Agency, in the absence of national guidelines. These standards (MOE 2011) were developed to assist owners of contaminated sites in decision-making regarding the quality of the soil and/or groundwater, and the assessment of the environmental condition of the land to determine the need for implementation of remediation measures. They consider the land use type, present or future, namely if the occupation is agricultural, residential, industrial or commercial.

In order to assess the presence of elements of interest (heavy metals) and study the relationship with the natural geological environment, it was proposed the determination of the parameters in laboratory, which being potential contaminants in the soil, also enables checking if they are of natural origin, particularly associated with the presence or influence of the "Lisbon Volcanic Complex" (in particular basalts).

The investigation plan allowed sampling at several sites with different types of occupation (preferably not industrial) in a 2 km area around the Ecopark (Table 1).

The location of the samples is shown in Fig. 4.

Table 1 Geological formations and number of samples around Ecopark (EGIAMB 2014)

Geological formation	Area occupied by the geological formation	Number of locations	Number of samples
	Ecopark (%)		
Lisbon Volcanic Complex	47.0	7	14
Bica Formation	45.2	4	8
Caneças Formation	7.5	4	8

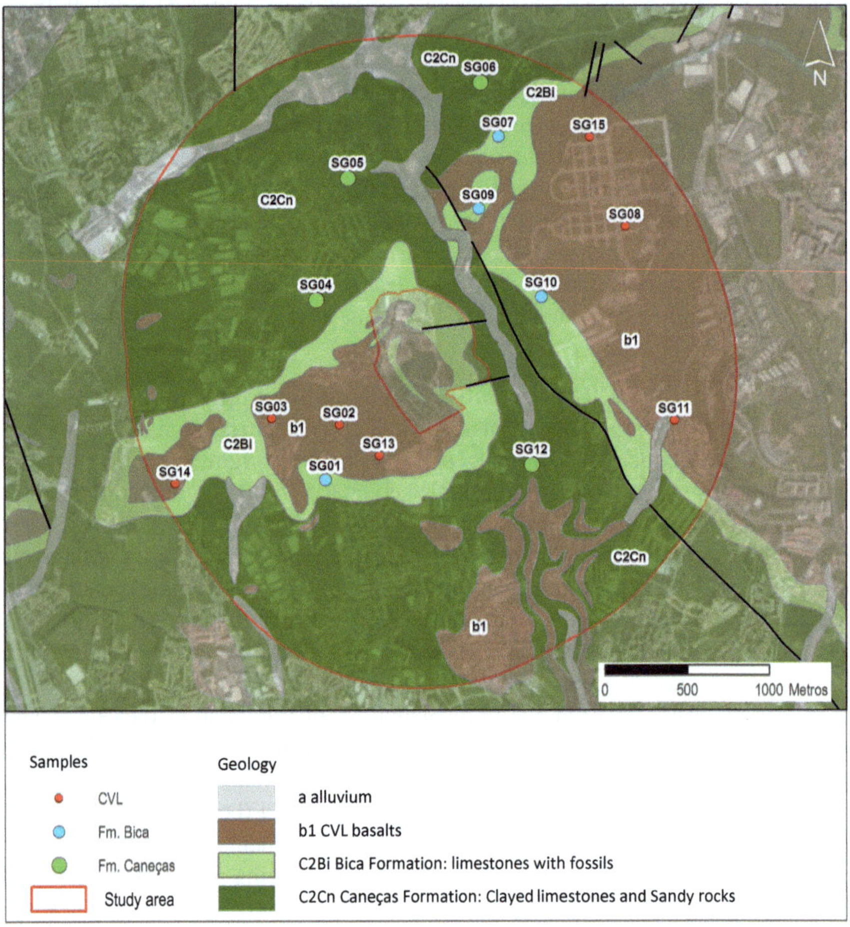

Fig. 4 Sampling plan (EGIAMB 2014)

For the design of this plan a number of factors were considered, such as the assessment of soils in situ, with representative substrates of different geological types occurring in the region and a statistically valid population for which it was considered a minimum of 30 samples.

Step 3: Laboratory Analysis
The analytical results of the 30 collected samples at the Trajouce Ecopark neighborhood are summarized with basic statistics and presented in Table 2.

The present study also considered the analytical results of the samples collected inside the Ecopark area under the rehabilitation process, as well as data collected in other studies performed by the authors in the Ecopark vicinity with the same geological context (Pérez 1985; Silva et al. 1998; Costa 2010).

Step 4: Statistical Analysis and Comparison to Reference Standards
According to Galuzka and Migaszewski (2011), Cram-Heydrich et al. (2010), Pérez (1985), Silva et al. (1998) and Costa (2010) there is a relationship between the metals concentration in soil and rock mass from which they originate. Incidentally, this is precisely the reason that different regions have different background metals concentrations, in particular heavy metals.

Table 3 presents the comparison of analytical results from Ecopark samples with the Ontario standards (OME, Table 1).

Figure 5 illustrate, for each geological unit, the comparison of the observed analytical results from Ecopark samples with the Ontario reference levels (OME, Table 1).

As it could be analysed from Table 2 and Fig. 5, samples collected at CVL Formation present the highest average concentration of vanadium, chromium and cobalt, exceeding the Ontario reference levels.

In the case of Trajouce Ecopark surroundings the distribution of the metal content, in particular vanadium, also follows the geological origin.

Figure 6 shows the vanadium content observed in the whole samples collected from the CVL geological formation in the previous studies (Pérez 1985; Silva et al. 1998; Costa 2010).

Figure 7 shows the vanadium content observed in the whole samples collected from the Caneças geological formation in the previous studies (Pérez 1985; Silva et al. 1998; Costa 2010).

Figure 8 presents the vanadium content observed in the whole samples collected from the Bica geological formation in the previous studies (Pérez 1985; Silva et al. 1998; Costa 2010).

The analysis of the previous figures allows identifying a trend for nickel, chromium and vanadium, where the average of the samples collected inside the Ecopark is clearly below that of the total population of samples collected in the Ecopark surroundings. This is justified by the increasing influence of carbonate formations in the samples collected inside the Ecopark.

It also allows identifying, in terms of geological types, a similar trend between the Bica Formation and the CVL. This trend is particularly evident in metals such as barium, chrome, nickel, titanium and vanadium. This may be related to the

Table 2 Basic statistics of elements observed in Ecopark soil samples for all geological units (EGIAMB 2014)

(mg/kg)	As	Ba	Be	Pb	Co	Cu	Cr	Fe	Mg	Ni	Ti	V	Zn
Avg	17.4	137.9	1.5	16.1	25.9	25.6	71.9	57,030	16,176	58.7	2378.0	97.3	79.9
Std dev (σ)	20.9	119.5	1.0	8.7	20.6	14.6	80.7	29,883	16,122	55.0	1893.0	92.2	36.9
Max	100.0	600.0	5.0	43.0	95.0	61.0	420.0	130,000	76,000	240.0	5900.0	520.0	170.0
Min	4.0	20.0	0.5	3.5	1.6	5.0	10.0	5800	2000	5.2	47.0	5.8	16.2
Avg − σ	0.0	18.5	0.5	7.5	5.3	11.0	−8.8	27,146	54	3.6	485.0	5.1	43.0
Avg + σ	38.2	257.5	2.4	24.8	46.5	40.2	152.5	86,913	32,299	113.7	4272.0	189.4	116.8

Table 3 Average metal grades soil samples versus geological units (EGIAMB 2014)

Reference levels		As	Ba	Be	Pb	Co	Cu	Cr	Ni	V	Zn	Geological units
OME	Table 1	18	220	2.5	120	21	92	70	82	86	290	
Average 14 samples (mg/kg)		13	106	2	14	27	22	70	44	114	71	*Fm. Bica*
Average 8 samples (mg/kg)		18	59	2	12	10	10	22	21	40	49	*Fm. Caneças*
Average 8 samples (mg/kg)		2	229	1	<10	51	39	194	118	231	95	*CVL*

Values in Red are above reference values

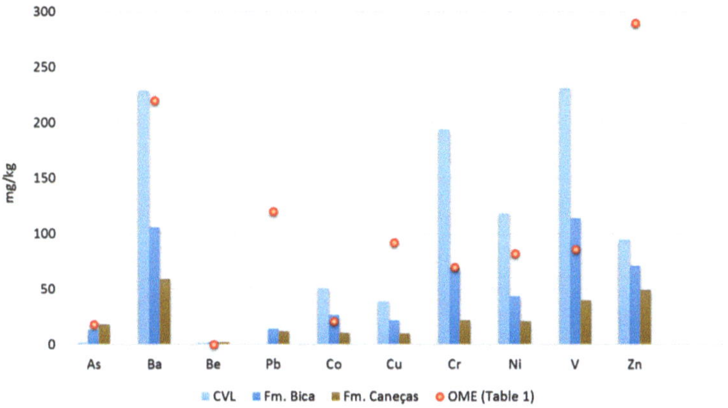

Fig. 5 Comparison of chemical elements from the geological units with the Ontario Standards (OME, Table 1) (EGIAMB 2014)

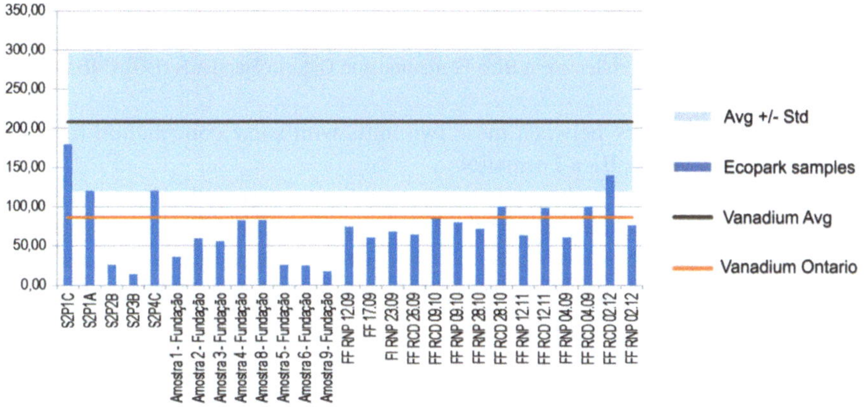

Fig. 6 Vanadium values collected in CVL Formation and Ontario Standards (EGIAMB 2014)

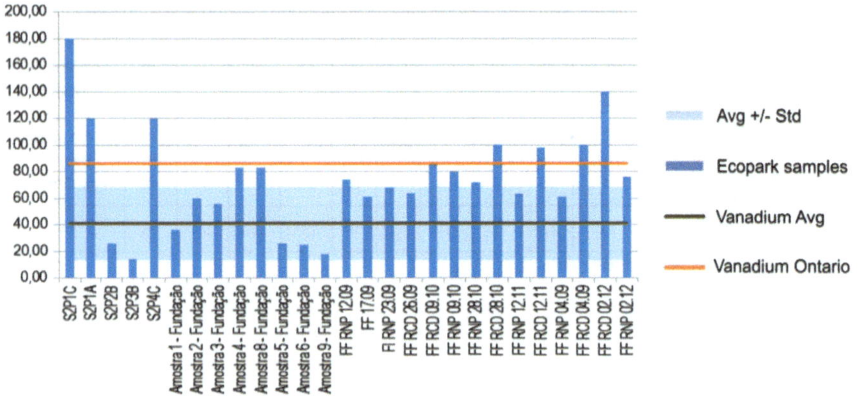

Fig. 7 Vanadium values collected in Caneças Formation and Ontario Standards (EGIAMB 2014)

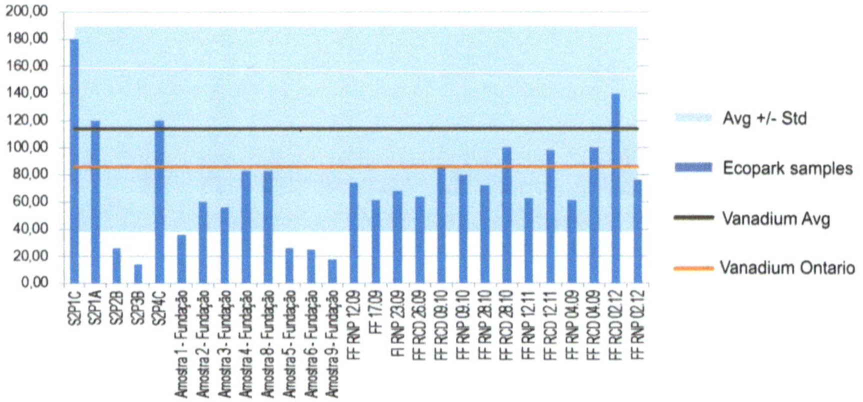

Fig. 8 Vanadium values collected in Bica Formation and Ontario Standards (EGIAMB 2014)

geographical proximity between these two units with some contribution from the CVL in soils covering Bica Formation.

The analysis of the correlation matrices identifies the association between cobalt, copper, chromium, nickel and vanadium. This same relationship was obtainable in Literature Review. This contributes to consider the presence of these metals in the Ecopark as naturally occurring associated with the Lisbon Volcanic Complex metallogenesis and also to the carbonate rocks.

4 Conclusions

Trajouce Ecopark is located on a geological complex mostly of Cretaceous origin, composed of carbonate (Bica and Caneças Formations) and volcanic formations (Lisbon Volcanic Complex). The CVL is in this region, consisting mainly of basaltic or peri-basaltic rocks.

Geochemical analysis of CVL basaltic materials allowed the assessment of the presence of relatively high levels of various heavy metals, such as chromium, nickel, titanium or vanadium. The presence of these metals in the bedrock that supports the soils of Ecopark region is considered a possible source of the metals found in soils.

The study of the distribution of various heavy metals in the Lisbon region shows that the materials of the CVL as bedrock, particularly when composed of basalts, led to the presence of heavy metals in the soils, mainly for chromium, nickel, titanium and vanadium.

A research plan to assess the background levels of heavy metals in the region was performed with the collection of 30 samples in the Ecopark surroundings.

Data interpretation focused not only on the results obtained from these samples, but also included data from a study with the same scope at Cabra Figa (2 km north) as well as the samples collected inside the Ecopark.

Results interpretation was made from three perspectives: Ontario Standards, geological and also statistics.

Regarding to Ontario Standards, if industrial/commercial use is considered there were exceedances for arsenic, cobalt, chromium and vanadium. But if the background is considered, then exceedances were observed for arsenic, barium, beryllium, cobalt, chromium, nickel and vanadium.

The geological interpretation showed that the distribution of heavy metal content in the soil has a close relationship with the geological formation that generates it. Finally, statistical analysis identifies the correlation between metals such as cobalt, copper, chromium, nickel and vanadium, which is consistent with the relationship between the same metals identified in the literature for the CVL.

In summary, the investigation, sampling and results interpretation confirms the existence of a relationship between the presence of vanadium at the Trajouce Ecopark and the geological background of the region, proving the hypothesis stated at the beginning of this study and allowing the use of those soils as not contaminated for site rehabilitation.

Acknowledgements The author Graça Brito acknowledge the support of the sabbatical grant number SFRH/BSAB/128454/2017 Ref. CRM:0061819.

References

APA (2015) Projecto legislativo relativo à Prevenção da Contaminação e Remediação dos Solos (PRoSolos). For consultation from 04.09.2015 till 04.11.2015. http://participa.pt/consulta.jsp?loadP=820. Accessed Oct 2015

Costa CSA (2010) Mineralogia dos Solos de Lisboa e sua Influência na Fixação de Elementos Tóxicos. Dissertação para a obtenção do grau de Mestre, Universidade de Aveiro, p 234

Cram-Heydrich S, Ponce de Léon C, Fernández-Lomelín P, Sommer-Cervantes I (2010) Chapter 2: Geogenic and anthropogenic differentiation of potentially toxic elements in

superficial sediments from Cuitzeo Lake in México. In: Ramsey G, McHugh S (eds) River sediments. Nova Science, New York, NY, pp 37–75. ISBN 971-1-60741-437-7
EGIAMB (2014) Estudo das Concentrações Naturais de Vanádio na Envolvência do Ecoparque de Trajouce. Final Report. R2013101A02. June 2014
Galuzka A, Migaszewski Z (2011) Geochemical background – an environmental perspective. Mineralogia 42(1):7–17. https://doi.org/10.2478/v10002-011-0002-y.; www.Mineralogia.pl
MOE (2011) Soil, ground water and sediment standards for use under Part XV.1 of the Environmental Protection Act., April 15, 2011. Ministry of the Environment
Pérez T (1985) Petrologia do Complexo Vulcânico de Lisboa. Dissertação para a obtenção do grau de Doutor, Universidade de Lisboa, p 260
Ramalho MM, Rey J, Zbyszewski G, Matos Alves CA, Palácios T, Moitinho de Alemeida F, Costa C, Kullberg MC (2001) Notícia Explicativa da Folha 34-C, Cascais. Instituto Geológico Mineiro, Lisboa, p 104
Silva MCR, Almeida, C, Munhá EJ (1998) Contaminação antropogénica por metais pesados nos solos da Várzea de Loures. Actas do V Congresso Nacional de Geologia, Comunicações dos Serv. Geológicos de Portugal. Tomo 84(2):E-98–E-101. Lisboa
USEPA (1989) Risk assessment guidance for superfund (RAGS), Volume I – Human health evaluation manual (Part B – Development of risk-based preliminary remediation goals). Office of Solid Waste and Emergency Response. Washington, DC. EPA/540/1-89/002. December

Graça Brito, Assistant Professor at NOVA University of Lisbon, Faculty of Sciences and Technology, Earth Sciences Department (DCT/FCT/UNL), for the Geological Engineering Graduation and Masters. Has a PhD degree in Geoenvironmental engineering, focused on the evaluation of contaminated industrial sites and risk analysis. Since the last 15 years she has collaborated in several Geoenvironmental R&D projects, mainly related to the evaluation of contaminated industrial sites, soils and groundwater modelling, risk assessment and environmental restoration. Its main areas of research are human health risk assessment for organic and inorganic pollutants, fate and transport of contaminants in soils and groundwater, environmental impact assessment and landfills monitoring. Is member of the Geobiotec UNL research centre and of the Professional National Engineering Association.

Carlos Costa, Expert in Geology, Geotechnics and Environment since 1980 in the implementation of more than a hundred studies design and engineering works, including for large hydroelectric plants and road infrastructure, in Portugal, Angola, Mozambique, Algeria and Turkey. Participated in several major environmental impact assessment (EIA) studies, including for the Sines LNG Terminal, the expansion of the Port of Aveiro, the Sines and Oporto refineries, the High Speed Railway Network and in the Alcantara (Lisbon) Container Terminal. Co-ordinated Strategic Environmental Assessment (SEA) studies for Urban Redevelopment in derelict industrial areas, e.g. Quimiparque (Barreiro) and Matinha (Lisbon). Co-ordinated demolition works, multi-material waste recovery and environmental rehabilitation of large industrial infrastructures. Since 1998 its main focus is environmental due diligence (EDD) primarily assessing soil and groundwater contamination in Industrial Parks, Power Plants, Chemical Industry, Steel, Metallurgy and Metal-mechanics, Shipyards, Explosives, Aeronautics and MSW and Industrial Waste.

Daniel Vendas, Expert in Geology and Environment since 2000. The main experience consists within soil and groundwater contamination and Human Health and Ecological Risk Assessment in Industrial Parks, Power Plants, Chemical Industry, Steel, Metallurgy and Metal-mechanics, Shipyards, Explosives, Aeronautics and MSW and Industrial Waste. In the last years has participated in EDD studies e.g. Highway Concession, Non-Hazardous Industrial Landfill, Recycled Paper Mill, MSW system, Agrochemical and Chemical Industries and Shopping Centres, in several environmental impact assessment (EIA) studies, including the expansion of the Alcântara (Lisbon) Container Terminal, High Speed Railway Network, Highways, Shipyards, Shopping Centres, and Strategic Environmental Assessment (SEA) studies for Urban Redevelopment in derelict industrial areas, e.g. Quimiparque (Barreiro) and Matinha (Lisbon).

Susana Dias, holds a degree of Advanced Studies in Sanitary Engineering and Integrated Waste Management of Environmental Engineering (FCT, Portugal, 2011) and a degree of Environmental Engineering (ULHT, Portugal, 2005) being ever since working in the waste management industry, namely at Tratolixo—Tratamento de Resíduos Sólidos EIM SA. Susana Dias works at Tratolixo as Monitoring Coordinator since 2009, with responsibility over the quality control of the process and products and environmental monitoring. Speaker at conferences held by Apemeta—Associação Portuguesa De Empresas De Tecnologias Ambientais (2014).

Soil Contamination: Case Study on Environmental Rehabilitation of the Trajouce Ecopark

João Dias Coelho, Joana Frazão, and Susana Dias

Abstract Soil contamination can be defined as the presence of pollutants above certain levels which cause the deterioration of its environmental functions (European Commission. http://esdac.jrc.ec.europa.eu/themes/soil-contamination. Accessed 30 Sept 2015). This contamination may be due to industrial chemicals or changes on the soil's environmental features. Between 2003 and 2005 there was an improper waste disposal on Trajouce's Ecopark. In 2012, Tratolixo began the recovery and rehabilitation of those areas through the "Environmental Rehabilitation Plan of Trajouce's Ecopark". In 2012, Tratolixo started removing and sending the residues from the "Waste Deposit" to its final destination. In June 2013, the company tested a new solution—sorting the materials and reusing them "in situ" maximizing the reuse principle. This new method allowed for a substantial reduction on amounts of waste sent to the appropriate destination, and it also reduced the initially estimated costs of the Rehabilitation Plan by approximately 85%.

This chapter describes a real case of soil contamination and its demanding process of remediation.

Keywords Soil • Rehabilitation • TRATOLIXO • Waste deposit

1 Introduction

Municipal and industrial waste disposal and treatment processes cause around a third of Europe's soil contamination problems. It is estimated that there may be around 2.5 million contaminated sites across Europe that need analysis and a deeper investigation. Of these 2.5 million, it is expected that approximately 14% are contaminated and require intervention through the application of damage remediation techniques. It is estimated that one third of these contaminated sites are known and are identified, 15% of which were subjected to such damage remediation

J.D. Coelho (✉) • J. Frazão • S. Dias
Tratolixo, São Domingos de Rana, Portugal
e-mail: dias.coelho@tratolixo.pt; joana.frazao@tratolixo.pt; susana.dias@tratolixo.pt

A. Dias et al. (eds.), *Modeling Innovation Sustainability and Technologies*, Springer Proceedings in Business and Economics,
https://doi.org/10.1007/978-3-319-67101-7_21

techniques. The most common traditional technique involves the excavation of contaminated soil and deposition at suitable locations (European Environment Agency—http://www.eea.europa.eu/highlights/soil-contamination-widespread-in-europe).

In 2008, at Trajouce Ecopark, some situations that could cause certain environmental problems, including soil contamination, were detected. This environmental contamination risk was detected in sequence of the study that aimed for the restructuring and rehabilitation of Trajouce Ecopark in order to improve operating conditions.

Knowing the situation, TRATOLIXO decided to suspend the implementation of the plan and draw up a diagnostic study, which aimed to get more concise information on the detected environmental problems and their potential risks.

According to the diagnostic study (DHV 2009b, c), between 2003 and 2005, several amounts and types of waste were deposited in an existing, non-waterproofed place, in the central area of the facility, and labeled "undifferentiated waste deposit". This deposit was made up of varied nature materials, namely of green waste origin, demolition waste, soils, rocks, light and heavy rejected, by-product of tuning the compost and other unspecified materials (Fig. 1).

After completion of the initial diagnostic study, and having a need for a more precise characterization, given the complexity of the site, TRATOLIXO moved to a new phase of the study, with the aim of gauging the limits of "undifferentiated waste deposit" and classifying the deposited materials, according to the waste acceptance criteria in landfills and assessment of their hazard degree.

Digging some inspection wells during the design of this study allowed us to identify the areas (Fig. 2) in which these wastes were dumped and these prospections allowed us to divide the area into four different sectors:

- Sector 1: Construction and demolition waste (CDW), with casual occurrence of cleaning wastes and other various materials (46,700 m^3);
- Sector 2: CDW, mixed with landfill of calcareous materials (10,600 m^3);
- Sector 3: CDW, mixed with cleaning waste, green waste, monsters and more (40,000 m^3);
- Sector 4: By-products of the composting process, light and heavy rejected, with compost (17,800 m^3).

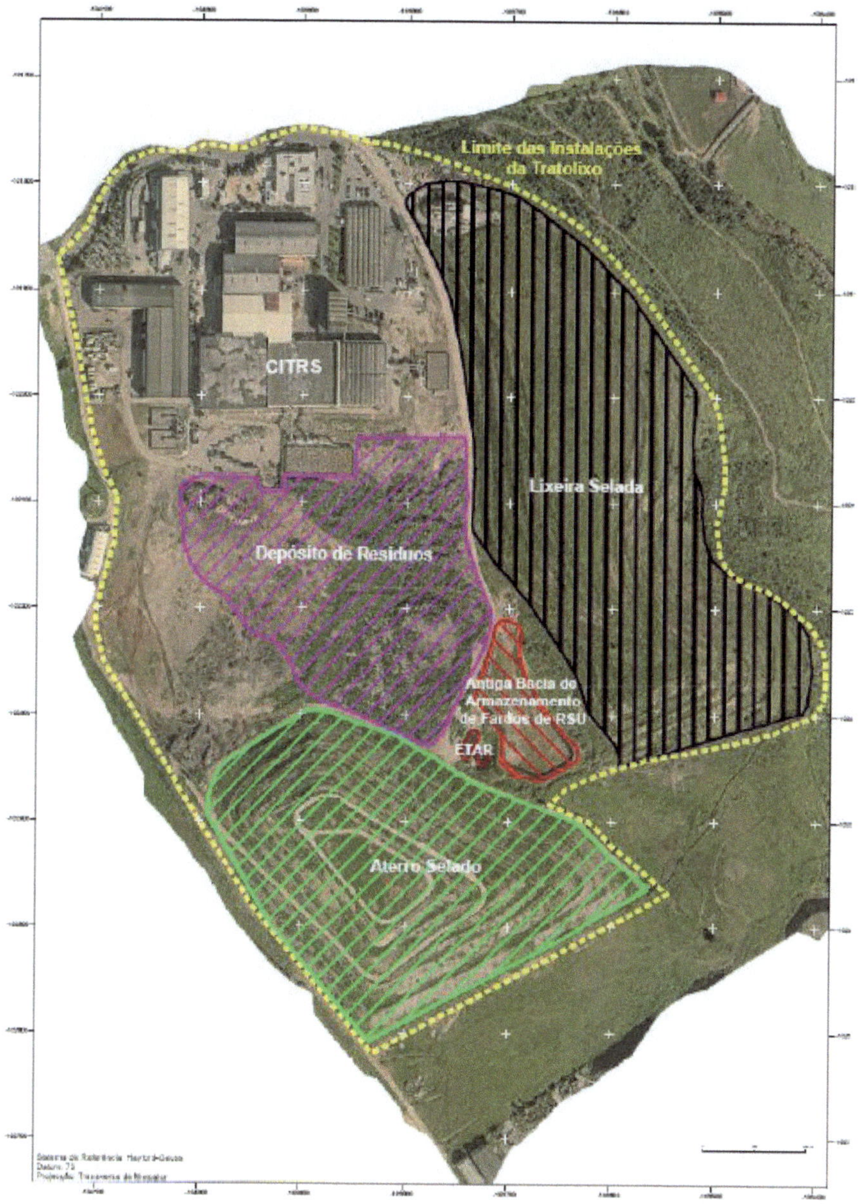

Fig. 1 Waste disposal location (DHV 2009c)

Fig. 2 Delimitation of the waste disposal area (DHV 2009b, c)

2 Methods

In September 2012, TRATOLIXO began the treatment of waste improperly dumped in Trajouce Ecopark between 2003 and 2005, marking the start of the rehabilitation plan called PRAET (DHV 2009c). The treatment of "undifferentiated waste dump" conducted in 2012 began in sector 1 and consisted on its removal and referral to an appropriate final destination.

In June 2013, a new methodology, based on the principle of maximizing the reuse in situ, was tested. This methodology consisted on a full screening of all the

Table 1 Amount of waste sent to their final destination (2012–2014)

Waste typology	2012	2013	2014
Hazardous waste	–	–	44.36
Non-hazardous waste	11,372.04	2525.68	2367.16
Inert waste	5831.40	857.32	1857.00

residues. Considering the results obtained in the environmental rehabilitation testing phase, all works, developed in 2013 and 2014, were conducted following this methodology.

The methodology consists in conducting a screening on the waste, which obtains three distinct fractions, carefully analyzed through the physical and chemical point of view, and then carried to an appropriate destination or reused locally.

3 Results

During the work of the PRAET, during 2012–2014, 24,854.96 t of waste were sent to their final destination (Table 1).

The comparison of the different types of quantitative waste sent to its final destination shows that the methodology adopted in 2013–2014 allows a significant reduction in the amount of waste going to its final destination, in turn allowing a considerable reduction in the costs of the environmental rehabilitation of the Trajouce Ecopark, currently estimated at 85% compared to the originally planned costs.

4 Conclusion

Soil contamination is a very relevant issue to the environment and economics, and may have implications for landowners, municipalities, territorial planning authorities, government and local communities (EPA – AU—http: //www.epa.sa.gov.au/environmental_info/site_contamination).

In short, the change in the methodology of waste treatment (2012/2013) allowed for a substantial reduction in waste amounts sent to an appropriate destination, as well as costs linked to the environmental rehabilitation of the Trajouce Ecopark. This reduction in costs compared to the predicted values is around 85%, which brings us to the conclusion that the methodology of comprehensive waste sorting translates into significant economic and environmental benefits for TRATOLIXO.

"Acting to Safeguard the Future", TRATOLIXO engaged in problem solving associated with the contamination of the Trajouce Ecopark soils, always promoting the use of environmentally friendly methods, while never forgetting its commitment to the reduction of costs associated with these operations.

References

DHV (2009a) Diagnóstico Sectorial das Instalações da Tratolixo em Trajouce, Unpublished Report E25340, January 2009, Algés

DHV (2009b) Diagnóstico Sectorial das Instalações da Tratolixo em Trajouce – Fase II, Unpublished Report E25360, March 2009, Algés

DHV (2009c) Plano de Reabilitação Ambiental das Instalações da Tratolixo em Trajouce, Unpublished Report E25360A, May 2009, Algés

EPA-AU. http://www.epa.sa.gov.au/environmental_info/site_contamination. Accessed 30 Sept 2015

European Commission. http://esdac.jrc.ec.europa.eu/themes/soil-contamination. Accessed 30 Sept 2015

The European Environment Agency. http://www.eea.europa.eu/highlights/soil-contamination-widespread-in-europe. Accessed 30 Sept 2015

João Dias Coelho is currently and since 2014 the Chairman of Board of Tratolixo, and member of the Board since 2010. Between 2007–2010 he was team management coordinator of the construction of Wastewater Treatment Plant (WWTP Guia). Previously (2003–2007) he was executive Member of the Board of the Multimunicipal Water and Sanitation Services of the Municipality's of Amadora, Cascais, Oeiras and Sintra (SANEST–ADP Group) and Member of the Board and co-founder of the Municipal Energy Agency—"Cascais Energy Agency".

João Dias Coelho (born in 1963) holds a Degree in Law (UML-Lisbon University-Portugal) and holds Post graduation degrees, one in Law Planning, Urbanism and Environment (FD-Lisbon University-Portugal) and the other in Labor Law (Lusiada University-Portugal).

He also attended the Advanced Studies in Environment, Spatial Planning and Sustainable Development, FCT UNL and Advanced Management in Human Resources, by the Faculty of Economics and Business Sciences of the Catholic University of Lisbon.

Joana Frazão holds the degree of Specialization in Environmental Quality (CET—Technological Specialization Courses of Universidade Católica Portuguesa, Portugal, 2005) and started her career in the year 2003 at Visteon—Palmela Plant, an American leading global automotive supplier, being responsible for preparing and implementing the monitoring plan for the Air Compressed Network Unit. In 2005, started to work in the waste management industry, first in Integrated System Management and a few years later in the Process and Product Control Department which has three main activities: Environmental Monitoring, Process and Product Control (Waste Characterization).

Speaker at MIST-Modelling Innovation Sustainability and Technologies (2015) and other seminars and workshops (Instituto Superior de Agronomia, Ordem dos Engenheiros, Apemeta-Associação Portuguesa de Empresas de Tecnologias Ambientais).

Susana Dias holds a degree of Advanced Studies in Sanitary Engineering and Integrated Waste Management of Environmental Engineering (FCT, Portugal, 2011) and a degree of Environmental Engineering (ULHT, Portugal, 2005) being ever since working in the waste management industry, namely at Tratolixo—Tratamento de Resíduos Sólidos EIM SA. Susana Dias works at Tratolixo as Monitoring Coordinator since 2009, with responsibility over the quality control of the process and products and environmental monitoring.

Speaker at conferences held by Apemeta—Associação Portuguesa De Empresas De Tecnologias Ambientais (2014).

Anaerobic Digestion of MSW: Challenges of a High Cost Technology

Inês Moura and David Pereira

Abstract In Portugal, as in Europe, there are several different technologies to sort and treat the different components of municipal solid waste (MSW). For treatment of the organic fraction of MSW (OFMSW) there are also several technological options. The Anaerobic Digestion (AD) technology is a well-known solution in Portugal, with several AD plants working at their full capacity. However, the technology between plants differs, and with it, its operation, maintenance and biogas production results.

To control and operate an anaerobic digestion process a lot of factors have to be taken into account. As with all biological processes, the optimum conditions are essential. The AD process has a rapid response to any changes, such as changes in the feeding, temperature and O2 addition (sometimes needed in small quantities (less than 1%) to decrease the H2S amount in the digesters). If the feeding stops, for any maintenance of the pumping system or an unforeseen problem, the pH decreases, the volatile fatty acids increase and the biogas production abruptly decreases, also decreasing, if the normal feeding is not restored in a short period, the methane content and increasing the H2S content. This kind of instability, as common as it can be in a plant operation, is highly unfavorable for the process and increases the operating costs by reducing plant revenue.

Other factors, mainly in the pretreatment, have to be considered, such as a good system of glass and grit removal especially in a wet treatment, which will prevent damaged pumps, clogging, and extremely high maintenance costs.

If the AD technology is based on a low solids content feeding, the plant will have a lot of piping and pumps, so the right project calculations for size and capacity are mandatory to decrease the costs. Arranging the different equipment in the plant to decrease the pumping distance will prevent capital costs in pumps and piping and will also prevent difficult clogging situations. To plan and design a simple piping

I. Moura (✉)
Engineering Department, Adductio, Lda, Lisbon, Portugal
e-mail: inescroftmoura@gmail.com

D. Pereira
Department of Environmental Sciences and Engineering, Faculdade de Ciências e Tecnologia, New University of Lisbon, Lisbon, Portugal
e-mail: djp@fct.unl.pt

A. Dias et al. (eds.), *Modeling Innovation Sustainability and Technologies*,
Springer Proceedings in Business and Economics,
https://doi.org/10.1007/978-3-319-67101-7_22

system, with as little curves and short piping as possible, will prevent high maintenance costs and a lot of plant stops due to clogging.

Keywords Municipal solid waste • Anaerobic digestion • Profit • Cost • Operation

1 Introduction

The construction of Anaerobic Digestion (AD) plants in Portugal comes within the scope of the Decree-Law 178/2006 establishing the PERSU II and Decree-Law n. ° 152/2002 of 23 May, which in Art.7. sets targets for the reduction of biodegradable municipal waste disposed in landfill. PERSU 2020 set a higher target to landfill biodegradable waste deviation with the Decree-Law n° 183/2009, increasing the need for treatment solutions that reduce, reuse and recycle the waste. Fig. 1 explains in a more detailed way the Portuguese situation in 2015.

In Portugal, as in Europe, there are several different technologies to sort and treat the different components of the municipal solid waste (MSW). For the treatment of the organic fraction of MSW (OFMSW) there are also several technological options. The AD technology is a well known solution in Portugal, with several AD Plants working at their full capacity. However, the technology between plants

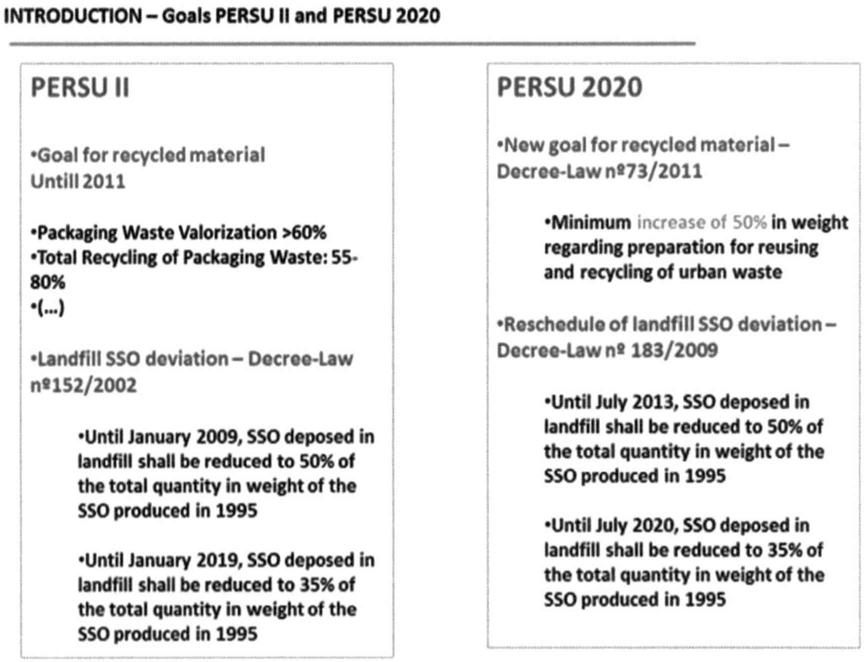

Fig. 1 Portuguese goals for recycled material according to PERSU II and PERSU 2020

differs, and with it, it's operation, maintenance, and biogas production results as well.

2 Control Ans Operation of an AD Process

To control and operate an anaerobic digestion process a lot of factors have to be taken into account. As with all biological processes, the optimum conditions are essential. The AD process has a fast response to any changes, such as changes in the feeding, temperature, O_2 addition (sometimes needed in small quantities (less than 1%) to decrease the H_2S amount in the digesters). If the feeding stops, for any maintenance of the pumping system or an unpredicted problem, the pH decreases, the volatile fatty acids increase and the biogas production abruptly decreases, consequently decreasing the methane content and increasing the H2S content, if the normal feeding is not restored in a short period. This kind of instability, very common in a plant operation, is highly unfavorable for the process and increases the costs of operation by reducing plant revenue.

Other factors, mainly in the pre-treatment have to be considered, namely a good system for the glass and grit removal, especially in the context of wet treatment, which will prevent damaged pumps, clogging, and extremely high maintenance costs.

Additionally, if the AD technology is based on a low solids content feeding, the plant will have a lot of piping and pumps, and being so, the right project calculations for size and capacity are mandatory to decrease the costs. Arranging the several different equipments in the plant to decrease the pumping distance will prevent capital costs in pumps and piping and will also prevent difficult clogging situations. To plan and design a simple piping system, with as little curves and short piping as possible, will prevent high maintenance costs and a lot of plant stopping due to clogging.

Figure 2 shows the Biogas Yeld as function of the Organic Loading Rate (OLR), temperature and different pre-treatments. As observed, plants that treat food waste and source separated MSW show high values of Biogas Yeld in the Anaerobic Digestion Treatment. The thermophilic reactors have also tendency to produce more biogas than the mesophilic digesters, for a similar OLR.

If the AD technology is based in a high solids content feeding, the digesters can be smaller but the efficiency of the process may then decrease. Reports about this systems point to a biogas yeld of 0.103–0.147 m^3/kg of Volatile Solid (VS) fed. For the cases where AD technology is used as wet systems reports of a biogas yeld of 0.4–0.5 m^3/kg VS were observed (see Fig. 2). However, the technologies are completely different and both have several advantages and disadvantages.

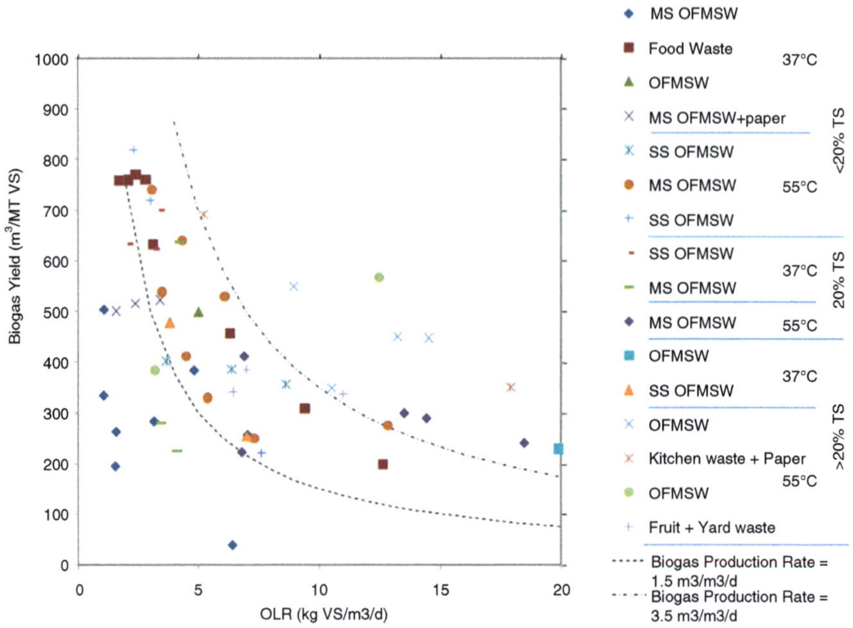

Fig. 2 Biogas Yeld as function of OLR, Temperature and Pre-treatments (H. Hartmann, B. K. Ahring, (2006))

3 The Multi-Staged Anaerobic Digestion

The multi-staged anaerobic digestion must be taken in understanding because of the goal of using more than one reactor. Sometimes these digesters operate in parallel (and are not multi-staged reactors) but single-stage digesters that can exist because of tank size limitations or just to simplify the management. A lot of economical studies have to be made to understand the viability of these systems in each real case study.

The first high-solids content digesters (dry systems) were a type of small landfills, so that resulted in batch systems. This type of system has some advantages, namely a higher possibility of controlling the digesters conditions, with higher biogas production rates. The main disadvantage is the uneven biogas production and the lack of stability of the biomass.

Another very important factor to take into account is the digesters temperature. Two main range of temperatures, thermophilic and mesophilic can be observed in almost every AD Systems. The mesophilic microbes are adapted to roughly 30–40 °C and the thermophilic microbes to 50–60°. The thermophilic system has higher costs of operation, especially a higher energy consumption, but the records found were of a higher biogas yeld content (see Fig. 2), so, the possible revenue of the energy production would be higher, albeit with higher costs of operation and energy consumption.

4 The AD Plant Profits

The AD Plant profits are based on the selling of electrical energy and recycles and on the gate fee. The profits may also include the reduction of landfill deposition, because there is no need to spend due to the waste deviation. In the future, the carbon offset credits should also be taken into consideration. However, an AD Plant has several costs to be taken into account, namely, the electrical energy of the equipment in the plant, the landfill deposition of part of the waste input, the labor costs, the water, the wastewater disposal, the maintenance costs and the financial costs if applicable (Faria T et al. 2013). The thermal energy used in the heating can also be considered a cost because it is energy not sold. The capital cost should also be considered and all the chemical products used in the operation.

The Digesters performance and cost depends on its configuration and OFMSW source. The biogas yeld of the different technologies vary and there are some factors to take into account, such as the percentage of methane in the biogas (the 50–70% of methane range present in the biogas can mean a totally different energy potential of a biogas volume unit) and the digestibility of the OFMSW. To compare biogas yeld of different technological solutions there is a need for standardized methods of characterization and accounting of organic matter input to the plants and digesters, to understand the efficiency of the pre-treatment and the quality of the process. Fig. 3 explains how the definition of organic matter is important on the overall process.

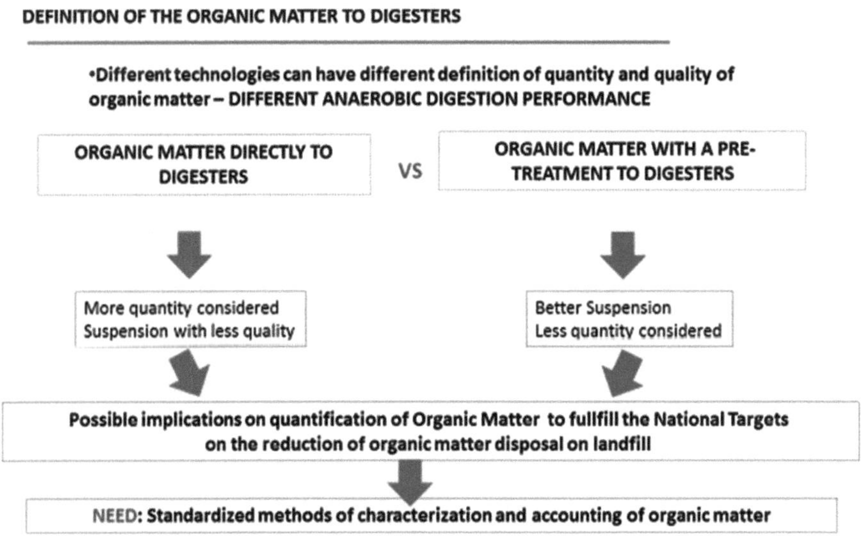

Fig. 3 Definition of the organic matter to digesters

5 Conclusion

Regarding the state of the art and the potential difficulties coming from the practical applications side, one may conclude that Anaerobic Digestion can be a solution to reduce the biodegradable municipal waste disposed in landfill. However, the costs of its implementation and operation have to be taken into account. The economic viability of these solutions is highly sustained in gate fees, the decrease of landfill deposition cost per ton and energy and recyclables selling market price. So, it's very important to study its viability before investments are decided, because effects from the behavior of the dynamic market, potentially fluctuating across time and geography. An example: for a developing country, an MT solution can be a solution because it does not need environmental awareness since it treats MSW and there is no investment cost in the waste collection. On the other hand, in a developed country, an organic valorization plant with an upstream automatic sorting can be a valuable solution due to the conditions of the energy selling market since the landfill deposition is expensive.

References

Faria T et al. (2013) Anaerobic digestion of MSW – opportunities and challenges. VIII Jornadas Técnicas Internacionais de Resíduos. Lisboa.

PERSU 2020, Plano Estratégico para os Resíduos Sólidos Urbanos, 2014, MAORT

PERSU II, Plano Estratégico para os Resíduos Sólidos Urbanos, 2007–2016, MAORT, 2007, ISBN: 978-989-8097-01-9

Hartmann, H, Ahring, BK (2006) Strategies for the anaerobic digestion of the organic fraction of municipal solid waste: an overview. Water Science and Technology 53(8):7–22

Inês Croft de Moura, born in 1985, is an BSc environmental engineer (2009) specialized in sanitary engineering (MSc, 2011) with, substantial experience in waste management, including fieldwork, and wastewater treatment design. She started her experience in Portugal being now working in Ireland.

David Pereira, born in 1956, is a Civil and Sanitary Engineer, with a PhD degree, with a professional career in very practical technology application and author of bestseller Portuguese innovative technical software, since 1977. He had a parallel career in both the academic and the industry sides, being partner of a consulting engineering conglomerate and an environmental sustainability company. Core competencies include water, wastewater and solid waste systems modeling (simulation and optimization), technology application in real life (water/wastewater transport and treatment, solid waste management), and technical and economic planning of those systems.

The Potential for Electricity Generation in Anaerobic Digestion of Municipal Solid Waste: The Real Case of TRATOLIXO

Ricardo Castro, Maria João Alves, and João Dias Coelho

Abstract The anaerobic digestion (AD) of Municipal Solid Waste (MSW) is a subject approached in different ways across Europe, as a result of the various national waste management and energy policies and other issues such as the typology of the waste. The Southern European countries, like Portugal, are among those that bet the most on AD for treating MSW from undifferentiated collection, instead of biowaste, the selective collection of the organic fraction of MSW, typical of Northern European countries.

The difficult pretreatment of this waste, which allows the separation of organic matter from all other fractions to AD, as well as the inert elements, that cause abrasion in the process, make the operation/maintenance costs high. Therefore, the AD of MSW is losing popularity in comparison with the lower operation costs and higher electrical production (most times) of landfills, still the dominant infrastructure in MSW management in Portugal.

It is therefore important to look beyond the obvious environmental benefits of this technology (reduction of greenhouse gases and landfills, renewable power generation and compost production), if it could represent an important contribution to the national grid. What is the true potential of organic residues from the undifferentiated MSW to generate biogas by AD and then electricity? Is it an economic asset? The plants operating in Portugal have been attaining their full capacity in recent years, therefore it is now possible to deliver concrete results that will help understand how much production is liable to be yielded.

This chapter intends to present actual data on the AD of MSW, taking as a case study the intercity system of AMTRES (Association of Municipalities of Cascais, Oeiras, Mafra and Sintra) and the AD plant (ADP) of Abrunheira, in Mafra, managed by Tratolixo.

Despite the difficulties associated with the operation of these plants, this work demonstrates that it is possible to carry out improvements in AD technologies, adapting them to the reality of waste management and the characteristics of

R. Castro (✉) • M.J. Alves • J.D. Coelho
Tratolixo, São Domingos de Rana, Portugal
e-mail: ricardo.castro@tratolixo.pt; maria.alves@tratolixo.pt; dias.coelho@tratolixo.pt

A. Dias et al. (eds.), *Modeling Innovation Sustainability and Technologies*, Springer Proceedings in Business and Economics,
https://doi.org/10.1007/978-3-319-67101-7_23

geographical areas which they are a part of, reaching beyond the projected production.

Keywords Anaerobic digestion • Biogas • Biowaste • Renewable energy • MSW • TRATOLIXO

1 Introduction

The need to meet community targets for diverting biodegradable waste (Biowaste) from landfill sites, led the municipal solid waste (MSW) managing bodies in Portugal, to invest in new mechanical and biological treatment (MBT) plants, investments which were co-financed by Community funds. Initially, these plants had composting as biological treatment method of reference, this second generation adopted the anaerobic digestion (AD), with subsequent composting, as preferred technology, due to the increase of revenues from the production and sale of electricity.

The AD of the organic fraction of MSW has some decades of development and is widely implemented in Northern Europe countries (EUNOMIA 2001), but in Southern Europe countries this technology was tentatively implemented and only by virtue of Community rules. To a large extent, this is due to several constraints associated with the collection of this Biowaste. While in northern countries of Europe a network of separate collection of Biowaste was set in place, in the countries of the South, this collection has been avoided, either for technical or economic reasons (Chang and Pires 2011).

In the case of Portugal, to a large extent, the Biowaste lost the opportunity to have a separate collecting because they were the last stream of waste to have an economical value. First, the separate collection of packaging emerged,—a system that has been adopted almost everywhere in the country and which added three flows to the collection of MSW: packaging (plastic and metal), paper/cardboard and glass. Later on, other flows of separate collection were being added for operational and specially for economic reasons: batteries, tyres, electrical and electronic equipment, monsters (large objects, such as furniture and mattresses), garden waste, among others. This philosophy of collection at national level prevented that, for example, a collection of only two streams, dry and wet, could be implemented with either gains in treating Biowaste and MSW collection costs.

The sum up of all these collection circuits led to a significant increase in costs that, according to the World Bank, in developed countries such as Portugal, can vary from 85 to 250 USD/ton. These costs exceed largely the treatment costs, estimated between 40 and 100 USD/ton for Landfills, 35–90 USD/ton for composting systems, or 65–150 USD/ton for anaerobic digestion (World Bank 2012).

Biowaste is the largest fraction of MSW, from 35 to 50% (EUNOMIA 2002), and one with a more significant impact on collective deposition in street containers, system adopted in Portugal, where the climate and the rapid leaching of this waste causes odors and contamination not compatible with high demands of citizens as regards the hygiene of public space. Importantly, the MSW collection in urban areas in Portugal is mostly done in daily circuits, or at worst, every other day. For these reasons, municipalities, directly or indirectly responsible for collection of MSW, would not bear the financial costs and political risks of selective collection of Biowaste. Therefore, only some collection circuits from large producers of Biowaste (industrial and commercial), have been implemented in the country.

This way, when the second generation of MBT plants were built, they did not provide the treatment of Biowaste, but MSW from undifferentiated collection. The few of those which predicted the Biowaste collection, later had to be adapted to the operation with MSW, as was the case of the ADP of TRATOLIXO, which we'll discuss as our case study.

Since various AD technologies originating from northern Europe were designed to the reality of Biowaste treatment, when the opportunity for their adaptation to the MSW collection systems arose, there have been several cases of failure. The pretreatment system of MSW passes by the choice of the smallest fraction, usually below 60 mm, where the contamination of inert elements (sand, stones and glass) is high. This contamination is responsible for high abrasion problems of equipment and in some cases, settling in anaerobic digesters, or intermediate tanks. These problems have caused several stops of MBT plants and maintenance costs have far exceeded the income from the sale of electricity, so the technology was viewed negatively by most waste management companies.

This study aims to demonstrate that despite these problems, it is possible to carry out adjustments during operation that allow to overcome these difficulties and, increase the reliability of these facilities. It is also an objective to show that the dry AD, is the most appropriate for the treatment of MSW to the typology found in Portugal, allowing electrical production results well above expectations, with a positive impact on the economy.

2 Methods

This study has been prepared based on the industrial-scale results of the Anaerobic Digestion Plant (ADP), of TRATOLIXO, located in Abrunheira, municipality of Mafra. This unit was designed to render 160,000 tons/year of MSW that after pre-treatment, would ensure 50,000 tons/year of organic waste and the filling of two digesters and; 40,000 tons/year of Biowaste that, after pre-treatment, would ensure 25,000 tons/year of organic waste and the filling of one digester. This plant was designed as a mesophilic dry process plant, with batch feeding, single-stage, according to Valorga technology.

The process of inoculation and start-up of the two MSW digesters took place at the end of 2012, having the plant operated only with these digesters during 2013. Due to the reduced amount of separate collected Biowaste on AMTRES system (about 2000 tons/year, only), insufficient to operate the Biowaste digester exclusively with this material, in early 2014, the decision was made to operate the third and last digester also with undifferentiated MSW. Since March of 2014 that the ADP operates with three digesters at full load, with MSW. This study aims to present the operating results of the ADP, with regard to the main production ratios of the digestion process:

- The amount of biogas per ton of sorted waste (SW) in the plant and sent to the digesters (Nm^3 of biogas/ton SW);
- The amount of electrical power produced per ton of MSW (kWh/ton MSW);
- The quantity of electrical power per ton of sorted waste in the plant and sent to the digesters (kWh/ton SW).

These ratios should also be compared with what was envisaged in the project, in order to understand what changes were made to achieve these results and what gains were obtained. Finally, we will evaluate the impact that such production figures have into the electricity network of these municipalities, and since the ADP has the ability to process only 65% of the MSW of AMTRES system, what would be the impact if this technical solution would be adopted for the treatment of all MSW generated by householders.

Chart 1 shows the evolution of the ADP performance since 2013, when the plant operation reached a rate close to the nominal in two of three digesters available. Full load in the three digesters was only reached in the end of 2014s first semester.

It should further be noted that from mid-2013, ADP now operates with MSW pre-screened to 80 mm in the Industrial Center of Solid Waste Treatment (ISWTP) of Trajouce, for logistics optimization reasons. For this reason, the amounts of

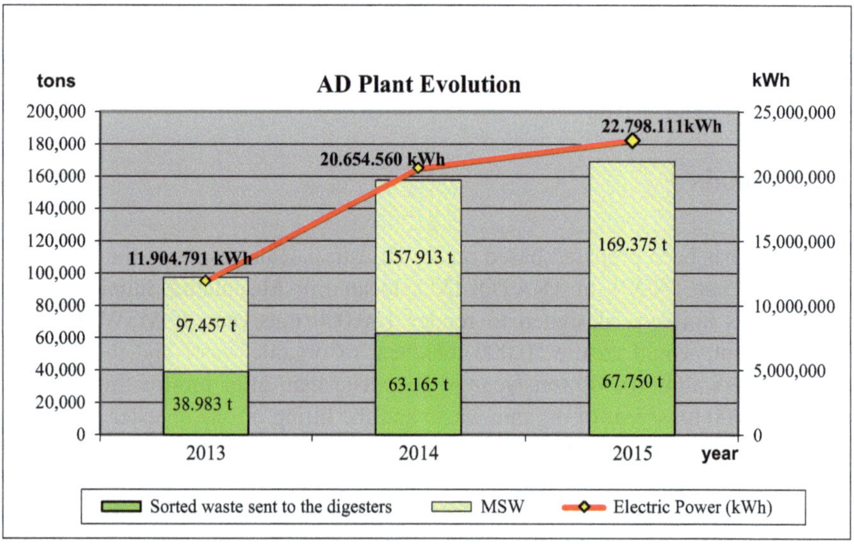

Chart 1 MSW treated and electric power produced in the ADP

MSW displayed are those that have been pre-processed at the ISWTP, and then sent to the ADP, to achieve amount of organic waste introduced into the digesters.

3 Results

In view of the several problems associated with these plants, some of them experienced since the start-up of the ADP, improvements have been implemented in the treatment process. In the commissioning phase, two substantial changes were made to the operation of the ADP by the technology contractor (Urbaser):

- Addition to the process of AD of the less than 15 mm fraction (previously rejected);
- Grinding to a particle size of approximately 20 mm from the fraction lying between 15 and 60 mm, for the AD;

After commissioning, see Table 1, TRATOLIXO has implemented a number of improvements aimed at two main goals: increasing the technical availability of the plant and, the improvement of the AD performance (reduction of contaminants to the digesters, increased productivity of biogas, reduction of the amount of biogas sent to the emergency flare). Listed below are some of these changes and their impact (Table 1):

These improvements allowed to surpass all expectations of biogas and electricity production, as shown by the following charts:

Chart 2 presents the amount of biogas produced per ton of SW that is introduced in the digesters. The project target was 147 Nm^3/ton, value close to reality in the

Table 1 Main changes implemented during ADP exploitation by Tratolixo

Improvements	Increasing technical availability of the plant	Improving AD performance
Processing pre-screened <80 mm MSW	✓	✓
Glass manual sorting		✓
Manual sorting of contaminants before AD (Q&C)	✓	✓
Ballistic system for rejection of inerts in the fraction <15 mm		✓
Implementation of technical cleaning team	✓	
Increasing AD temperature from 38 to 40 °C		✓
Improvements in agitation network (high pressure biogas)	✓	✓
Construction of Bypass to the dewatering unit	✓	
Schedulling of technical maintenance stops to avoid deviation/losses of biogas and electrical power production	✓	
Increase of dry matter content, from 25 to 28%		✓

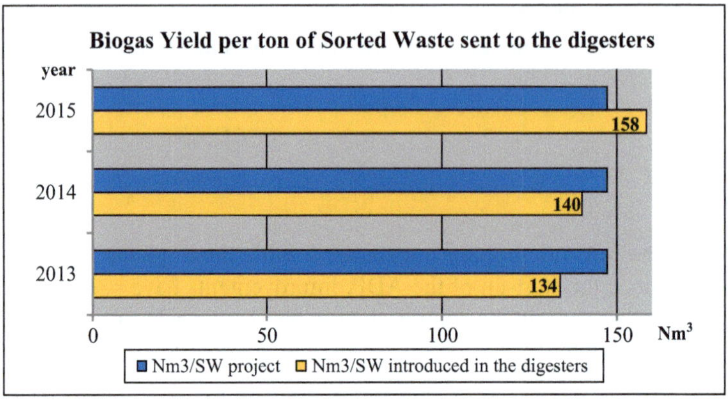

Chart 2 Biogas yield evolution in ADP

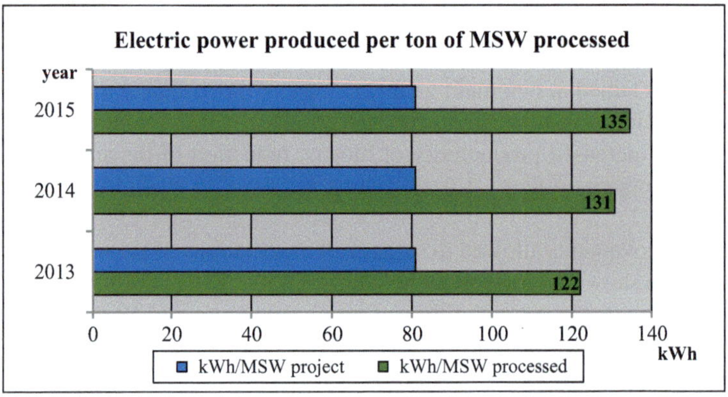

Chart 3 Evolution of electricity production performance per MSW treated

years 2013 and 2014 but still below expected. In 2015 this value has been exceeded for the first time, with 158 Nm^3/ton.

Regarding the electricity produced from biogas, gains on technical availability of critical equipments allowed to spend only 2% of the biogas produced, in the plant's emergency flare, increasing ratios of electrical production per MSW and SW as the Charts 3 and 4 show.

Considering that the project foresaw the production of only 81 kWh/ton of MSW processed, the results are quite better. In part, this improvement is related to the increase in the ratio of the SW for the digesters, which went from 37 to 40%.

In terms of electricity production for every ton of waste actually introduced in the digesters, this also overcomes the project in about 30%, which estimated a ratio of only 259 kWh/ton RT.

It is also important to clarify that these values of electrical power produced, refer to the amount actually injected into the national grid. There are still about 5% of

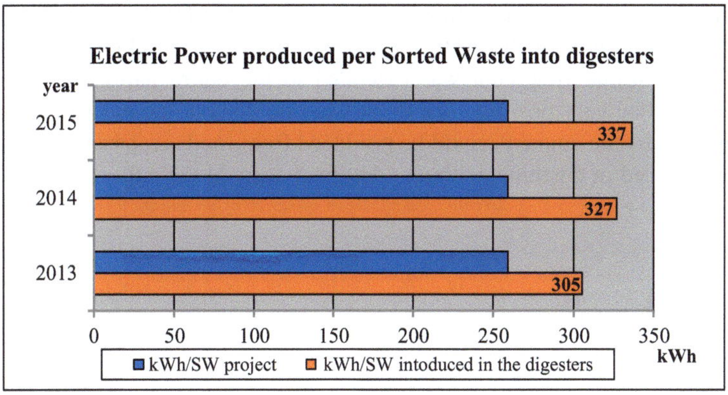

Chart 4 Electric power produced per ton of sorted waste into digesters

losses between the production and the injection on the network, already considered in all these results. The ADP does not use any electricity for self-consumption.

4 Conclusions

The results presented here, demonstrate a positive reality of a technology that is often viewed in a reticent manner as a solution for MSW treatment. It is possible to notice that despite the problems associated with the treatment of this highly heterogeneous stream of waste, the potential of the organic fraction contained therein allows to achieve very satisfactory electrical production ratios, without further burdening Municipalities with new separate collection systems.

There are, in fact, numerous technological and operational improvements that are changing the reality of these plants. Regarding to the impact that a plant like this can have in the community, in addition to the obvious environmental benefits that should be emphasized—renewable energy production, the significant reduction of greenhouse gas emissions and the potential of organic recovery for incorporation in the soil in the form of compost—there is a very positive impact on the power grid.

Considering that the production of MSW in the municipalities of AMTRES is 0.346 ton/inhab.year (TRATOLIXO 2013), knowing that for every ton of this waste in ADP is possible to produce 135 kWh, we can calculate that for each inhabitant, it is possible to produce in a year, 47 kWh of electrical power from its MSW. Taking into account the electrical consumption per capita (PORDATA 2015) in Portugal, divided by different components, we know that domestic consumption is 1173 kWh a year and the public street lightning is equivalent to 141 kWh. If all AMTRES's MSW was treated in the ADP, the total electrical production would be enough to guarantee 33% of the total public street lightning or, 4% of the total domestic power consumption.

Although unknown to most, this is a reality in the AMTRES system, where with the installed capacity of the ADP it is already possible to achieve 65% of these results. If a similar treatment was extended to the entire national territory, the impact would be very positive, just think that one third of the electricity for street lightning across the country, could be produced from food scraps and other organic materials placed in dustbins, without additional costs in the collection.

References

Chang NB, Martinho PA (2011) Solid waste management in European countries: a review of systems analysis techniques. J Environ Manag 92:1033–1050

EUNOMIA (2001) Costs for municipal waste management in the EU Final Report to Directorate General Environment, European Commission. http://ec.europa.eu/environment/waste/studies/pdf/eucostwaste.pdf

EUNOMIA (2002) Economic analysis of options for managing biodegradable municipal waste – Final Report to the European Commission. http://ec.europa.eu/environment/waste/compost/pdf/econanalysis_finalreport.pdf

PORDATA (2015) Base de Dados Portugal Contemporâneo. http://www.pordata.pt/

TRATOLIXO (2013) Relatório e Contas e Sustentabilidade. http://www.tratolixo.pt/Comunicacao/Documents/Relatório%202013%2021%20de%20abril%202014_web%20final.pdf

World Bank (2012) Urban Development & Local Government Unit, what a waste – a global review of solid waste management

Ricardo Castro holds the degree of Environmental Engineering (ULHT, Portugal, 2005), being ever since working in the waste management industry, namely, at Tratolixo—Tratamento de Resíduos Sólidos EIM. Between 2007 and 2011 he was part of the team managing the construction of the Anaerobic Digestion Plant (for Municipal Solid Waste), in Abrunheira, Mafra, the biggest of its kind in the country, with a capacity of 200,000 tons/year. Previous to commissioning he has done on the job training in similar plants in Spain (UTE EBRO, Zaragoza) and France (Everé, Marseille). Since then until 2016, he was the plant's Process Engineer (Chemical/Biological), with responsibility over the laboratory and methanization areas, including biogas production control, biogas treatment (sulphur remotion), wastewater pre-treatment and analysis. Also supervised several students' internships, and their thesis to obtain the Masters of Science degree.

Since January 2017 he is the Plant Manager, overseeing all the activity of the company in terms of waste treatment.

Speaker at two conferences of the project GR3—Grass to green gas (co-funded by the Intelligent Energy Europe Programme of the EU), held by LNEG (Lisbon, 2014, 2016). Speaker at MIST—Modelling Innovation Sustainability and Technologies (Cascais, 2015).

Has been part of a national working group related to biowaste and another related to biogas regulation, he is now a member of a national working group regarding the problems of the refused stream of MSW and the way to address European recycling goals.

Maria João Alves holds the degree of Chemical and Biological Engineering (ISEL, 2009). In 2010 started working in the waste management area in Tratolixo as a trainee. Tratolixo EIM—Inter-Municipal Company-is responsible for the management and treatment of Municipal Solid Waste produced by Oeiras, Mafra, Cascais and Sintra municipalities. Began working in the process and product control department, in environmental monitoring, process and product control and waste characterization. In 2011 was part of the team that did the startup of the Anaerobic

Digestion Plant, in Mafra. From 2011 until now, works in the plant's laboratory, monitoring the anaerobic digestion (mechanical and biological treatment) and supporting processes (MSW pre-treatment, biogas and wastewater) managing the laboratory and the development of new analysis methods.

Had a participation in the conference Modeling Innovation Sustainability and Technologies—1st Edition MIST 2015–"The potential for electricity generation in anaerobic digestion of MSW—the case of TRATOLIXO".

João Dias Coelho is currently and since 2014 the Chairman of Board of Tratolixo, and member of the Board since 2010. Between 2007–2010 he was team management coordinator of the construction of Wastewater Treatment Plant (WWTP Guia). Previously (2003–2007) he was executive Member of the Board of the Multimunicipal Water and Sanitation Services of the Municipality's of Amadora, Cascais, Oeiras and Sintra (SANEST–ADP Group) and Member of the Board and co-founder of the Municipal Energy Agency—"Cascais Energy Agency".

João Dias Coelho (born in 1963) holds a Degree in Law (UML-Lisbon University-Portugal) and holds Post graduation degrees, one in Law Planning, Urbanism and Environment (FD-Lisbon University-Portugal) and the other in Labor Law (Lusiada University-Portugal).

He also attended the Advanced Studies in Environment, Spatial Planning and Sustainable Development, FCT UNL and Advanced Management in Human Resources, by the Faculty of Economics and Business Sciences of the Catholic University of Lisbon.

Wastewater Reuse: Case Study of Abrunheira's Industrial Water Treatment Plant

Ana Barbosa, Lúcia Bonifácio, and Susana Dias

Abstract Increasing pressure on water resources and technological development of water treatment systems, coupled with savings from water reuse, have prompted the urgency of wastewater recycling and reuse.

Based on this premise, Tratolixo has built an Industrial Wastewater Treatment Plant in Abrunheira (IWTP)—Mafra, a system that reuses 60% of effluent for industrial purposes. The final effluent shows a significant reduction on its pollutant parameters, leading to a reuse in the Anaerobic Digestion Plant.

Since it started working the Abrunheira's IWTP has saved more than 25,000 m^3 of water that was reintroduced into the industrial process. This chapter describes a real case of successful investment on wastewater reuse.

Keywords Tratolixo • Reuse • Industrial water

1 Introduction

It has been estimated that the water saving potential in Europe stands at 40%. This fact illustrates the importance of improving water efficiency across a variety of sectors as one aspect of tackling water scarcity and droughts (http://ec.europa.eu/environment/water/quantity/water_efficiency.htm).

This water scarcity leads us to reflect on drinking water applications, prompting the urgency of wastewater recycling and reuse, and finding alternative water resources (Asano 1998). In Portugal, there are already some standards and legislation on reusing wastewater, specially conducted by ERSAR—the water and waste regulatory entity (Monte and Albuquerque 2010).

There's already a technological development on wastewater treatment systems. At an industrial level, there can occur a closed-loop system, where the industrial treated wastewater can be used on the same process (Fig. 1), thereby increasing

A. Barbosa (✉) • L. Bonifácio • S. Dias
Tratolixo, São Domingos de Rana, Portugal
e-mail: ana.barbosa@tratolixo.pt; lucia.bonifacio@tratolixo.pt; susana.dias@tratolixo.pt

A. Dias et al. (eds.), *Modeling Innovation Sustainability and Technologies*,
Springer Proceedings in Business and Economics,
https://doi.org/10.1007/978-3-319-67101-7_24

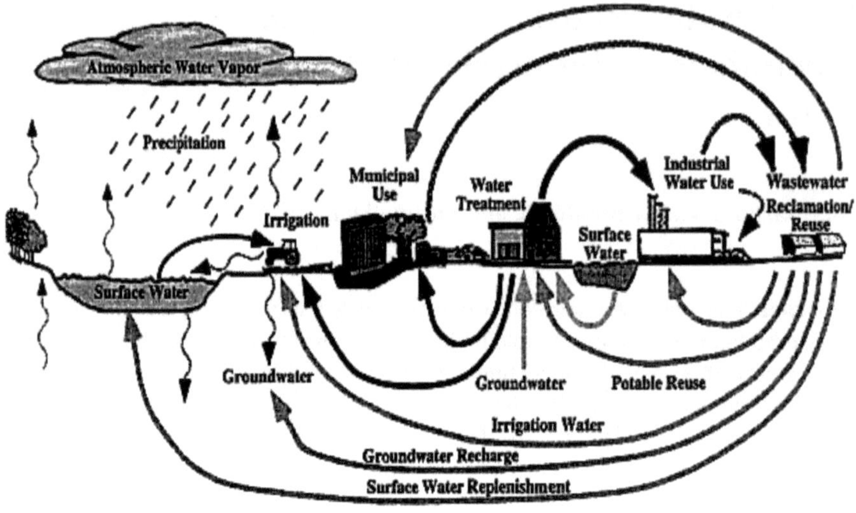

Fig. 1 Engineered treatment and reuse facilities in the cycling of water through the hydrologic cycle (Asano 1998)

water savings. About 25% of worldwide water demand is related to industrial applications.

The cost-effectiveness of reusing water for industrial purposes depends on the distance the water must be transported, between the reclamation facility and the point of use (Asano 1998).

In Abrunheira's Ecopark—Mafra, TRATOLIXO has built an industrial wastewater treatment plant (IWTP) 300 m distant from the Anaerobic Digestion Plant (ADP). This is a new technological facility for waste treatment and has three digesters that use industrial water into the organic process. This water may come from the IWTP, making the whole process economically and environmentally more respectful.

This work reflects on treated wastewater reuse (final water/effluent reuse), specially taking as an example the matching between these two Abrunheira's—Mafra facilities: ADP—as the industrial wastewater production and final water consumer and IWTP—as the treatment unit and final water producer (this facility may also consume some of the treated effluent).

2 Case Study

In 2012, Tratolixo started a new technological solution for waste treatment—an Anaerobic Digestion Plant in Abrunheira's Ecopark—Mafra, an innovative solution, strategic in terms of organic waste deviation from landfills, enhancing organic and energetic valorization.

This anaerobic digestion process, despite being called a dry process, needs water/moisture for the microorganisms, around 90 m^3/day. The water that results from this digestion process, around 113 m^3/day, has a high organic content.

This leachate together with the wastewater from social facilities (and from the future landfill) of the Ecopark, go straight into the Industrial Wastewater Treatment Plant, that began working in May 2014, with a capacity to receive 245.5 m^3/day. This infrastructure was specially designed to receive and recover waters with a high and complex organic content and has a high level of technical know-how. The final effluent achieves a quality that allows re into the industrial circuit, minimizing potable water consumption (Fig. 2).

Since IWTP started working, the potable water consumption reduced (Fig. 3). Late results from September 2015 have shown that almost half of the total water consumption in the Ecopark is industrial water (44%), leading to savings in drinking/potable water. TRATOLIXO already saved around 25.796 m^3 of potable water since the IWTP started.

Fig. 2 Production, treatment and reuse cycle of industrial waters in Abrunheira

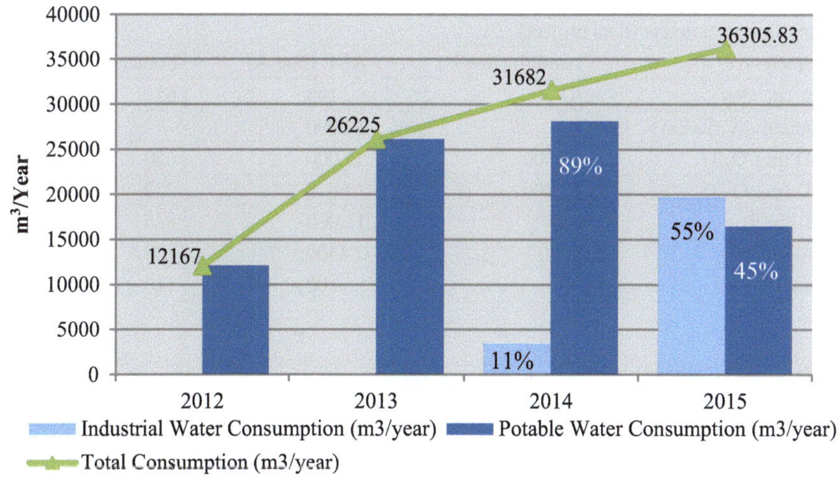

Fig. 3 Potable water consumption and industrial water consumption on ADP process

3 Treatment Process

The differences on quality and quantity of the leachate have made it necessary to build two distinct lines, that just converge on the tertiary treatment. The IWTP was dimensioned to receive wastewater from three different origins, described on Table 1 (Socamex 2013).

Except for BOD, it was already expected that ADP wastewater would be the most heavily charged. Until now, the landfill origin has no influence because it is still under construction, and the IWTP only works with the wastewater from the ADP and Social Areas, with a flow of 113 m^3/day. This flow has the following characteristics (Table 2):

Not only does ADP wastewater have a strong organic level but also a high conductivity and solid content.

The treatment process has three different phases. The final effluent achieves such a quality that it can be reused in the ADP industrial process:

1. Primary Treatment: coarse and perforated screens, equalization tanks for flow regulation and two treatment lines:

   - Line A: Less significant flows—ADP
   - Line B: Significant flows—Landfill

Table 1 Parameters of the three different wastewater origins

Admission parameters (data project)			
Parameters	Landfill (line B)	ADP (line A)	Social areas (line A)
Flow (m^3/day)	67	196	65
Conductivity (µs/cm)	–	30,000	–
COD (mg O_2/l)	20,000	35,000	650
BOD (mg O_2/l)	10,000	6875	250
TSS (mg/l)	500	13,000	325
N (mg/l)	3000	4500	100
P (mg/l)	30	100	11

Table 2 IWTP admission parameters (project and real data)

Admission parameters		
Parameters	Project	Real
Flow (m^3/day)	327	113
Conductivity (µs/cm)	30,000	35,000
COD (mg O_2/l)	35,000	28,000
BOD (mg O_2/l)	20,000	6000
TSS (mg/l)	13,000	14,000
N (mg/l)	3000–6000	4500
P (mg/l)	100	40

Table 3 Final effluent parameters (project and real data)

Final effluent parameters		
Parameters	Project	Real
Flow (m^3/day)	196	68
Conductivity (µs/cm)	approx. 200	approx. 300
COD (mg O_2/l)	<100	<16
BOD (mg O_2/l)	40	<5
TSS (mg/l)	<80	<5
NH_4 (mg/l)	<15	<3
Cl (mg/l)	<200	50

2. Secondary Treatment: Biological treatment (Line A and Line B) and MBR—Membrane Bio Reactor;
3. Tertiary Treatment: reverse osmosis by "Cross-Flow" with a pore of <0.001 microns;
4. Solid phase treatment: one treatment line with rotary drum thickener, for the sludge removed from the biological treatment and MBR.

The main advantage of the final phase with the tertiary treatment is the high quality of the effluent that can be reused on the industrial process of the ADP and also in the IWTP.

Based on this premise the final parameters were defined (Table 3).

At the present moment, except conductivity (not so relevant on the ADP process), every other final parameter meets all the expected values for reusing: a medium flow of 68 m^3/day and a conductivity of <300 µs/cm, COD < 16 mg O_2/l and BOD < 3 mg O_2/l (Table 3).

Since ADP reached its full capacity, in 2014, the volume of wastewater naturally increased. However, in 2014 and comparing with 2013, at an economic level, TRATOLIXO saved around 17% on the wastewater treatment costs (wastewater was previously sent out to external treatment).

Between 2014 and 2015 (until the present moment) there was a 9% saving and the amount of effluent treated raised from 583.2 to 682 m^3/week (median values).

4 Conclusion

Water as a production factor has an undoubted importance in a country such as Portugal, due to its edaphic and climatic constraints. This fact, together with domestic and industrial demands, and comparing raising demand and known water resources, renders inevitable its proper and sustainable use (Lencastre and Franco 2003).

"Acting to Safeguard the Future", Tratolixo is strongly committed in water saving and reusing the effluent of the waste treatment process. In Abrunheira's Ecopark—Mafra an Industrial Wastewater Treatment Plant was conceived to reclaim and treat the ADP, Social Areas and landfill effluents. Nowadays, the major contribution comes from the ADP with highly organic effluent containing 35,000 µs/cm of conductivity, 28,000 mg O_2/l of COD, 6000 mg O_2/l of BOD and 14,000 mg/l of TSS.

The IWTP has a phased treatment, with a final step reverse osmosis. This infrastructure's performance leads to a final effluent with a very high quality: conductivity < 300 µs/ cm, COD < 16 mg O_2/l and BOD < 3 mg O_2/l allowing final water reusing.

This fact has already allowed savings of around 23,267.1 m^3 of potable water since May 2014 when IWTP started working.

The reusing solution of the high quality final effluent in Abrunheira's Ecopark proves to be the best economic option but mainly the best environmental solution.

References

Asano T (1998) Wastewater reclamation and reuse. CRC Press LLC. https://books.google.pt/books?hl=ptPT&lr=&id=9vAGkleP234C&oi=fnd&pg=PR29&dq= water+reuse+water+treatment+system&ots=_nP46qMTNb&sig=dJlCIWjZZIGjR-mIwSSEXKo6I#v=onepage&q&f=true. Accessed 23 Sept 2015

Lencastre A, Franco FM (2003) Lições de Hidrologia. Fundação Armando Lencastre (Eds). Caparica

Monte H, Albuquerque A (2010) Reutilização de Águas Residuais. Série Guias Técnicos 14. Entidade Reguladora dos Serviços de Águas e Resíduos, Instituto Superior de Engenharia de Lisboa (Eds)

Socamex (2013) Memória Descritiva de Funcionamento da ETAL

Online Resources

European Commission – Environment. http://ec.europa.eu/environment/water/quantity/water_efficiency.htm. Accessed 21 Sept 2015

Ana Barbosa holds the degree of Environmental Engineering and Master of Bioenergy (UNL, Portugal, 2008), being ever since working in the waste management industry, namely, at Tratolixo—Tratamento de Resíduos Sólidos EIM. Between 2010 and 2014 she joined the strategy and innovation department, developing a refused derived fuel (RDF) project in Tratolixo together with IST, CIMPOR, CITRI and LNEG.

Between 2014 and 2016 managed the Industrial Waste Water Plant in Trajouce—Cascais, based on a new technical process, which included water phase changes.

Currently working as Production Chief Department.

Speaker at MIST—Modelling Innovation Sustainability and Technologies (2015).

Lúcia Bonifácio, degree in Environmental Management (pre-bologna) (New Atlantic University-Lisbon Portugal, 2004).

Final Degree Project in Economic Assessment Area of the Final Products of Wastewater Treatment Plants (2004). In 2005 started her career at COLEU Company (Cascais-Lisbon) in the development of urban solid waste collection services' projects. In 2008 joined the team of the Tratolixo as an Environment superior Technician—Operation and process management in the Water Treatment plant.

Lúcia Bonifácio is the Security delegate of Anaerobic Digestion Central Ecoparque of Abrunheira.

Susana Dias holds a degree of Advanced Studies in Sanitary Engineering and Integrated Waste Management of Environmental Engineering (FCT, Portugal, 2011) and a degree of Environmental Engineering (ULHT, Portugal, 2005) being ever since working in the waste management industry, namely at Tratolixo—Tratamento de Resíduos Sólidos EIM SA. Susana Dias works at Tratolixo as Monitoring Coordinator since 2009, with responsibility over the quality control of the process and products and environmental monitoring.

Speaker at conferences held by Apemeta—Associação Portuguesa De Empresas De Tecnologias Ambientais (2014).

Choosing the Appropriate Technology for Wastewater Treatment Regarding Energy Sustainability

David Pereira and Inês Moura

Abstract This paper intends to present a study about the costing of construction and operation of a particular branch of wastewater treatment: the use of activated sludge systems. There are three classical options of this process taken into consideration: extended aeration, conventional aeration with standard rate sludge digestion and conventional aeration with high rate sludge digestion and energy valorization.

In a large range of served population the treatment alternatives were evaluated, designed according to the current parameters.

The economic study of the treatment alternative used a geometrical pre-design, followed by measured work and quotation. This technical-economical modeling allows for the obtainment of interesting conclusions for the future design of wastewater treatment plants, whether they be new facilities or installations that need enlargement and/or rehabilitation.

Based on practical assumptions there's a definition of the served population which brings advantages to the use of one or other alternatives, considering almost exclusively economical and financial criteria. The designer must have the responsibility to consider other factors that are discussed widely.

The purpose of the present study, being an academic dissertation, was to offer a general tool for decision. Since it was based on actual average market conditions, it may present deviations when applied to real situations, requiring adaptation from qualified technicians.

As it is well-known a wider range of contributing equivalent population should be served, with advantages, by a treatment that incorporates biogas use from mesophilic anaerobic digestion, to produce heat and/ or electricity. On the other hand, smaller equivalent populations use to their advantage extended aeration systems,

D. Pereira (✉)
Department of Environmental Sciences and Engineering, Faculdade de Ciências e Tecnologia, New University of Lisbon, Lisbon, Portugal
e-mail: djp@fct.unl.pt

I. Moura
Engineering Department, Adductio, Lda, Lisbon, Portugal
e-mail: inescroftmoura@gmail.com

A. Dias et al. (eds.), *Modeling Innovation Sustainability and Technologies*,
Springer Proceedings in Business and Economics,
https://doi.org/10.1007/978-3-319-67101-7_25

without separated digestion. The question to answer was for what population there was no doubt about the decision to take without making too many design exercises.

In this study the range of populations where one or other system is obviously a good option are divided by a middle range, called "the doubt region", that has to be studied with more accuracy, due to local and particular characteristics. This "doubt region" includes activated sludge systems with conventional aeration using standard rate anaerobic digestion or treated digestion or extended aeration ones, competing according to actual local conditions.

These qualitative results were expected, being the present dissertation a contribution to delimit these results in a quantitative way. It was concluded that the conventional aerated systems with standard rate digestion start to compensate, comparing with the extended aeration systems costs, in 15 years, for served equivalent populations higher than 10,000 inhabitants. The activated sludge conventional aeration system with heated and mixed digesters and energy production from biogas becomes economically viable for the same project duration, from about 35,000 inhabitants served (according to this study, from 33,500 inhabitants), regarding average Portuguese conditions.

In between, deep studies must be done, wherein there is the possibility of using separate digestion of primary and secondary sludge in standard rate digesters, a type of solution that is falling into disuse, because of the exploration´s disadvantages, but whose study should be performed to reduce the inconveniences, decrease the energetic consumptions in comparison with extended aeration systems, for a range of equivalent population where the implementation of high rate digestion with energy production is not yet justified.

Keywords Wastewater treatment • Activated sludge process • Anaerobic digestion • Cost minimization • Energy balance

1 Introduction

Municipal wastewater treatment is a costly process which needs to be economically and financially optimized. The present paper intends to present a study about the costing of construction, maintenance and operation of a particular branch of wastewater treatment, the use of activated sludge systems, with three classical options of this process: (i) extended aeration, (ii) conventional aeration with standard-rate sludge digestion and (iii) conventional aeration with high-rate sludge digestion and energy valorization.

This technical & economical modeling allows the obtainment of interesting conclusions for the future design of wastewater treatment plants, whether for new facilities or installations that need enlargement and/or rehabilitation.

The purpose of the present study, being an academic one (de Moura et al. 2012), was to offer a general screening tool for decision help between alternatives, limiting the hard work which is the study of a larger set of alternatives.

2 General Comparison of Activated Sludge Processes

Briefly the main activated sludge systems types may be compared in the following way:

(a) EXTENDED AERATION

- Lower investment
- Greater energy consumption (electricity)
- Easy operation and maintenance
- More reliable
- Shock loading relatively resistant
- Stabilized sludge
- Less amount of sludge for deposition
- More difficulties in optimized layout under hilly (less flat) ground conditions

(b) CONVENTIONAL AERATION

- Greater investment (greater number of unit operations)
- Less electricity consumption in oxygenation
- More complexity in equipment
- More dependent of incoming quality conditions, especially industrial toxicity
- Separate sludge treatment with heat consumption
- Potential for gas & electricity production

3 Methodology

In a wide range of served population, the treatment alternatives were evaluated, designed according to the current parameters. Classical litterature was followed as Eckenfelder, Novotny 1989; Pires 2009; Qasim 1999; Spellman 2003; Vesilind 2003; U.S. Environmental Protection Agency 1974.

The economic study of the treatment alternatives used a geometrical pre-design, followed by quantity measurement and quotation using cost functions. Sensitivity analysis was an important point in the study.

Based on practical assumptions there's a definition of the served population which brings advantages to use one or other alternative, considering almost exclusively economical & financial criteria. The designer must have the responsibility to consider other factors that were discussed.

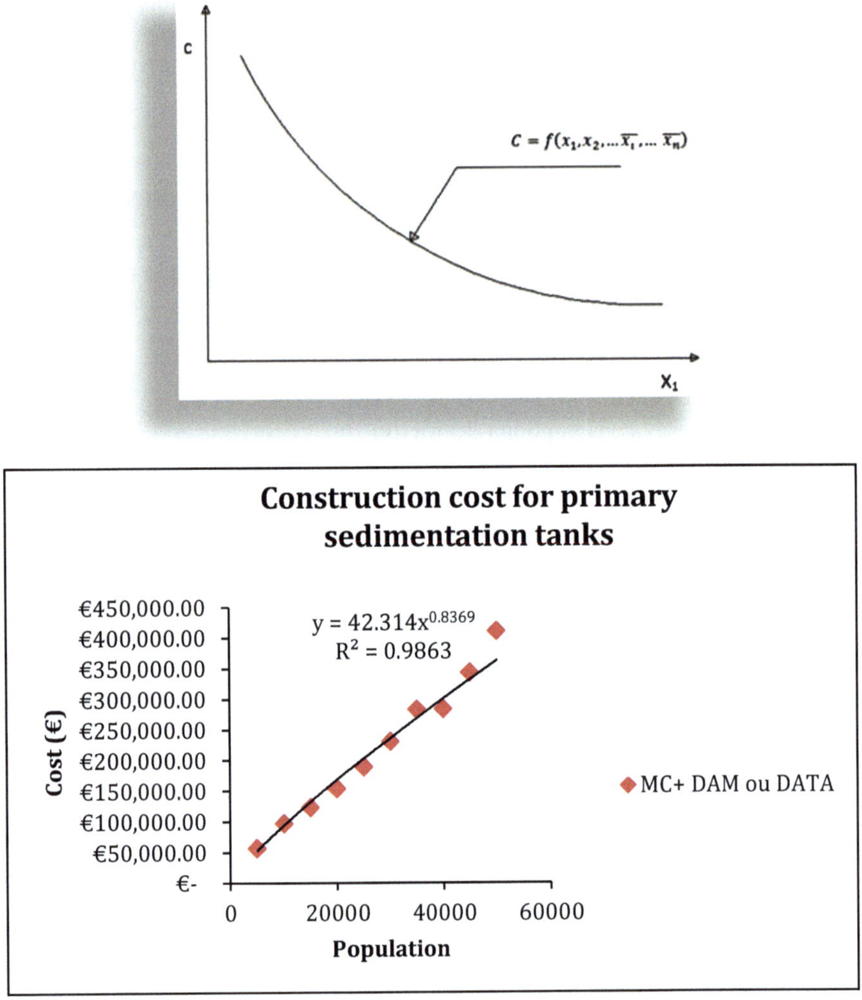

Fig. 1 Typical cost function and a synthetic example with relative costs (only for comparison with parallel assumptions, not absolute values)

As it is well known, wider ranges of contributing equivalent population better justify incorporation of biogas use from mesophilic anaerobic digestion, to produce heat and/or electricity. The question to answer was for what population there was no doubt about the decision to take without making too many design exercises.

The defined target was to obtain a middle range, called "doubt region", that has to be studied with more accuracy, due to local and particular characteristics, being relatively safe to define the extreme areas.

In Fig. 1 we may observe a general cost function, which shows the economy of scales, i.e., the decrease of unit costs with the increase of dimension. This example

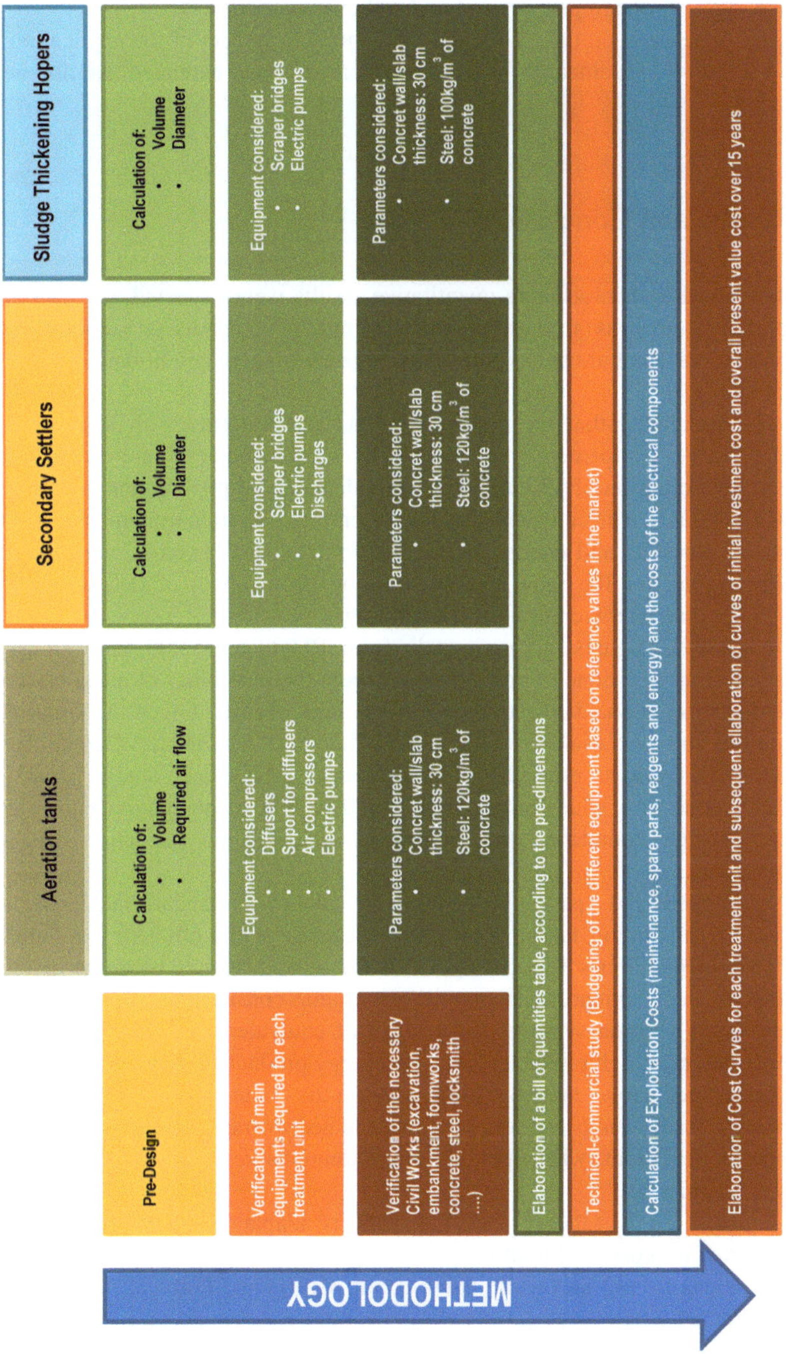

Fig. 2 General methodology flowsheet—example for the extended aeration activated sludge alternative

does not give absolute values, since there were a set of assumptions regarding, for instance, type of topographic ground and soil characteristics.

Figure 2 presents, on the other side, the general flowsheet of the application of the methodology, specifically for the simpler alternative: extended aeration activated sludge.

4 Results and Conclusions

Qualitative rather than accurate quantitative results were expected and obtained. The present study was a contribution to delimit these results in a approximate quantitative way, that are only valid for average Portuguese conditions on the year it was built.

It was concluded that the conventional aerated systems with standard-rate digestion start to compensate, compared with the extended aeration systems costs, in the project duration of 15 years, for equivalent served populations higher than 10,000 inhabitants. The activated sludge conventional aeration system with heated and mixed digesters and energy production from biogas becomes economically probably viable, for the same project duration, from about 35,000 inhabitants served, regarding average Portuguese conditions.

Between these figures (approximately from 10,000 to 35,000 equiv. inhab.), deep studies must be done, if one does not accept the possibility of using separate sludge digestion in standard-rate digesters, without heating. Given the simplified assumptions and the variations on the energy market (Comissão das Comunidades Europeias 2001; Monte et al. 2010; Moreira et al. 2015), deep studies shoud be preferably done in a wider range, from 7,500 to 50,000 population equivalent. Outside of this boundaries the probability to exist an exception will be extremely low.

In developed and, partially, developing countries standard-rate unheated digesters are a solution falling into disuse, because of the efficiency, reliability and environmental issues disadvantages, but whose study should be performed too, reducing its inconveniences: decrease the energetic consumptions in comparison with extended aeration systems, for a range of equivalent population where the implementation of high rate digestion with energy production is not yet justified.

As underlined above, there are limitations on the obtained results.

Since the study was based on probably outdated average market conditions, it may present deviations, when applied to real situations, requiring adjustment from qualified technicians.

Resulting from Portuguese data, the study is adaptable for other conditions, different in aims, costs and tradition.

Conclusions are not universal! Think about Algeria and Portugal, for instance! And the future?

Limitations will not allow absolute boundary answers but will fix the region of the answer. When doubts appear, additional studies are necessary.

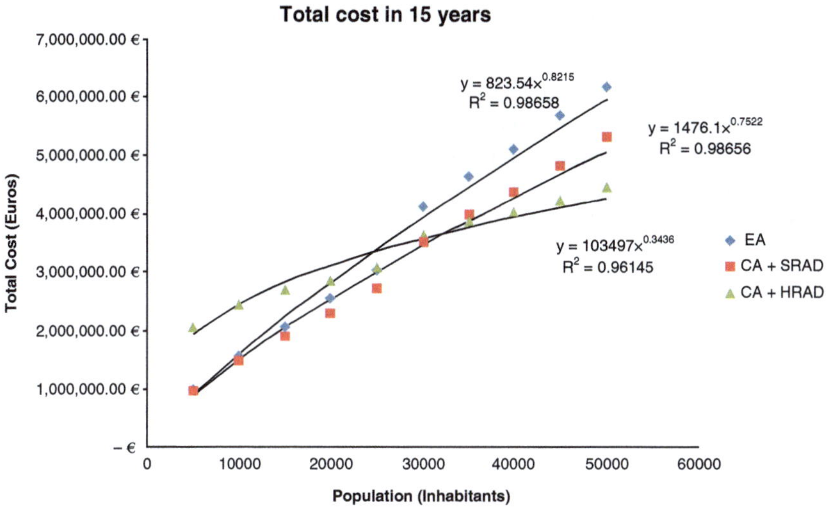

Fig. 3 Example—scenario of comparison costs for the three alternatives

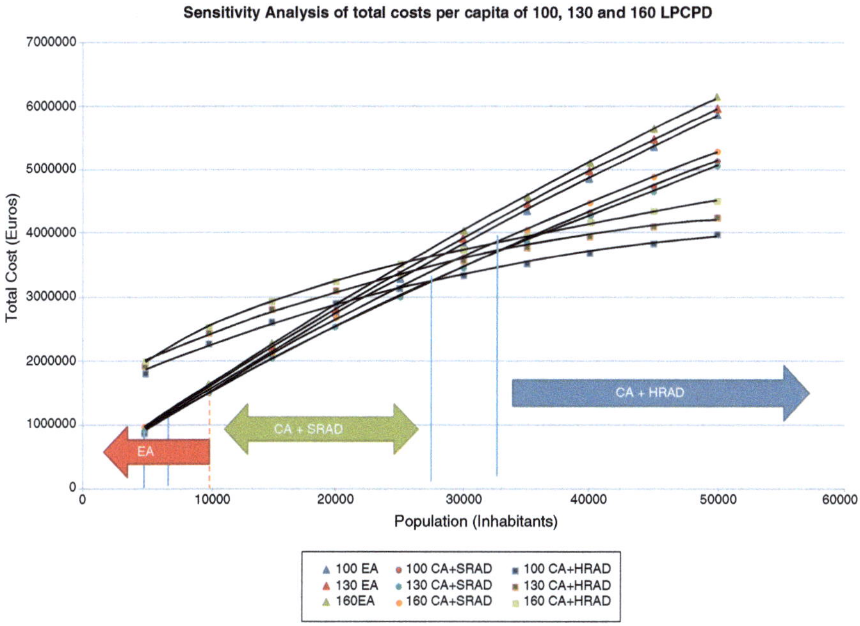

Fig. 4 Example of sensitivity analysis exercise, dependent on specific flow

Figures 3 and 4 shows some results of the methodology. One may follow these figures with the next key:

EA	Extended aeration
CA	Conventional aeration
SRAD	Standard rate anaerobic digestion
HRAD	High rate anaerobic digestion
LPCPD	Liters per capita per day

References

Comissão das Comunidades Europeias (2001) Towards a European strategy for the security of energy supply, Bruxelas

de Moura IC, Pereira D (2012) Opções de tratamento de águas residuais por sistemas clássicos de lamas activadas numa perspectiva de minimização de recursos aplicados. Master of Science Thesis. Faculdade de Ciências e Tecnologia da Universidade Nova de Lisboa

Eckenfelder WW (1992) Activated sludge treatment – industrial wastewater. Taylor & Francis, London

Monte M, Mano A (2010) Contributo para o Estudo da Valorização Energética de Biogás em Estações de Tratamento de Águas Residuais, Master of Science Thesis. Faculdade de Ciências e Tecnologia da Universidade Nova de Lisboa

Moreira RFC, Bessa R (2015) Previsão Probabilística dos Preços de Energia Elétrica do Mercado Ibérico de Eletricidade. MSc Thesis. Universidade do Porto. Porto. Portugal

Novotny V, Imhoff K (1989) Karl Imhoff's handbook of urban drainage and wastewater disposal. Wiley, New York

Pires A (2009) Contribuição para o Estudo de Avaliação do Desempenho de Um sistema de estabilização anaeróbia e utilização do gás produzido

Qasim SR (1999) Wastewater treatment plants, planning, design and operation. CRC Press, Boca Raton

Spellman FR (2003) Handbook of water and wastewater treatment plant operations. Lewis, London

U.S. Environmental Protection Agency (1974) Process design manual for sludge treatment and disposal

Vesilind PA (2003) Wastewater Treatment Plant Design. IWA, London

David Pereira, born in 1956, is a Civil and Sanitary Engineer, with a PhD degree,with a professional career in very practical technology application and author of bestseller Portuguese innovative technical software, since 1977. He had a parallel career in both the academic and the industry sides, being partner of a consulting engineering conglomerate and an environmental sustainability company. Core competencies include water, wastewater and solid waste systems modeling (simulation and optimization), technology application in real life (water/wastewater transport and treatment, solid waste management), and technical and economic planning of those systems.

Inês Croft de Moura, born in 1985, is an environmental engineer (BSc, 2009) specialized in sanitary engineering (MSc, 2011) with substantial experience in waste management, including fieldwork, and wastewater treatment design. She started her experience in Portugal being now working in Ireland.

Sustainabilty of Large Real Estate Projects: Case Study of Vila Nova de Santo Estêvão

David Pereira and Susete Mestre

Abstract This article analyses the contribution to the systematic assessment of sustainability in major real estate projects, both at the environmental level and its interactions with the various infrastructures that constitute the project or in which they are integrated. Hence, the adoption of sustainability policies as an instrument of paradigm change, enabling better use of new technologies and allowing significant waste reduction is reducing the ecological impact.

Aiming to assess qualitative and quantitative impacts and to propose a set of measures that can reduce these ecological impacts it is increasingly urgent to consider the actual and potential impacts associated with the construction environment of large buildings . The analysis of the situation should be held in a preliminary phase, in order to facilitate the identification of the adequate measures that minimize the impacts and, if possible, its complete elimination.

The creation of more sustainable environments where urban green spaces play a key role in the citizens' quality of life is a common concern and priority for all stakeholders, including not only industry and governments but also citizens.

This Vila Nova de Santo Estêvão enterprise (VNSE) is a real case project whose impacts' assessment has been done for the infrastructures of the golf course, the WWTP (Wastewater Treatment Plant) and the reservoir, according to an adaptation of the LiderA classification, widely used in Portugal.

Keywords Sustainability • Environmental evaluation • LiderA certification system

D. Pereira (✉)
Department of Environmental Sciences and Engineering, Faculdade de Ciências e Tecnologia, New University of Lisbon, Lisbon, Portugal
e-mail: djp@fct.unl.pt

S. Mestre
Engineering Division, Empreendimento Vila Nova de Santo Estêvão, Adductio, Lda, Lisbon, Portugal
e-mail: susetemestre@gmail.com

A. Dias et al. (eds.), *Modeling Innovation Sustainability and Technologies*, Springer Proceedings in Business and Economics,
https://doi.org/10.1007/978-3-319-67101-7_26

1 Introduction

1.1 Framing

The modern society has developed based on technological advancement and the assumption of inexhaustibility of natural resources, which caused the devastation of the systems that support life on the planet, endangering the well-being and survival of mankind. As a result, today we live a huge environmental crisis.

There is a growing awareness about these phenomena and a consequent proactive attitude, which gave rise to a model of sustainable growth. This recognizes the complexity of the relationship between development and the environment, realizing that this is a relationship of interdependence, and cannot be treated separately.

According to this definition present in the Brundtland Report-Our Common Future of 1987, the concept of sustainable development is about meeting the needs of the current generation and not compromising the future generations, taking into account the rational use of resources and preserving both the species and their habitats.

The green spaces are an urban component essential to the sustainability of cities, benefiting them ecologically, socially and aesthetically. However, these also must be sustainable. The sustainability of a green space results from the balance of the various elements that compose it. As such, if this balance is compromised it can become unbearable. It becomes, then, important to know these elements and their dynamics, not only in space but also in the environment that surrounds it, and realize how we can intervene, to achieve sustainability.

The efficient and conscious use of available natural resources, such as energy or water, is a central issue in sustainable development and is one of the most visible aspects in the pursuit of sustainability.

The water is a limited resource, which is necessary to protect and preserve. Its efficient use is an environmental imperative in any country in the world.

In Portugal, climate change can significantly affect water availability, in short to medium term, which is why it is necessary to develop measures in different sectors, to ensure an increase in the efficiency of water use. The need for an efficient use of water has been recognized as a national priority, through the publication of the resolution of the Council of Ministers of Portugal n° 113/2005, which approved the National Program for the Efficient Use of Water (PNUEA - Implementation from 2012 to 2020).

Also, the energy sector has been the subject of management policies aimed at sustainability, as for example the extinction of non-renewable or fossil energy sources. Renewable energy sources are all those that are inexhaustible and can draw on a permanent basis, such as solar, hydro, wind, biomass, tidal, wave and geothermal energy. They cause null or reduced impact on greenhouse gas emissions, which is one of the causes of the greenhouse effect.

Non-renewable energies, as the name indicates, use limited reserves, which decrease as they are consumed. Examples include coal, natural gas, oil and uranium. The cost of these types of energy increases according to their scarcity and cause environmental impacts (ADENE 2010).

1.2 Scope and Objectives

The environment, the economy and the society sensitize us to view the world not just as parts of a whole. The social, economic and environmental aspects are essential for achieving sustainability.

In the context of the introduction of the concept of sustainability in enterprises, the need arises to apply a model of performance evaluation, which gives importance to factors affecting the nature and social welfare measures that minimize the environmental, economic and social impact: improvements in water management, ecosystem recovery, reduction of carbon footprint and energy consumption, among others in order to achieve sustainability.

The overall objective is to propose an assessment for the project under study, through criteria that define the environmental sustainability, proposing for further implementation measures on the sustainable use of VNSE (Vila Nova de Santo Estêvão) venture through a model to evaluate, through classes, its behavior.

2 Sustainability

Global awareness about the need for a better relationship with the planet forces us to be more aware of protecting natural resources and the environment.

Throughout History, various scientific publications and events were held, with the aim of alerting society for negative impacts on the environment from development without limits, whose continuity would compromise future generations.

It is necessary to point out that in this article, the concept of negative impact as a loss, cost or fail and of the positive impact as a benefit, advantage or opportunity (PARTIDÁRIO 2011) is adopted.

The data demonstrates the Nature as an economic object of uncontrolled consumption, which caused a series of negative impacts on human health and the planet, such as global warming, environmental imbalances, social inequalities, increased pollution and related diseases and extreme poverty.

In 1984 Canada was the first nation to launch environmental management legislation—the Responsible Action Program (SOUSA 2010).

In the year of 1987, with the release of the Brundtland Report (Our Common Future), from the World Commission on Environment and Development, from the United Nations Organization (UNO), the concept of sustainable development began to be employed for the first time in History.

According to the definition, the needs of the poorest people should receive top priority, and there must be a limit of technology and social organization, so that future generations can meet their real needs (UNO 1987).

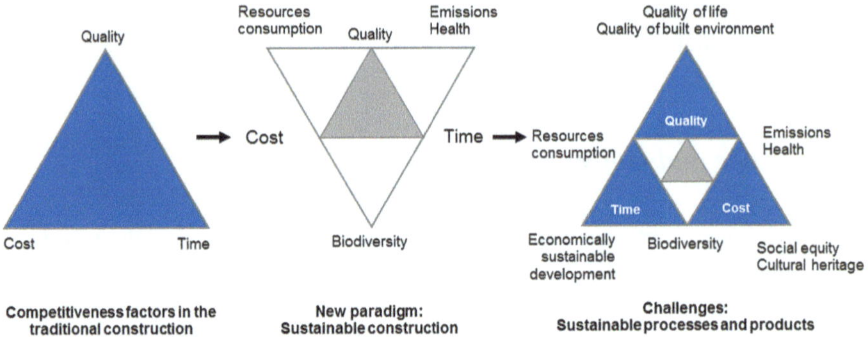

Fig. 1 Evolution of the paradigm of sustainable construction (PINHEIRO 2006)

The participation of society in sustainable development (Borrego 2002) is essential. The great pillars of sustainability (environmental, social and economic) are not sufficient, if these were not to serve stakeholders. Therefore, the intervention of each of us in an individual way is necessary, and through governmental and non-governmental organizations, business associations and associations of citizens, among others.

At the beginning of the twenty-first century, sustainability, which was restricted to intellectuals and politicians, became more democratic and linked to the media as a way of encouragement, or even pressure, so that governments and corporations incorporate sustainable development concepts.

Sustainable construction is a new approach to the traditional construction process, in addition to the issues related to the cost, time and quality, it seeks to give emphasis, in a first phase, to ecological aspects such as emissions, consumption of resources and biodiversity, passing then to broaden the scope, into quality of life, social equity and sustainable economic development. The figure below reflects the evolution of this paradigm of construction (PINHEIRO 2006) (Fig. 1).

3 LiderA: Portuguese National System for Evaluation and Certification of Sustainability

The system LiderA supports, evaluates and certifies the sustainability, integrating the various phases: design, construction and operation. This standard protects six principles to achieve sustainability: valuing the local dynamics and promote adequate integration; promote efficiency in the use of resources; reduce the impact of loads (either in value or toxicity), ensure the quality of the environment, focused on the environmental comfort; promote the sustainable socio-economic experiences; ensure the best sustainable use of built environments, through the environmental management and innovation.

Classification LiderA structured principles for achieving sustainability in six aspects, with 22 areas and 43 criteria:

1. Local integration, with regard to the soil, to natural ecosystems and the landscape and heritage;
2. Resources, covering energy, water, materials and food;
3. Environmental loads, involving the effluents, atmospheric emissions, waste, outside noise and thermal-lighting pollution;
4. Environmental comfort, in the areas of air quality, thermal comfort and lighting and acoustics;
5. Socioeconomic experience, which includes access to all, life-cycle costs, economic diversity, amenities and social interaction, participation and control;
6. Conditions for sustainable use integrating environmental and innovation management.

Each of these strands has a weight, which allows your evaluation, and the sum of the various chapters will match 100% (see Table 1).

After obtaining the value of each criterion, the global efficiency and environmental sustainability can be evaluated. When linking the performances of the criteria, the total performance in each area is obtained. On the other hand, weighing each area the positioning in each strand can be found and to perform the weighing for each strand is found the class of sustainability of the solutions considered, through the scale presented in Fig. 2.

Table 1 Weight of the sorting sheds LiderA

Strands	%	Criteria	% Each criterion
Local integration	14	6	2.33
Resources	32	9	3.56
Environmental loads	12	8	1.50
Environmental comfort	15	4	3.75
Socio-economic experiences	19	13	1.46
Sustainable use	8	3	2.67
Total (%)	100		

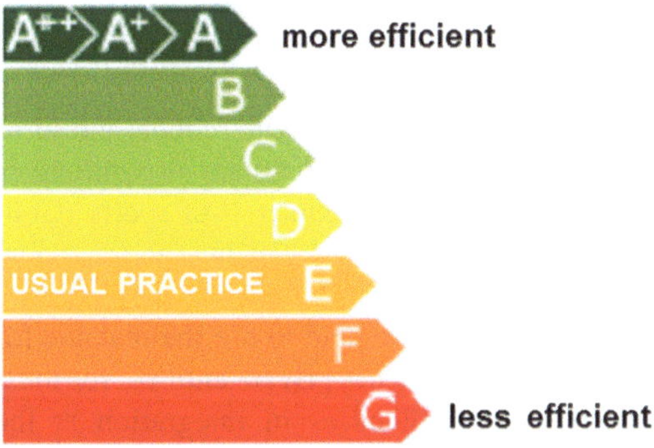

Fig. 2 Performance Classes assigned the criteria, according to LiderA (Pinheiro 2013)

4 Application of Integration and Evaluation Model

It was decided to apply the model LiderA, to the three main infrastructures of the enterprise of VNSE, in order to ascertain the final class of each performance, in terms of sustainability.

4.1 Golf Course

The evaluation rating of the infrastructure of the golf course is 54%, which is integrated in the class A^+ according to the classification LiderA.

These are the following concerns regarding the golf course:

- In terms of local landscape there is concern of select species for restocking, according to the soil and climate characteristics, mainly *Quercus suber* and *Pinus Pinea.*
- There is an old gothic-style building, which served to guard the cattle—OVIL—with historic heritage value; in the designed commercial galleries was provided for the maintenance of the beautiful facade.
- Buildings in support of the golf course have many windows, to ensure natural lighting; given that the facilities are almost exclusively active during working hours (day) that helps to reduce the energy consumption.
- Carbon intensity—there wasn't a direct reflection on the subject, but the overall environmental concern related with energy consumption tended to value this component indirectly.
- About the quality of water greater efficiency in the reservoir of hydrocarbon retention and reuse of washing water and rain are required. The irrigation water that the golf course uses is managed in a rational way, through a computer program—Site Pro, which allows to define, in real time, the water needed in the different zones.
- Most of the solid waste produced on the golf course is handed over to the competent units, for recycling.
- Whenever possible, one opts for battery equipment to avoid noises that disturb the residents.
- The Club House possesses the infrastructure adapted to people with disabilities.

There are partnerships with schools and surrounding towns to enjoy the facilities of Santo Estevão golf. There are also initiatives of open days, when all the people can, free of charge, contact with sports (Golf). There is a picnic park to enjoy Nature, a physical maintenance inside course and a magnificent terrace for relaxation.

Tournaments are held to use the revenue for charities. There is also a collaboration with public and private entities, for the accomplishment of traineeships for people with difficulties in integration at the work market, both with motor deficiencies or economic difficulties, through the official network of employment agencies.

4.2 Natural Water Reservoir

With regard to the natural water reservoir, the weight of each strand was calculated using the number of criteria, i.e. each criterion shows the same weight of importance in the calculation of sustainability. Being the welfare and social interactions of greater importance (25%), followed by local integration (20%) and resources, environmental management and awareness and communication, into% each. Finally, it positions itself to lower weight: environmental charges (10%).

The dam was built primarily for leisure and wellness in the venture to create and if needed to respond in emergency water supply in case of fire. In this way, you can check that matches the aspect of greater weight: welfare and social interactions. The reservoir is integrated within the complex and surrounded by local fauna and flora. It is a valued and protected place.

Every day, except on weekends, the storage level range is read, which rises through the rainfall, runoff, stormwater network and treated wastewater from the existing WWTP.

The dam produces no noise that disrupts the population. It is accompanied by a quarterly report delivered to the Portuguese Environmental Agency (APA), on the behavior of the dam.

It is a stunning location accessible to the community and with a line of road transport. This dam is built to support the water reserve for fire fighting, fire training divers and recreational purposes.

The assessment of the natural water reservoir infrastructure took a rating equal to 55%, which is integrated in the class A+, as the golf course had too, according to the classification LiderA.

4.3 Wastewater Treatment Plant (WWTP)

Without more details, which may be seen in the full document that serves as base for this paper (Mestre and Pereira 2014) it is presented the evaluation rating for the WWTP infrastructure (31.25%), which is integrated in the class B, according to the classification LiderA.

5 Discussion of the Results Obtained by Applying the Model and Recommendations

The major goal of this study is the analysis of the environmental perspective of infrastructure implementation of the considered type, contributing for future benefits planning. The conclusions may be addressed not only to promoters of these ventures, but also to entities that regulate and oversee them. It is also important to consider strengthening the awareness and education of players, users and residents, to the problems of the environment, in order to be able to establish an informed and active participation of those involved, in the protection of ecosystems and habitats present in the venture, to safeguard the rational use of natural resources.

The objectives of this study were attained, because it was possible to determine the assessment of the sustainability of the enterprise, with the evaluation of major infrastructure, despite the difficulties in evaluating certain parameters and of subjectivity. It was observed that, in accordance with the procedures, both the reservoir and the golf infrastructure are integrated in class A+, according to the LiderA system and the WWTP has a lower class, i.e., is integrated in class B, due to the small amount of effluent that enters for treatment, which does not allow a great efficiency of the system.

6 Conclusions

In this study, it was possible to define, critically and consistently, the current situation of the development VNSE, as well as recommend a set of actions that can contribute to increase the sustainability of the environment as a whole.

The major conclusion was to show that, through simple methods one may contribute to mitigate the impacts caused by the deployment of large enterprises. Despite it being a particular case study, it can offer generalized conclusions about typed ventures, both in Portugal and in other locations.

This work aimed to propose a general evaluation for the enterprise, through criteria that define environmental sustainability. Proposed measures to the sustainable use of VNSE venture through a model that allowed evaluating through classes, the behavior of the enterprise, based on the concept of sustainability.

There was difficulty in defining the weights of some factors, in the LiderA method, for the infrastructure of the project, because, often, the order of magnitude is not easily quantifiable. But, on the other hand, this method allowed a simple and fast evaluation with reduced resources.

In this sense, the result obtained for the sustainability of the enterprise is not absolute, but the results obtained with the methodology adopted for the evaluation corresponded to the objectives proposed, having been transmitted in a clear and objective way.

An overall assessment study, simple and succinct, but pragmatic and objective of sustainability from the overall complex, was carried out. The importance of environmental management systems was verified, which contributes to a continued improvement, a role more responsible towards society and ecology. Also, through the simple, individual or collective voluntary behaviors, the value chain may be optimized, ensuring sustainability, to improve the permanent social welfare.

The importance of an environmental manager is extreme, because he/she is the coordinator of all environmental policy, supporting and motivating the team and all stakeholders by setting the strategies and investments in the area of Environmental Protection. Finally, this work has been one step ahead, to internally evolve this company toward a greater efficiency of environmental sustainability, serving as example and leverage that sector typified. Through the implementation of environmental policies, all actors in this process will contribute to overall positive developments, understand the motivating aspects that influence organizations to invest in actions accountable to the environment and society.

References

ADENE (2010) Energy efficiency guide, Agency for energy, Lisbon

Borrego C (2002) 21 Environment Magazine n° 6, year I, Lisbon

Mestre SG, Pereira DP (2014) Sustainability of real estate ventures of large size: study case of Vila Nova de Santo Estêvão, Dissertation submitted to obtain the degree of Master of Science in Engineering and Water Management in New University of Lisbon

Partidário MR (2011) AIA and sustainability: loop IST, concepts, sustentabilidade.pdf. https://dspace.ist.utl.pt/bitstream/2295/711702/1/2

Pinheiro MD (2006) Environment and sustainable construction. Environmental Institute, Amadora

Pinheiro MD (2013) Leading tourism enterprises: voluntary system for the sustainability of built environments, voluntary Portuguese standard produced by the LiderA. LiderA, Portugal

Sousa (2010) The evolution of environmental policy in 20th century, Brazil, BC. http://www.achegas.net/numero/vinteeseis/ana_sousa_26.htm

UNO (1987) United Nations Organization – Report of the World Commission on Environment and Development: our common future. New York. http://www.un-documents.net/wced-ocf.htm

David Pereira Born in 1956, is a Civil and Sanitary Engineer, with a PhD, with a professional career in very practical technology application and author of bestseller Portuguese innovative technical software, since 1977. He had a parallel career in both the academic and the industry sides, being partner of a consulting engineering conglomerate and an environmental sustainability company. Core competencies include water, wastewater and solid waste systems modeling (simulation and optimization), technology application in real life (water/wastewater transport and treatment, solid waste management), and technical and economic planning of those systems.

Susete Mestre Born in 1980, is an agronomist since 2004, with a master's degree in Water Management and Engineering (2014). Her professional career focuses on the environmental management and maintenance of golf courses, always following the principle of sustainability and economic development.

MIX
Papier aus verantwortungsvollen Quellen
Paper from responsible sources
FSC® C105338

If you have any concerns about our products,
you can contact us on
ProductSafety@springernature.com

In case Publisher is established outside the EU,
the EU authorized representative is:
Springer Nature Customer Service Center GmbH
Europaplatz 3, 69115 Heidelberg, Germany

Printed by Libri Plureos GmbH
in Hamburg, Germany